Environmental Issues
in the 1990s

'Ceux qui viennent trops tard ne trouvent que des os'

'For those who come late only the bones are left'

French proverb

Environmental Issues in the 1990s

Edited by

A. M. Mannion *and* **S. R. Bowlby**
Department of Geography,
University of Reading, UK

JOHN WILEY & SONS
Chichester · New York · Brisbane · Toronto · Singapore

Published 1992 by John Wiley & Sons Ltd,
Baffins Lane, Chichester,
West Sussex PO19 1UD, England

Reprinted December 1992

Other Wiley Editorial Offices

John Wiley & Sons, Inc., 605 Third Avenue,
New York, NY 10158-0012, USA

Jacaranda Wiley Ltd, G.P.O. Box 859, Brisbane,
Queensland 4001, Australia

John Wiley & Sons (Canada) Ltd, 22 Worcester Road,
Rexdale, Ontario M9W 1L1, Canada

John Wiley & Sons (SEA) Pte Ltd, 37 Jalan Pemimpin #05-04,
Block B, Union Industrial Building, Singapore 2057

Library of Congress Cataloging-in-Publication Data:

Environmental issues in the 1990s / edited by A. M. Mannion and S. R.
   Bowlby.
      p.   cm.
   Includes bibliographical references and index.
   ISBN 0-471-93326-0
   1. Human ecology.   2. Environmental protection.   3. Economic
development—Environmental aspects.   I. Mannion, Antoinette M.
II. Bowlby, S. R. (Sophie R.)
GF41.E53   1992
363.7—dc20   91–36821
                                                                      CIP

British Library Cataloguing in Publication Data:

A catalogue record for this book is
available from the British Library.

ISBN 0-471-93326-0

Typeset in 10/12pt times by Cambridge Composing (UK) Ltd, Cambridge
Printed and bound in Great Britain by Dotesios Ltd., Trowbridge, Wilts.

To

JOHN B. WHITTOW

Contents

About the Contributors

All the contributing authors are past or present members of the Department of Geography at the University of Reading, UK.

SOPHIA R. BOWLBY graduated from the Geography Department at Cambridge University before going to Northwestern University, USA to do her PhD. She is a senior lecturer at Reading where she teaches courses in human resources, social geography and retailing and marketing. Her main research activities are in the fields of feminist geography, retailing and access and mobility.

MICHAEL J. BREHENY is professor of applied geography at Reading. He is a chartered town planner, having worked in local government before taking up an academic career. He has research interests in planning theory and practice, regional development, retailing and environmental issues. He is chairman of the British Section of the Regional Science Association, a council member of the Town and Country Planning Association and a Fellow of the Royal Society of Arts.

ERLET CATER graduated from the Geography Department, University of Exeter, before undertaking her MSc at the London School of Economics and her PhD at Reading where she now lectures on Third World development. Tourism in the Third World is one of her main research interests and she has travelled extensively throughout the Third World and has witnessed many changes wrought by international tourism over the past twenty years.

BRIAN GOODALL is professor of geography and now Dean of the Faculty of Urban and Regional Studies, having previously been head of the Reading Department of Geography from 1980 to 1990. For the last 20 years his research and teaching interests have concentrated on environmental management, tourism and recreation.

PETER HALL is now emeritus professor of geography at Reading having been professor in the department from 1968 to 1989. On leaving Reading he became director of the Institute of Urban and Regional Development at the University of California, Berkeley, and he now holds a concurrent position as professor of planning at The Bartlett, University College, London University. He is the author of over 20 books and numerous articles on planning related subjects.

JOCELYNE HUGHES graduated in geography from Cambridge University and gained her PhD from the University of Tasmania. She has worked on the ecology and conservation of wetland environments for the past ten years in Australia, Antarctica, New Zealand, Tunisia and Guatemala. She joined Reading in 1990 where she lectures in physical geography and wetland ecology.

MICHELLE LOWE has been a lecturer and researcher at Reading for the past four years and teaches courses in human resources and industrial and regional development. Prior to this she was a PhD student in geography at Cambridge University and an undergraduate at the University of Birmingham. Her main research interests are in economic and social geography, in particular, local economic development and feminist geography.

ANTOINETTE M. MANNION graduated from the universities of Liverpool and Bristol. Her research interests are Quaternary studies and environmental archaeology and their relevance to global environmental change. She teaches biogeography/ecology and biotic resources for undergraduate years one and two and a specialist option on environmental change at year three.

ANDREW MILLINGTON is a reader in physical geography at Reading, having previously lectured at Fourah Bay College, University of Sierra Leone. He has interests in soil erosion and conservation, woody biomass assessment and environmental degradation in developing countries, particularly drylands.

JOHN G. SOUSSAN did both his BA and PhD in geography at the University of Leeds and subsequently taught at the universities of Leeds, Liverpool and Lancaster before coming to Reading. His principal research and teaching interests are in the fields of resource appraisal and energy and biomass studies in the Third World.

DAVID SPENCER read geography at Sheffield University and did a M. Soc. Sci. degree in urban and regional studies at Birmingham University. After a period in local government and a research post at the Centre for Urban and Regional Studies in Birmingham, he lectured in urban studies at Bulmershe College of Higher Education, Reading. He returned to the university sector in 1990, and now lectures in human geography, specializing in population, housing, and planning. His current research examines the counterurbanization phenomenon and its impact on rural communities.

RUSSELL THOMPSON has studied climatology over the past 30 years with research projects in many parts of the world, including both polar regions, the humid and arid tropics and the temperate lands of New Zealand, Canada and England. He has held lectureships in geography departments at the University of New England, New South Wales; Massey University, New Zealand; the University of Guelph, Ontario; the University of Western Australia; the University of the South Pacific and (currently) at Reading.

JANE WELLENS is a postgraduate student in the geography department. She graduated from the University of East Anglia, where she obtained a BSc in development studies (natural resources). Her research involves using satellite imagery to monitor and model biomass resources in Tunisia. Her other interests include desertification, common property resources and environmental degradation especially in Nepal.

Preface

Books on environmental issues are not a novelty: all the issues presented in this text have received widespread attention in the popular and academic media. However, with the increasing number of new courses in institutes of higher education that cover these issues and with their inclusion in 'A' level syllabuses, we believe that an introductory text such as this, which covers a wide variety of social and physical issues, is necessary. We have aimed to provide a thorough introduction to each topic. The text should thus be of value to sixth formers and first-year undergraduates as a foundation for the pursuit of further study.

The editors of and contributors to this book are all university teachers and researchers in geography. This is a subject that has long acknowledged the importance of studying the relationships between the physical and human environment. The discussions in this book of both the physical and social processes which affect the environment emphasize the need to focus on these relationships when evaluating the environmental issues of the 1990s.

The text is divided into three parts. The first discusses existing frameworks for examining people–environment relationships. The concept of sustainable development is also given prominence because it currently commands wide acceptance among policy makers, although there are many misconceptions about its meaning and implications. The second part covers global issues, including a scene-setting chapter on the environmental and cultural changes of the last two to three million years. Discussions of climatic change, deforestation and marine pollution follow. The relationship between population change and environmental degradation and the current and future patterns of energy production and consumption are discussed in the ensuing two chapters. Changing political attitudes and technological developments in biotechnology and genetic engineering are considered as representative of society's changing understanding of and response to environmental problems through political policies and technological applications.

The third section of the book is concerned with the local environmental impacts of resource use and misuse, notably the impact of industry and fossil fuel energy consumption, which cause atmospheric, aquatic and terrestrial pollution. The environmental issues created by agricultural practices, including wetland destruction, eutrophication, soil erosion and desertification, are also discussed. Local issues that relate to transport, urban sustainability and quality and tourism are addressed. Finally, a brief perspective and prospective is presented which highlights some key themes from the book and discusses their implications for sustainable development.

Our overall policy has been to provide an easily digestible set of individual chapters

with cross referencing to indicate the overlap that exists between issues relating to
the physical environment and to economic and social issues. Each chapter also
includes recommendations for further study, which will be useful to the student,
teacher, policy maker and, indeed, any interested reader.

This book has been written at a time when environmental issues are high on the
political and popular agenda. There is much gloomy talk of global environmental
damage, but at the same time there is some optimism that human beings can find a
better way of living within the global ecosystem of which they are a part. We wonder
how a similar text will read as the 21st century opens a decade hence, and hope that
when that century ends such a text will prove unnecessary.

This book is dedicated, by all its contributors, to our colleague, Dr John Whittow.
We wish to acknowledge his very considerable contribution to the Department of
Geography at the University of Reading and to the subject of geography. In
particular, we wish to pay tribute to his ability, as a researcher and teacher, to
integrate the physical and human elements of this discipline in the study of
environmental issues. For many years he has directed the human and physical
geography degree course at Reading and has shown generations of students the
excitement and enjoyment to be gained from bringing the two sides of the subject
together. His numerous books and papers have also made a major contribution to
the understanding of environmental issues. He has placed a particular value on and
has a rare talent for communicating knowledge of environmental processes to a wider
public than the academic community alone.

A. M. MANNION and S. R. BOWLBY
Reading
August 1991

Acknowledgements

We acknowledge the help of Sheila Dance and Heather Browning of the Drawing Office, Department of Geography, University of Reading, for the provision of diagrams, and of Jacqueline Wicks, Sarah McQuillan and Pat Wylie for typing the manuscript. We also record our thanks to the following colleagues at Reading and elsewhere who contributed to this text by commenting on drafts of chapters: Dr K. Crabtree, Professor I. R. Gordon, Professor R. J. Gurney, Dr J. R. Hardy, Dr S. Nortcliff, Mary A. Parry, Mr R. B. Parry, Mr P. J. G. Pearson, Mr M. J. Stabler, Dr G. Wadge and Dr J. B. Whittow. We are also indebted to Dr M. D. Turnbull for compiling the index.

SECTION I

IDEAS AND CONCEPTS

CHAPTER 1

Introduction

A. M. Mannion and S. R. Bowlby

1.1 PRELUDE

This book presents discussions of current environmental issues which underpin the environmental debates of the 1990s. Analysing these issues requires an understanding of the interactions between human society and the physical environment. Some of the chapters are concerned primarily with the physical processes involved in environmental degradation, whereas others focus on analyses of the social and economic processes which give rise to specific environmental problems. Whatever the topic, all the chapters examine the ways in which human action affects the physical environment and the impact that this has on society. As this relationship is not always straightforward, there are many difficulties in analysing the relationships between people and the environment. These arise because there is no appropriate conceptual framework within which such relationships can be examined.

In Western intellectualism there has been a tradition of analysing society separately from the environment. This dualistic approach has its origins in the work of Strabo (64 BC to AD 20), a Roman scholar whose 17 volume work entitled *Geographica* drew distinctions between people and places. The tradition was maintained by such scholars as Bacon (1561–1626), Varenius (1622–50) and Kant (1724–1804), who also argued that human beings have the ability and right to exert power over the natural world in a way that other species cannot. As discussed in Chapter 10, attitudes to the natural world in the West have centred on the subjugation of nature for human benefit. This attitude has recently been criticized and in the 1990s alternative ideas about the relationships between human beings and the natural world are becoming more widespread. These alternatives suggest that human beings should be considered as part of the natural world, rather than as separate entities, with a responsibility to care for it, either for their own long-term benefit or for the benefit of other organisms. Despite this shift in attitudes, there is still no satisfactory conceptualization of the relationships between human societies and the physical environment.

Humans undoubtedly exploit the physical environment, creating the physical environmental problems discussed in this book. Such exploitation has also led to the congregation of people in urban settlements, which nevertheless remain reliant on

Environmental Issues in the 1990s. Edited by A. M. Mannion and S. R. Bowlby
© 1992 John Wiley & Sons Ltd

the wider environment for resources and waste disposal (Fig. 1.1). These centres of population have created their own impacts upon the environment. In the 1990s there is a growing realization that the long-term subjugation of nature by society has begun to create a less accommodating milieu for the sustainability of society. This involves the impairment of so-called 'open access resources' which, at the global level, constitute the global ecosystem.

The separation of physical and social environmental problems is reflected in this text. It is arranged in two main sections which distinguish between environmental changes of global significance and changes which, although they may occur in a variety of spatial contexts, have only a local impact. This local impact may, however, be of great importance to those affected. Within these sections the issues are further divided into those which emphasize change in the physical environment, whether or not that change has a 'physical' or 'human' origin, and those where the focus of concern is social and economic change.

This separation of physical and social environmental issues should not be taken to imply that there is no need to find better ways of examining people–environment relationships. It is not satisfactory to polarize the efforts of scientists and social scientists. Current conceptualizations of the people–environment relationship deriving from the physical sciences make use of the notion of a *system* and imply that social and physical processes should be examined as interacting parts of an environmental system. Social science conceptualizations draw on the *political economy* approach and suggest that human beings should be considered as acting within an environmental, political and economic structure which constrains their actions in the short term but which can be modified, or radically changed, in the long term. From this perspective the characteristics of the physical environment act as constraints on human activity even though these may be modified by technological and social organization.

In the next section, existing ideas and approaches which might be used to explore further the interaction of social and physical processes are discussed. The first three subsections examine systems theory, the ecosystem concept and the Gaia hypothesis. These are unified by the fundamental flows of food and fuel energy that power all earth surface processes, including human activity, and which, in turn, affect global biogeochemical cycles. The last subsection discusses ideas from human geography and political economy.

1.2 THEORETICAL FRAMEWORKS: SYSTEMS AND SOCIETY

The need to reconcile the analysis of the environment and society has become more pressing in the last 20 years as environmental problems have emerged as potential threats to society's fabric and existence. For too long, Western society has taken the environment for granted, utilizing resources in such a way that they are wasted rather than conserved. A greater understanding of society and the physical environment is essential if the prodigal resource use that characterizes pre-21st century society is to be translated into post-21st century sustainable development (Chapter 2). Sustainable resource use requires that resources should be used in such a way that their value for future generations is preserved. Achieving this will require a better understanding of processes such as energy flows and biogeochemical cycles and how they can be manipulated to satisfy society's needs in the long term. Inevitably, new ways of achieving sustainable development will emerge as science and technology advance but, in contrast to past and present resource use, society will also need to change its aspirations and attitudes. A new symbiotic relationship between people and environment must develop if many of the issues discussed in this text are to be solved. A first step in reaching such a goal is to establish appropriate frameworks within which the people–environment relationship can be analysed.

1.2.1 Systems theory

Systems theory is a useful starting point for a discussion of holism because, in an environmental context, it attempts to define the interrelationships that occur in nature. The idea was initiated by a biologist, Ludwig von Bertalanffy, in the 1920s, as a means of assessing the laws governing the life of individual organisms. It was later adopted in 1949 by the newly developing subject of cybernetics, which is the study of regulating and self-regulating mechanisms in nature and technology. In essence the systems approach encompasses the view that changes in one component of an interlocking complex will promote changes in all of the components to some degree.

In the early 1970s this theory was formally proposed by Chorley and Kennedy (1971) as an appropriate way of analysing geographical phenomena. It was presented as a means of segmenting a complex whole, i.e. the environment, into recognizable units or subsystems. In principle, some of these units could relate to physical interactions and others to social interactions. It thus facilitates the analysis of subsystems which follow very different rules of interaction, but it maintains an holistic approach by emphasizing the relationships between such subsystems.

Systems theory is a widely used concept in the physical sciences because it provides a framework for identifying and quantifying the components, processes and interactions of environmental systems, thereby facilitating prediction. Virtually all environmental systems are open systems in which energy and matter are imported and exported. In nature all such systems are in a state of dynamic equilibrium with a balance between the components, processes, inputs and outputs. In common with cybernetic systems, this balance is maintained by internal controls known as negative feedback loops. For example, the uneven distribution of global heat is counteracted by atmospheric circulation which carries heat away from the equatorial zone to the polar regions. Positive feedback, however, has the opposite effect and is a major agent of environmental change. For example, the loss or reduction of its protective vegetation cover will render soil susceptible to erosion which could, in turn, prevent revegetation. Unless events are catastrophic, e.g. volcanic eruptions and earthquakes, the impact of positive feedback usually occurs gradually. This is because environmental systems have some degree of resistance to change so that time elapses between the stimulus, or input, and the output, or response.

Over geological time the major stimulus causing positive feedback in environmental systems has been climatic change. This is discussed in relation to the most recent geological period, which has witnessed the extension and contraction of polar ice caps, in Chapter 3. Climatic change has also been responsible for many of the environmental changes of the present interglacial period, but in the last 10 000 years the most significant progenitor of positive feedback has been human activity. Of the many different types of environmental systems that exist, virtually none has remained unaffected by anthropogenic activity. Much of this has been and continues to be deliberate, such as the past deforestation of Europe (Chapter 3) and the current deforestation of the tropics (Chapter 5). Much of it is caused inadvertently; the ecological impact of fossil fuel consumption has recently become manifest as acid-stressed ecosystems (Chapter 11). Only now is society beginning to contemplate the potential global warming that might ensue from the release of carbon dioxide and other so-called 'greenhouse gases' into the atmosphere (Chapters 3 and 4).

These issues, and many others, are a result of the disruption of regulatory mechanisms the earth surface systems. They have become issues because positive feedback has been sufficiently intense or prolonged to overcome resilience and create observable and often detrimental environmental change. Systems theory thus provides a practical framework within which to develop solutions to extant problems, and to prevent new issues developing, although this requires a thorough investigation of the interrelationships within and between earth surface systems. The identification of thresholds would go a long way to preventing further resource degradation and is vital for the adoption of sustainable practices. For example, how much vegetation cover can be removed before soil erosion becomes significant? Or, what sort of crop can be grown which affords a particular soil the greatest protection against erosion?

Systems theory has also been promoted as an approach to understanding social interactions. It has not proved to be particularly appropriate for the analysis of social and economic interrelationships for two reasons. Firstly, although social scientists have made progress in understanding the structure and behaviour of some individual sets of social, political and economic relationships, their ability to specify the interrelationships between such social subsystems is too limited for the application of general systems theory. This does not mean that systems theory is necessarily unsuitable in principle, but that it is not a practicable method of analysis at present.

A more fundamental criticism is that systems theory finds difficulty in dealing with the inventiveness of human society. Despite inadequacies in understanding societies, many of the social and political relationships and structures that human beings create can be analysed to provide some understanding and prediction of the future. This is clear, for example, in the case of economic relationships for which successful short-term predictions can be made about people's reactions to economic changes. However, human beings can produce novel ways of organizing themselves and can change their values and political organization. Such changes can mean that the old analyses no longer work and that new approaches to understanding and predicting social change are required. Although systems theory is able to analyse dynamic environmental systems, it cannot readily cope with the innovation that is part of social systems.

Despite enthusiasm for it among social scientists in the 1950s and 1960s, systems theory has not so far generated powerful insights into human society. Nor has it been possible to show that the systems approach to understanding society, through examination of the simultaneous interaction of complex processes, is more effective for analysing social change than existing methodologies. This is exemplified by the development in the late 1960s of a model based on 'the scientific method, systems analysis and the modern computer' (Meadows *et al.*, 1972). This attempted to predict the effect on the environment and on human society of the then existing global trends in population growth, industrial production, food production, use of non-renewable resources and pollution. This was the famous Club of Rome model, which predicted exponential growth of the world system followed by collapse. The model is now regarded as seriously flawed because it adopted an inappropriate approach to the analysis of social interactions (McCutcheon, 1979).

Ironically, despite this lack of success, there is a re-emergence of interest in constructing such models. This has arisen because of renewed concern about global environmental problems and the need to make forecasts of likely future scenarios given different policy actions. There is a risk that if such models are used uncritically, they will produce misleading but superficially convincing results which may suggest inappropriate courses of action. This is not to suggest that modelling complex interactions is not worth doing. Rather, it implies that such modelling can only provide partial analyses at present and that its results should be viewed with caution.

1.2.2 The ecosystem concept

The ecosystem concept is not divorced from systems theory insofar as it represents a type of environmental system that includes living organisms. In some respects the concept is better suited than systems theory to provide a basis for a holistic framework to unite environment and society. This is because people are functional components of ecosystems, as are plants and animals. From a temporal perspective human beings have, for a large proportion of their history, been integral components, rather than controllers, of ecosystems (Chapter 3).

The term 'ecosystem' is a contraction of the phrase 'ecological system', which derives from the study of ecology. The latter was defined by a German zoologist, Ernst Haekel, in 1869 as 'the entire science of the relations of the organisms to the surrounding exterior world, to which relations we can count in the broader sense all the conditions of existence. These are partly of organic, partly of inorganic nature'. Ecology thus emphasizes interrelationships between biotic and abiotic environmental components, which are described by the term ecosystem that was coined by the British ecologist, Sir Arthur Tansley, in 1935. Ecosystems, like general systems, consist of components between which there are exchanges (or processes) that exist in a state of dynamic equilibrium if undisturbed. This state is maintained by negative feedback and is altered by positive feedback. Again, human activity is responsible for much of the positive feedback which creates environmental change if thresholds are transgressed.

Central to the processes that occur in ecosystems are energy flows. Firstly, solar energy fuels the atmospheric system, which then determines global climates; these, in turn, impose constraints on plant growth, or, conversely, climatic conditions are among the most important environmental factors to which plants are adapted. In addition, green plants are the means by which the earth's biota (all plants and animals, including humans) is sustained. In photosynthesis, carbon dioxide from the atmosphere and water from the soil are combined using solar energy to produce carbohydrates. Solar energy is thus converted to food energy and green plants are described as autotrophs (self-feeding) because of their ability to transform one type of energy to another. The organic matter so produced is gross primary productivity. The food energy that remains after taking account of the energy used by the plants themselves, for metabolic processes such as respiration, is net primary productivity. This provides a food energy source for heterotrophic organisms that cannot produce their own energy. Primary productivity is thus the basis of all food chains and food webs, i.e. the trophic relationships that characterize all ecosystems at any scale. All animals and humans depend on this ability of green plants to provide food energy. Science has not yet discovered how to improve the process, although this may be possible via genetic engineering (Chapter 9), or indeed even to mimic it. Moreover, many of the environmental issues discussed in this book, notably deforestation (Chapter 5), soil erosion (Chapter 14) and desertification (Chapter 15), are the direct result of society's attempts to channel energy flows by replacing natural ecosystems with agricultural systems.

The importance of the ecosystem concept to an examination of the people–environment relationship is exemplified by the work of a US ecologist, Eugene P. Odum, who has emphasized the central role of energy flows (Odum, 1975). This approach is illustrated in Fig. 1.1, which shows that the fuel-powered urban–industrial systems (cities and societies) are energy consumers and waste producers. The energy consumed derives from natural ecosystems, agricultural systems and the products (fossil fuels) of ancient ecosystems, emphasizing the dependence of society on its environment for food and fuel energy. The production and use of this food and fuel energy generates waste products that cause various types of environmental problems (e.g. Chapters 4, 11 and 12). These activities can also impair remaining natural ecosystems such as wetlands (Chapter 13).

The ecosystem concept provides a viable framework for dissecting the energy interdependencies of human societies and the environment. Chapter 2, in which sustainable agricultural production is discussed, also illustrates that it is valuable to examine the energy inputs and outputs of human production systems. However, it is difficult to use the ecosystem concept to examine people–environment interactions where an analysis of potential human reactions and adaptations is required. For some purposes it is helpful to consider human beings as just another species, but for other purposes it is necessary to recognize the power, variety and complexity of the values and motivations and social and economic structures of humans, and their ways of interacting with the environment. The problems of applying such concepts to the analysis of human behaviour have been pointed out in criticisms of the human ecology approach proposed by the Chicago school of urban sociologists (e.g. R. E. Park and E. W. Burgess). In the 1920s and 1930s they adapted ideas concerning the invasion and succession behaviour of plant communities to explain the changing pattern of human land use in the city. Whereas this work has proved valuable to urban analysts, attempts to draw analogies between human and plant behaviour have been abandoned, along with suggestions that humans are driven by similar competitive imperatives to those of plants and animals [see Robson (1969) for criticisms of the Chicago school].

1.2.3 The Gaia hypothesis

This hypothesis was postulated in the mid-1960s by James E. Lovelock, an independent UK scientist based in Cornwall. The hypothesis states that the temperature and

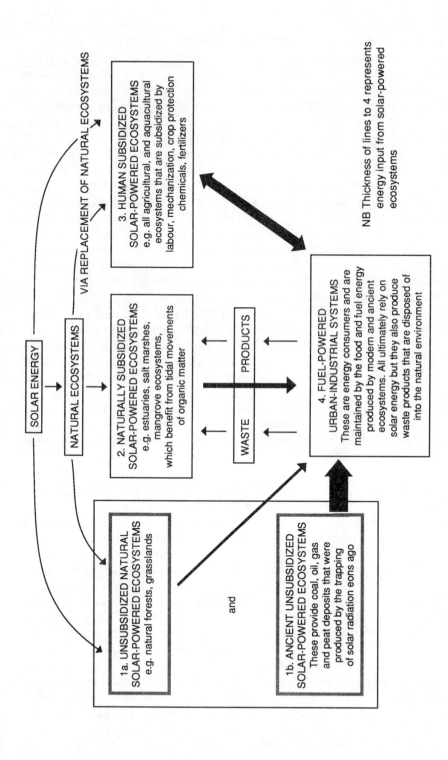

Figure 1.1   An interpretation of the energy flows of Odum (1975), juxtaposing natural ecosystems, agricultural systems and urban–industrial complexes.

composition of the earth's surface, including those of the atmosphere, are regulated by the biota, the evolution of which is influenced by these same factors that it regulates. In other words, the earth's biota acts as part of a control system to maintain conditions at the earth's surface that will support life. Since its initial publication, this hypothesis has aroused strong supporting and opposing debate. Criticisms of Gaia (the Greek name for the goddess of the earth) centre on Lovelock's original idea of Planet Earth as a living organism, apparent conflicts with Darwinian evolutionary theory, difficulties of testing the hypothesis *in toto* and its alleged teleological aspect, which suggests purposeful change and a predetermined end.

Lovelock's view of Planet Earth as a living organism is indeed difficult to countenance when such a large proportion of it consists of inorganic materials. Adherents of Gaia have, however, modified this view by expressing the idea of Planet Earth as a single system wherein the biotic components act as regulators of the interactions between the inorganic components. For this there is considerable evidence. Certain types of bacteria, for example, in soils and the root nodules of specific plants, fix nitrogen from the atmosphere and convert it to nitrates which can then be used by plants as a source of nitrogen. In addition, plants remove carbon dioxide from the atmosphere in the process of photosynthesis and return oxygen; they thus influence the composition of the atmosphere.

In relation to evolutionary theory, Lovelock (1990) argues that traditional theory is misguided because it ascribes a passive role to the biota throughout earth history. Both Darwinian evolutionary theory, which involves steady, gradual evolution, and punctuated evolutionary theory, which involves periods of rapid evolution separated by long periods of stability, suggest that evolution occurs as a response to the physical environment. Gaia, Lovelock believes, reconciles these views by suggesting that evolution has occurred by both processes. However, he also argues that both life and the environment evolve together, not that one controls the other but that they are coupled, with long periods of relative stability, or homeostasis, which are interrupted by periods of rapid change in both organisms and environment. Thus, the evolutionary paths of the environment and of organisms are interdependent and reciprocal. Under such circumstances, natural selection would occur just as Darwin suggested. Darwinism and Gaia are, therefore, not mutually exclusive.

The question of testability is more vexed. There are aspects of the Gaia hypothesis that can be tested by observation, but it is impossible to ascertain whether or not the coupling of organisms and environment is the most important factor that has determined the nature of life on earth and environmental characteristics. One aspect of this debate is the fact that the evolution of photosynthesizing green plants 2.5×10^9 years ago created an oxygen-rich atmosphere. This presumably allowed evolution to take a different course than it would otherwise have done had the atmosphere remained anaerobic. On a shorter time scale, the analysis of air encapsulated within polar ice cores (Chapter 3) indicates that glacial–interglacial cycles were characterized by changes in the concentrations of atmospheric carbon dioxide. Less atmospheric carbon dioxide in glacial periods may well have been due to a greater abundance of photosynthesizing organisms (as well as changes in oceanic circulation and the burial of organic material). The complexity of such interactions and the many different inputs and outputs involved over geological time, however, mean that a complete analysis is impossible. What is perhaps most crucial is whether society, as part of the biota, is changing, or will change, the environment, especially the atmosphere, to such an extent that it becomes unsuitable for humans, who, in the Darwinian sense, then become 'unfit'.

There are observable precedents which indicate that this is possible. For example, the succession of vegetation types that colonize newly formed substrates, e.g. glacial or volcanic deposits, cause physical and chemical transformations in those substrates to such an extent that a further range of species can then invade. The new arrivals compete so well with the original colonizers that the latter are either reduced to low populations or are relegated. Even this comparatively simple successional process is not easy to model or to analyse because of the many factors involved. This example does not necessarily involve evolution, but it illustrates the importance of the biota as an agent of environmental change and the ability of organisms to create an environment that is no longer suitable for their maintenance.

The final criticism of the Gaia hypothesis as teleological is also refuted by Lovelock (1990). He dismisses the idea of the earth and its biota as a sentient system with a predetermined path and a predetermined end. It is also unfortunate that Gaia's credibility has been tarnished by harnessing it to such a premise and this is a major reason why many scientists refuse to consider it as a feasible hypothesis and, equally unfortunately, why it has been adopted by the more extreme green movements. If Gaia is teleological then it should, once the predetermined pathway is established, be possible to predict future environmental changes. This is not so and the fact that the Gaia hypothesis cannot be used to predict future changes is a much more fundamental and important criticism.

What then, is the relevance of the Gaia hypothesis for providing a framework for describing and analysing the people–environment relationship? In this context, Gaia's greatest asset is holism, a concept that it shares with systems theory and the ecosystem concept (Sections 1.2.1 and 1.2.2). It not only suggests a close coupling of life and the environment, but emphasizes that this has exerted control on the development of the interrelationship. Control does not mean a one-way influence or dominance, but the operation of regulation within and between the life-support systems that maintain the earth and its biota. The examples given of biotic influence on atmospheric composition testify to this. The coupling of biota and environment in a non-passive way is also exemplified by the operation of biogeochemical cycles which, like energy flows, are vital processes for the maintenance of the global ecosystem. Biogeochemical cycles are the flows of materials that occur within the environment at all scales. Most of the naturally occurring elements (e.g. carbon, hydrogen, oxygen and nitrogen) have established pathways which involve the movement or flux of materials between reservoirs or stores. The latter are the atmosphere, biosphere and lithosphere, between which there are exchanges, or fluxes, brought about by agents of transfer that are often biotic.

Figure 1.2 illustrates the global carbon cycle, which is intimately associated with energy flows as carbon is a component of all organic molecules produced by photosynthesis. The biota, therefore, play a major role in regulating this global biogeochemical cycle, although it is not the only control as is the case in the biogeochemical cycles of other elements. Figure 1.2 illustrates, simplistically, a range of complex interactions that comprise the real world. The regulation of global biogeochemical cycles is, therefore, the embodiment of Gaia. Moreover, there is a

12

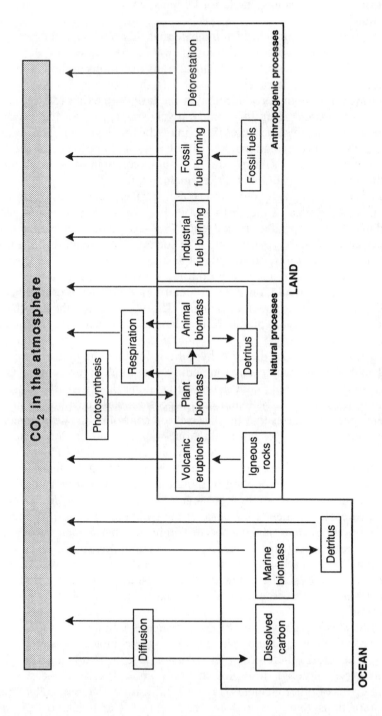

Figure 1.2   The global biogeochemical cycle of carbon.

considerable degree of control on the biogeochemical cycling of carbon by humans. This biogeochemical cycle is not only concerned with contemporary exchanges of carbon between reservoirs, but also with exchanges between ancient reservoirs of carbon laid down as fossil fuels (Fig. 1.1). Such fluxes would occur naturally by weathering but are accelerated by society's use of fossil fuels. Agriculture, both directly and indirectly, also influences the global carbon cycle. The replacement of natural ecosystems, especially forests, by agricultural systems has released and continues to release carbon from the biotic reservoir into the atmospheric reservoir. Human activity also impinges on many other biogeochemical cycles by exploiting mineral resources and producing waste products. The Gaia hypothesis is orthodox, in this context, because it formalizes this close coupling between environment and biota. It is especially relevant to geography, which has always attempted to reconcile holism with dualism. It is, however, no more or less ideal as a framework for examining the people–environment relationship than is systems theory or the ecosystem concept.

Systems theory suggests that any complex of human or physical relationships can be analysed through the identification of subsystems and their positive and negative feedbacks. The ecosystem approach suggests that the important elements to analyse are the relationships within and between biotic and abiotic components of the environment. It emphasizes the importance of energy flows between the elements of an ecosystem. It is not clear that Gaia offers any advance on these methods of approach. The Gaia hypothesis is valuable in stressing that human beings are only a small element of a complex system and an element that may well have only a limited life. However, it does not offer any advance over the systems approach or the ecosystem approach in its implications for the methods of analysis that should be used to disentangle the interactions between people and the environment. In particular, it offers few ideas on ways to approach the analysis of the impact of change in the non-human biota and the environment by human societies or vice versa. The strengths of Gaia are that it suggests new kinds of interrelationships to examine and it has certainly provoked interest, debate and rethinking of theories about the processes governing change in the earth's organic and inorganic components.

1.2.4 Environment and society

The discipline of geography has always focused on explaining the relationship between people and the environment. An early view of this relationship was that of *determinism*, that is, that the nature of society is determined by the environment. This approach drew on Darwinian evolutionary theory, which suggests that the environment is the major determinant of the path of evolution (Section 1.2.3), and concepts of social Darwinism, which suggested that those societies and groups which are best able to adapt to the environment will become dominant. The idea of determinism was put forward in geography by Friedrich Ratzel (1844–1904) and popularized among English-speaking geographers during the early 20th century. The approach was criticized for ignoring both the capacity of human beings to make diverse choices, given the same environmental conditions, and to modify those conditions.

Consequently, a new theory called *possibilism* was advanced to express this two-way relationship. It was proposed by the French geographer, Febvre (1878–1956), and advanced the view that although the environment presents limits to human activity, it also offers a range of possibilities within those constraints. The French geographer Vidal de la Blache (1845–1918) suggested that through adaptation to and modification of specific environments, people produce distinctive ways of life (*genres de vie*) in different environments. He used this approach to study the *pays* of France, which were then rather isolated, rural, agricultural regions. However, although his approach suggested convincing interpretations of the social and economic life of the peasantry within each *pays*, it proved impossible to apply to the study of industrial economies where the physical characteristics of an area might provide few clues as to the reasons for the activities carried out within it. Thus de la Blache's approach is not appropriate for the study of any society affected by industrialization or the economic relations of capitalism.

During the second half of the 20th century human geographers have therefore turned away from discussion of the relationships between people and the environment. They have preferred to explore either the ways in which spatial separation affects the organization of human activities or the ways in which social and economic relationships play themselves out over space and affect the societies of different places. During the 1980s the concept of locality has refocused attention on the distinctive characteristics of particular places, but in a different fashion from the approach of de la Blache.

The idea of locality owes much to the work of Massey (1984), who suggested that the economic and social characteristics of places resulted from the impacts of successive phases of economic organization. These phases are characterized by a distinctive 'spatial division of labour' between different places. For example, the pattern of industrial location that characterized the early Industrial Revolution in Britain, in which the north of England was the main site of manufacturing activity, was transformed in the latter part of the 19th century to one in which the south of England played a more important role in production activity. Each phase of economic organization assigns a particular economic role to different places, which then affects social relationships within each area. The effects of successive spatial divisions of labour will be superimposed within a particular place, resulting in a unique combination of social and economic characteristics.

However, places and their inhabitants are not merely passive recipients of the impacts of change. They react to and influence these impacts. Moreover, the way in which a new spatial division of labour develops will be influenced by the characteristics of places produced by preceding phases of economic organization. This approach says nothing about relationships between people and the environment. However, it could be extended to provide a limited framework for examining the impacts of general social, economic and environmental changes on the social and environmental relationships in a particular place.

A more appropriate framework may be based on the *political economy* approach that has been developed in geography over the last 20 years. In the early 1970s human geographers became interested in examining how power relationships influenced patterns of development and spatial organization within society. They turned to the work of sociologists and political scientists and especially to the work of Marx (1818–83) and Weber (1864–1920). Although Weber developed his ideas in opposition to those of Marx, both were concerned to elucidate the structures of power within society and, in particular, to explain how certain groups, or *classes*, were able to gain power over others. Each arrived at different answers, but their work and debates over class differences have formed the basis for the development of the political economy approach in human geography. The approach has been extended to analyse the role of racism in creating oppressed social groups and the original emphasis on class divisions has been changed by feminist work which argues for the significance of the divisions of power between women and men to any understanding of social relationships.

This political economy approach suggests that if the social relationships and processes of change within a given society are to be understood, then it is necessary to examine the nature of the economy and the power relationships that it sustains. In particular, it is necessary to examine the divisions of property ownership (both land and capital); the structure and conduct of relationships between employers and workers; the structure and conduct of relationships between women and men and between different ethnic groups; political groupings (from parties, and ethnic or locality based pressure groups to trade unions) and the organization of state power. The relationship of the values of the society to these power divisions must also be explored.

This approach suggests that to understand people–environment relationships it is necessary to examine how the social relationships of power, listed briefly above, relate to the control and use of environmental resources. Geographers who are concerned with issues of resource use and misuse in the developing world have already carried out studies in which a political economy approach is used to understand the interplay of environment and society. This is exemplified by the work of Blaikie (1985) on soil erosion (Chapter 14). The value of this approach can also be seen in Chapters 5, 7, 15 and, in a different context, in Chapters 18 and 19. Furthermore, the question of whether humanity can create sustainable global development rests on its ability to devise social structures and relationships which do not lead to the exploitation and impoverishment of the environment.

1.3   ABOUT THIS BOOK: TIME, PLACE AND SCALE

One of the common themes that emerges from the frameworks discussed above is the role of humans as part of the biota that interact with, and are dependent upon, the environment. Examining this interplay over time (Chapter 3), it is possible to compartmentalize the development of society into three stages. Before the inception of permanent agricultural systems 10 000 years ago, humans existed as integral components of the biota, manipulating natural resources without detrimental effect. They were constrained, by relatively low technological skills, in their ability to appreciably change the environment. With the onset of permanent cultivation came a new range of skills directed towards the manipulation of energy flows. Humans thus became controllers of their environment, a role that intensified as resource use and agriculture developed. Ideas that nature should be subjugated for the development of society accompanied these skills, and with industrialization in the 18th century came new technologies that allowed this subjugation to continue.

The 1980s witnessed the onset of a new age of awareness which acknowledges that such subjugation, manifest as environmental degradation, is no longer tenable for the long-term maintenance of society. Although this 'new age' remains, in the early 1990s, rhetoric rather than reality, it marks the beginning of a change in attitudes that is embodied in the concept of sustainable development. This is why, in Chapter 2, this idea is discussed in detail. Society can never return to pre-industrial or pre-agricultural conditions; even if this was possible it would not be desirable. The idea that sustainable development conveys is the need to embrace a more conservational attitude towards resources, including those resources that are impaired by the disposal of waste products, within existing technological ability.

The environmental impact of human communities is not globally uniform, but the sum total of this impact is creating change on a global scale that, in the spirit of the Gaia hypothesis, is altering the earth's life-support systems. To appreciate the magnitude of this change and the factors involved in it, this book is divided into two main sections. The first relates to global issues that pertain to the physical environment and to society, whereas the second part is concerned with more localized issues.

1.3.1 Global issues

In a temporal context (Chapter 3) human communities began to exert a major influence on global environments with the onset of permanent agriculture 10 000 years ago. Although this was not the 'revolution' that it has often been called, as it grew out of sophisticated hunter–gatherer strategies, it created environmental and cultural changes of an unprecedented nature. It developed in tandem with permanent settlement, facilitated the division of labour and marked the emergence of society as a significant force on the landscape. The onset of permanent agriculture also marked the first stage of global deforestation, a process that began in localized areas but which is now a global problem (Chapter 5). Deforestation together with fossil fuel energy production (Chapter 8) are among the most important issues of the 1990s, not least because of their effect on global climate via their impact on the biogeochemical cycle of carbon (Chapters 3 and 4). Global warming could influence all aspects of environmental and cultural processes, although it remains controversial because, as yet, there is no definitive evidence that it is becoming reality. Deforestation (Chapter 5) is also linked with species extinction and declining gene pools; both represent the diminution of resources, the potential of which has not been fully realized. What is perhaps most disconcerting about the current deforestation of tropical forests is the fact that many of the agricultural systems replacing them are also failing. For example, large-scale ranching in Central and South America has, in general, not provided the salvation to national economies that it initially promised.

Population growth and demographic trends are also global issues. Burgeoning populations in the developing world and changing demographic patterns in the developed world, are both a matter for concern. As discussed in Chapter 7, increasing populations inevitably mean greater pressure on resources, but the relationship with environmental damage is complex. Individuals in the developing world exert different pressures on resources than do those in the developed world, where income, age and family structures will influence environmental impacts in the 1990s. At the same time, there are scientific and technological developments that may offset some of the problems that affect humanity. Biotechnology and genetic engineering (Chapter 9) provide new opportunities for agriculture which, in principle, could lead to the tailoring of crops to the environment rather than vice versa, which has been the traditional approach. Biotechnology is providing new ways to treat waste, including the recycling of useful materials. In the longer term such developments should contribute to sustainable resource use.

Global energy consumption (Chapter 8) is likely to continue to create environmental problems. Although there are trends to reduce fossil fuel energy consumption in the developed world, there are counteracting trends in many developing nations as they industrialize. The enhanced greenhouse effect is thus unlikely to be appreciably mitigated, nor is the problem of acidification (Section 1.3.2). Nuclear power is not a viable alternative on a large scale, even if the problems of waste disposal could be solved. This is particularly so for the developing world because of the huge costs involved in constructing and maintaining nuclear reactors. Biomass fuels may be acceptable alternatives in some parts of the world, but in others the use of fuelwood is creating environmental degradation.

However, the last decade has witnessed the rise to prominence of environmental issues in local, national and global politics and in the media. This is important because it means that society is becoming better informed about such problems and, as a result, consumer preferences are beginning to change. In addition, individuals are more prepared now than a decade ago to participate in recycling programmes. In relation to politics, the Montreal Protocol of 1987, to curb chlorofluorocarbon emissions for the protection of the stratospheric ozone layer, was the first international agreement on an environmental issue. This is particularly important because it formalizes the recognition that environmental problems know no national boundaries. There has also been an increase in support for green political parties (Chapter 10). This is also a welcome trend, not least because it has prompted mainstream political parties to pay more than lip-service to environmental issues. Globally, such movements are increasingly likely to influence decision-making bodies.

1.3.2 Local issues

Most of the environmental impacts of human communities are manifest at least initially as local environmental changes, although the distinction between local and global issues is rather arbitrary. In relation to the physical environment, local issues include fossil fuel consumption which causes acidification (Chapter 11), and industrial processes and transport which release toxic materials (Chapter 12). Agricultural systems also create environmental issues, for example the diminution of habitats such as wetlands (Chapter 13), soil erosion (Chapter 14) and desertification (Chapter 15). The scale on which some of these processes now occur means that they are globally significant. Some issues are the result of a combination of factors; cultural eutrophication, for example, is the product of agriculture, domestic and industrial detergents and sewage output from urban concentrations of people. Urban centres also suffer from a range of environmental problems caused by waste generation, industrial activity and transport. Inevitably, these result from large concentrations of people and are thus localized. Social and cultural patterns and their temporal changes also influence the fabric of urban–industrial centres and generate a further range of issues, such as urban building decay and polluted urban atmospheres (Chapters 17, 18 and

19). Moreover, the wealth generation of such centres, especially in the developed world, has created a further range of environmental issues linked to tourism (Chapter 19).

Currently, acidification is mainly confined to the developed world in the northern hemisphere, as it is here that the Industrial Revolution of the 18th century began. Two centuries of fossil fuel use have led, almost imperceptibly, to the crossing of ecological thresholds in freshwater ecosystems on acid bedrock in Europe and North America, and possibly in forests in these areas as well (Chapter 11). The continued release of sulphur dioxide and nitrous oxides has given rise to acidified lakes with impoverished biotas as well as the decay of urban buildings. In the longer term, acidification is also likely to be a significant problem in the developing world as nations industrialize. Fossil fuel consumption has also caused increased concentrations of lead in the atmosphere, which may be detrimental to human health (Chapter 12).

Agricultural practices have also had a significant local impact. The need to expand the area of food, fibre and animal-product output has exacted a massive toll on environmental systems. The expansion of agricultural systems in both the developed and the developing worlds has caused the demise of natural habitats, notably forests (Chapter 5). Among the most significant environmental issues created by agriculture are soil erosion and desertification. Both are widespread due to agricultural practices which are mismatched with the environment. Such practices are adopted in response to a variety of social, economic and political forces, ranging from financial incentives that promote large-scale monoculture, as in parts of Europe and the USA, to population pressures and the need to generate income by cash cropping, which leads to the cultivation of marginal areas, as in parts of Africa. Solving such problems is not simply a matter of soil conservation programmes and improved cropping or grazing regimes. A more holistic approach is required to take account of social and economic relationships. Similar points can be made about local environmental issues in the developed world where the demands for increased living space in semi-rural environments and increased car use are creating acute local problems in the rural and urban environment (Chapters 16 and 18). These are not problems that can be solved merely by new methods of transportation and communication. They require more efficiently organized settlements designed to maintain the quality of life and the resources of the built and natural environments (Chapter 17).

1.4 CONCLUSIONS

Environmental issues must rank amongst the most important problems facing politicians and policy makers in the 1990s and beyond. All of the issues discussed in this text, and there are undoubtedly many others, require solutions in the next few decades. There are no sections of society that some, or indeed all, of these issues fail to touch. As Fig. 1.3 illustrates, both local and global issues relate to three major aspects of human activity: agriculture, industry and concentrations of population. All three conspire towards environmental degradation which may ultimately create an inhospitable environment for the progenitors of these activities. Human communities may be controllers within ecosystems, but deflections of energy flows and disruptions of biogeochemical cycles have had, and will continue to have, unforeseen, detrimental effects as well as positive benefits, as the following chapters illustrate.

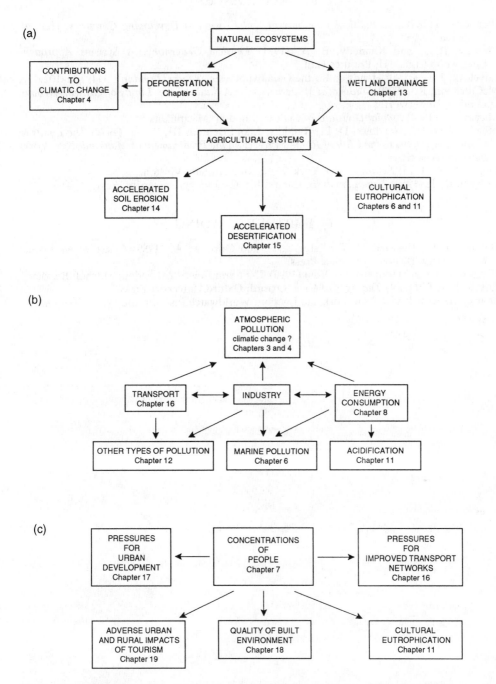

Figure 1.3 The impact on the environment of (a) agriculture, (b) industry, energy and transport and (c) concentrations of people.

1.5   REFERENCES

Blaikie, P. (1985) *The Political Economy of Soil Erosion in Developing Countries*. Harlow: Longman.

Chorley, R. J. and Kennedy, B. A. (1971) *Physical Geography: A Systems Approach*. Englewood Cliffs, NJ: Prentice Hall.

Lovelock, J. E. (1990) Hands up for the Gaia hypothesis. *Nature (London)*, **344**, 100–102.

McCutcheon, R. (1979) *Limits of a Modern World: A Study of the 'Limits of Growth' Debate*. London: Butterworths.

Massey, D. (1984) *Spatial Divisions of Labour*. London: Macmillan.

Meadows, D. H., Meadows, D. L., Randers, J. and Behrens III, W. W. (1972) *The Limits to Growth: A Report to the Club of Rome's Project on the Predicament of Mankind*. New York: Potomac Associates.

Odum, E. P. (1975) *Ecology*. New York: Holt, Rinehart and Winston.

Robson, B. T. (1969) *Urban Analysis*. Cambridge: Cambridge University Press.

1.6   FURTHER READING

Botkin, D. B., Caswell, M. F., Estes, J. E. and Orio, A. A. (1989) *Changing the Global Environment*. Boston: Academic Press.

Goldsmith, E. and Hildyard, N. (Eds) (1990) *The Earth Report 2*. London: Mitchell Beazley.

Lovelock, J. E. (1989) *The Ages of Gaia*. Oxford: Oxford University Press.

State of the World 1990. New York and London: Worldwatch Institute and W. W. Norton.

CHAPTER 2

Sustainable Development

J. G. Soussan

2.1 INTRODUCTION

This chapter explores ideas which draw together many of the themes found in this book. These ideas are explained in some depth, and like all concepts these may at times appear to be rather abstract, but the reader will find that the arguments set out here question some of the fundamental principles which underpin our way of life. The way people think about the environment, about economic development and about the links between the two is changing considerably. The end of the 1980s saw a radical reappraisal of our concerns over resource availability and use, the environmental consequences of resource exploitation and the relationship between the environment, poverty and economic change. This re-appraisal has given rise to a new approach to environment and development issues; an approach which seeks to reconcile human needs and the capacity of the environment to cope with the consequences of economic systems. This approach is called *sustainable development*. There are many problems with this approach, and these are addressed in this chapter. To date, it consists largely of broad goals, and the human institutions responsible for managing the planet are a long way from turning these goals into a clear programme of achievable actions.

Before the 1970s natural resources were seen as plentiful, and as there to be exploited as cheaply as possible in quantities which did not acknowledge resource frontiers. Little thought was given to the amounts of minerals used or forest cleared or the capacity of the air, land and water to absorb pollutants. In particular, development was seen as synonymous with economic growth (Chapter 12).

There were people who raised questions about this pattern of resource exploitation, but these tended to be voices on the margins. Authors such as Schumacher (1973), with his book *Small is Beautiful*, did have some impact, and concern with pollution and the environment led to the growth of the green movement in many developed countries and to more effective legislation to control environmental impacts in countries such as the USA and Sweden. In most cases environmental legislation was partial and responsive to specific problems (see Chapter 12, which discusses pollution

---

Environmental Issues in the 1990s. Edited by A. M. Mannion and S. R. Bowlby

and development), and nowhere were resource or environmental issues central policy concerns (Chapter 10).

There was very little fundamental questioning of the economic system producing these impacts until the late 1960s, when an environmental movement began to make itself felt in Europe and the USA. The world was then hit by the hammer blow of the first oil crisis in 1973 (see Chapter 8 on energy resources). This traumatic shake-up of the general complacent acceptance of resource abundance coincided with the publication of *The Limits to Growth* (Meadows *et al.*, 1972), a study which claimed to show that the existing pattern of resource use would lead to a collapse of the world system within the next century. These events created a widespread perception of resource scarcity. Commodities of all sorts were stockpiled, prices rose with fears of other producer cartels to rival the Organization of Petroleum Exporting Countries (OPEC), and it was generally believed that the world was faced with a future of scarcer, more expensive resources.

This viewpoint dominated throughout the 1970s and much of the 1980s, and led to a much wider acceptance of resource issues as central economic and political concerns. It was, however, based on a false premise; that resources (and especially mineral resources) were about to run out.

These pessimistic predictions have proved to be false. Resources such as minerals may be finite, but they are a long way from being exhausted. The amounts available for human use are now understood to be much greater than assumed in the past. This reflects a better knowledge of resource availability; knowledge which is itself largely a consequence of efforts made in response to scarcity fears. Higher prices, fears about the security of supplies, technological developments, more efficient resource use (especially for energy) and, above all, renewed exploration efforts, have meant that mineral resources are discovered faster than they are used, and that those consumed are used more efficiently. This, combined with the economic depression of much of the 1980s, produced a glut of most minerals on the world market. Commodity prices collapsed in the early 1980s, and are now cheaper than in the past (Fig. 2.1). Indeed, the problem for many countries (and in particular much of the developing world, which relies on commodity exports) is not that resource prices are too high, but that they are too low. The 'terms of trade' for mineral exporters (the prices of primary products relative to those of manufactured goods) have continually deteriorated. In the 1990s, one of the main economic problems facing the developing world is low demand for, and the low price of, their main exports. A combination of decreased revenue from primary exports and crippling debt burdens leads to declining exchange rates and acute foreign exchange shortages, which in turn leads to less money for investment and essential imports, declining government revenues, widespread unemployment and other factors which, for many countries, have destroyed prospects for economic growth. For much of Africa and Latin America, in particular, this has resulted in declining standards of living in the 1980s.

OPEC kept oil prices high for a while, but even they had to bow to the inevitable (see Chapter 8). The collapse of oil prices in 1986 was as dramatic (if less publicized) as the price rises of 1973–74 and 1979–80. The era of resource scarcity was over. At the same time, new concerns were emerging over the future of the global environment. These concerns centred not on the future supply of resources, but on the impact on the environment of using these resources. There is now less concern about

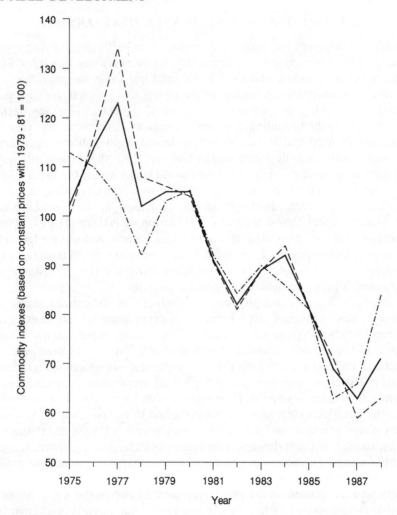

Figure 2.1 World commodity indices and prices, 1975–88. Adapted from World Resources Institute and International Institute for Environment and Development (1988), Table 14.3, pp. 240–241, 'World Commodity Indexes and Prices, 1960–1987', and World Resources Institute (1990), Table 15.4, p. 250, 'World Commodity Indexes, 1975–1988'. (–.–.–) Metals and minerals; (—) all non-fuel commodities; (– – –) agricultural commodities.

the amount of non-renewable resource stocks than about the quality of renewable resource flows. Pollution of the world's atmosphere, waterways and oceans, declining forests and woodlands and degrading land and soil quality are the resource concerns of the 1990s (these issues are explored in Chapters 4, 5, 6, 8, 11 and 12). Today there is little concern over the effects of resource scarcity on economic processes; rather the great challenge facing the world is to cope with the impact of economic growth on environmental processes. This change in emphasis is reflected in the approach which has become known as sustainable development.

2.2   ORIGINS OF SUSTAINABLE DEVELOPMENT

Sustainable development has emerged during the late 1980s as a unifying approach to concerns over the environment, economic development and the quality of life. It has meaning at global, national and local levels, and has value for providing a context within which to understand and influence the trajectory of growth and change in both developed and developing countries. As an idea, it needs to be approached with caution. It is rapidly becoming the new orthodoxy, and as is the case with all bandwagons it is used and abused to justify development policies which are clearly neither sustainable nor developmental. Despite this, the notion of sustainable development does offer a vision of a future world which meets the needs of all of the global community without undermining the integrity of the environment.

The idea of sustainable development was first used in the *World Conservation Strategy* [International Union for the Conservation of Nature (IUCN), 1980]. This first formulation stressed sustainability in ecological terms, and was far less concerned with economic development. It argued for three priorities to be built-in to development policies: (a) the maintenance of ecological processes; (b) the sustainable use of resources; and (c) the maintenance of genetic diversity.

The emphasis was on the physical environment in its current state, and this formulation was criticized for being anti-developmental. It saw economy–environment relationships simply in terms of the human impact on the environment as a static end-state, and tended to imply that any impact was negative. As such, it tended to attack symptoms, rather than causes, of environmental degradation. For example, it led to simple calls for stopping all tropical forest clearance without considering the forces leading to clearance, whether they could be controlled and, indeed, whether forest clearance was always a bad thing. As Chapter 5 shows, there are many places where some form of forest clearance is a logical and desirable use of resources. Indeed, without deforestation there would be no agriculture. It is better to attempt to manage the process in a sustainable way than to try to impose prohibitions which could never be enforced.

Poverty and the actions of the poor were seen as one of the main causes of non-sustainable development, rather than recognizing that poverty and environmental degradation are both consequences of existing development patterns (a theme developed in Chapter 7 on population). This lack of vision of the relationship between the economy and the environment led to a re-formulation of the concept of sustainable development to reflect concerns over what many commentators saw as an 'anti-poor' bias in the IUCN (1980) report.

This re-formulation led to the creation of the World Commission on Environment and Development (better known as the Brundtland Commission after its chair, Gro Brundtland of Norway) in 1984. The Commission initiated studies which culminated in the publication of *Our Common Future* (the Brundtland Report) in 1987; a report which has set the benchmark for all future discussions of sustainable development (World Commission on Environment and Development, 1987).

Our Common Future defines sustainable development as:

> Development that meets the needs of the present without compromising the ability of future generations to meet their needs.

In essence, the vision of sustainable development set out in the Brundtland Report is a call for policies which recognize the need for economic growth, and seek to maximize growth, but which do so in a way which does not jeopardize the position of vulnerable people or deplete the future viability of the resource base. It calls for a different attitude to economic development, in which the *quality* of growth is seen to be as important as the *quantity* of growth.

This theme is explored in Chapter 17 on sustainable urbanization, in which the value and dynamism of the city as a built environment is discussed in the context of the developed world. As this chapter shows, it is important not to view all forms of development or change as negative. Human intervention can, and in many cases does, create a more stable, productive and sustainable environment than that found in nature. One way of expressing this is that humankind can 'add value' to the environment by specific forms of management. This is often achieved at the cost of lower short-term economic returns, but emphasizes that good environmental practices and sound long-term economic management go hand-in-hand.

The Brundtland Report identifies two key concepts in sustainable development policies:

(a) The basic needs of all people must be met in a way which provides for their needs with security and dignity—in the world today, where the needs of so many are not met, this inevitably means giving the needs of the poor priority; as Chapter 7 shows, this is not just desirable in moral or equity terms, it is also good development practice.

(b) There are no absolute limits to development—development potential is a function of the present state of technology and social organization, combined with their impact on environmental resources; this is illustrated in the chapter on energy (Chapter 8), which shows that access to energy relates to the cost and availability of technologies to harness different energy resources.

The report argues that poverty, resource depletion and environmental stress arise from disparities in economic and political power. From this, it is argued that sustainable development at a global level can only be achieved through major changes in the ways in which the planet is managed.

In this approach, sustainable development is not seen as a fixed state, but rather is a process of change in which each nation achieves its full development potential, while at the same time building upon and enhancing the quality of the environmental resources on which development is based. This ambitious goal will require different forms of resource exploitation, investment patterns and decision-making processes, technological development and institutional change. The Brundtland Report advanced the following list of objectives for sustainable development policies:

(a) reviving economic growth
(b) changing the quality of growth
(c) meeting essential needs for jobs, food, energy, water and sanitation
(d) ensuring a sustainable level of population
(e) conserving and enhancing the resource base
(f) re-orienting technology and managing risk
(g) merging environment and economics in decision-making processes.

This is a long list of ambitious goals, many of which are imprecise and some of which are not mutually consistent. They reflect the problems of the mid-1980s, which were a period of economic stagnation or decline, of widening disparities between the income and prospects of the rich and poor and of increasing awareness that the form of economic development which characterized the modern world was undermining the sustainability of environmental processes.

The report advanced the following changes as needed to achieve these goals:

(a) a political system which allows effective citizen participation in decision-making processes
(b) an economic system that is able to generate surpluses and technical knowledge on a self-reliant and sustained basis
(c) a social system that provides solutions to the tensions arising from our present form of disharmonious development
(d) a production system that respects the obligation to preserve the ecological base for development
(e) a technological system that can search continuously for new solutions
(f) an international system that fosters sustainable patterns of trade and finance
(g) an administrative system that is flexible and has the capacity for self-correction.

The focus of the work of the World Commission for Environment and Development is global in scale, and the report inevitably has some of the compromises which emerge when seeking a consensus in a body of national governments. A fair criticism of the Brundtland Report is that it set a broad agenda for change without confronting the many barriers which exist to achieving these goals. As such, the report contains a series of fine statements which are impossible to disagree with, but which are too vague to be translated into concrete actions. Despite this, the conclusions are bold and ambitious, and have set the direction of the debate on the re-orientation of future development policies.

Governments and international agencies such as the World Bank can no longer argue for policies in terms of simple economic growth alone. They must account for the environmental and distributional consequences of their policies, and are increasingly recognizing that development is concerned with far more than simple economic growth. This reflects a significant change to the approaches of the past; changes which are due in no small part to the work of Gro Brundtland and her colleagues.

2.3   DEVELOPING THE CONCEPTS:
THE DILEMMAS OF SUSTAINABLE DEVELOPMENT

Despite the progress which has been made, the ideas listed in Section 2.2 remain abstract: the goals of sustainable development are defined, but mechanisms to bring about these changes are not. The challenge of the 1990s is to turn the principles of sustainable development into achievable policies which lead to concrete change. This move from principle to practice is far from easy, and poses a series of dilemmas for which there are no clear solutions.

The need for development is clear: in the modern world most of the global community live in circumstances which are morally unacceptable, while a small minority practise gross and wasteful consumerism. More than this, the development prospects for most of the world's poor are declining, while the security of their meagre standard of living is low. The poor are not just poor, they are increasingly vulnerable to different forms of economic and environmental disruption.

Central to attempts to give more meaning to sustainable development is the study of environmental economics. In a major study, *Blueprint for a Green Economy* (Pearce *et al.*, 1989), David Pearce and his colleagues at the London Environmental Economics Centre advance three precepts that lie at the heart of a sustainable development approach.

The first precept is the need to give a proper value to the environment. This includes an emphasis on the value of natural, built and cultural environments and the quality of life as a policy objective. The importance of this basic concept is clear. In the past the environment has been poorly used and abused because its true value has not been understood. The sustainable management of environmental resources is not possible until their worth is appreciated and an appropriate price is paid for the benefits received.

Valuing the environment is a main theme of environmental economics, but has led to a preoccupation with the technical problems of assigning monetary values to environmental properties or goods which do not have a market price. There are profound problems with valuing environmental resources which do not have a current commercial value: resources such as clean air and water, pleasant surroundings or even, in a developing world context, communal woodlands which are the source of subsistence products for peasant households. These problems are compounded by the uncertainties of assessing the seriousness of resource degradation. Too little is known about the extent and rate of change of degradation in such areas as tropical deforestation, soil erosion, ocean pollution and ozone depletion. This is even worse for a possible future threat, such as global warming or nuclear disasters such as Chernobyl. It is believed that these are a problem, but there are no certainties, and by the time clear evidence is available it will be too late. Expedient responses to extant problems, such as soil erosion, tend to be too little too late, and proactive responses to likely future threats are non-existent. The whole question of valuation is consequently beset with uncertainty, and any exercise to give values to resource problems is often little more than a quantification of the assumptions and prejudices of the person doing the valuation.

The second precept is the need to extend the time horizon over which development policies are viewed; the notion of futurity. Policies should be concerned with both the short- to medium-term effects of decisions (less than 10 years, as now) and the long-term impact of decisions, to include the inter-generational effects, on our grandchildren and beyond, of the way the resource base is managed. The key idea here is that the present generation should pass on at least as much 'capital' as they inherit, including environmental capital, or resource endowment and human knowledge (people as capital), as well as the physical and financial capital of conventional economics.

This again poses a series of dilemmas. The ability to predict needs, values and technical developments into the future is limited. Resources' values change over time, something which is clear for aesthetic resources such as the appreciation of landscapes, but also holds true for physical resources. For example, uranium had no resource value until the nuclear age, and is likely to have little value in the future as nuclear weapons and power diminish in importance. This leads to the key issue of who decides which trade-offs should be made (an issue explored in Chapter 20). Are people willing to sacrifice consumerism today for the welfare of future generations? What happens if some people gain and others lose from such policies? Over what sort of time horizon are trade-offs to be made? Is it reasonable to take risks if current resource exploitation may cause problems for the future? These and other questions need to be resolved before the inter-generational problem can be addressed.

The third precept Pearce et al. (1989) emphasize is the necessity to provide for the needs of the least advantaged in society, i.e. intra-generational equity. This will involve policies which require society to sacrifice economic growth to diminish the gaps between rich and poor. These ideas operate at all levels: within local communities, at a national level and between countries in the global community.

The problem is that everyone recognizes the undesirability of inequalities, but people differ on both the causes of and the solutions to these inequalities. The central dilemma is how, in a competitive world economy, can countries, firms and individuals survive to meet these goals? If the different actors adopt sustainable development policies in isolation, then this undermines their immediate competitiveness. At an international level, problems such as the debt trap force nations to take a short-term perspective even when it is clear that the actions needed to survive in the short term are not desirable in the long term. Similarly, many firms would undoubtedly be more willing to bear the cost of externalities if they did not undermine their competitiveness, or if they could be sure that other firms also had to bear these costs. In these and many other ways, reducing the gap between the haves and the have-nots in the global community is one of the intractable problems of the modern era, and clearly cannot be solved without some form of co-ordinated regulation and enforcement of standards.

Environmental economics are important, but there are some problems with the techniques which are emerging. These problems are partly technical and partly ideological. Central to this is the debate on the role of market mechanisms versus state regulations and enforced standards as mechanisms for environmental policies. Similarly, the approach to equity and meeting the needs of the poor, which is found in much of the literature on sustainable development is essentially welfare-oriented. It sees production and distribution as separate activities, with the needs of the poor somehow provided for through charitable mechanisms. In the short term this will undoubtedly be necessary in many places; the social democracies of Europe have shown how beneficial a welfare state can be in ameliorating the inequalities of the economic system. A really sustainable form of development will, however, do far more than this. Rather than simply providing for the needs of the poor, it will give the poor the capacity to provide for their own needs in a secure and dignified manner. This in turn means that sustainable development raises questions about the control of resources used in the development process and who has power over the decisions on which forms of development should take place (Chapter 20). This is essentially about giving people greater control over their own lives, and as such is about a form of development which creates democracy alongside growth and equity. This means democracy in its widest sense; not just voting to elect governments, but a process whereby people can have a say in the decisions which shape their destinies.

This, of course, is a political minefield, as it raises the whole question of which forms of political organization and economic ownership will best give people real control over their own lives. There is no easy answer to this, and indeed any attempt to create a prescriptive blueprint will deny the essence of the process, but the need to raise the question is clear. This is the central dilemma of sustainable development: what is the best way to confront and overcome the massively powerful vested interests that would feel threatened by structural changes to the status quo? As is true for other approaches to development, sustainable development is about power relationships, and the fine intentions of the approach will come to nothing until this central dilemma is recognized.

2.4 CHARACTERISTICS OF SUSTAINABLE DEVELOPMENT POLICIES

Despite the dilemmas raised in the preceding section, it is possible to define a number of common themes that should characterize sustainable development policies. Building a sustainable future will be a long and difficult journey, but the most difficult of journeys starts with a first step, and is most easily undertaken if broken down into a series of shorter, less daunting stages. Concerns over the environment and development are now at the centre of the political debate. Throughout the world politicians are responding to the fears and concerns of their constituents that the world cannot continue down the same development road. Politicians now ignore issues of the quality of life, the welfare of the disadvantaged and the preservation of the global environment at their peril (see Chapter 10).

These concerns are rarely translated into effective action. There have been some changes, such as differential taxes on unleaded petrol in the UK, legislation from the European Commission that all cars sold in the European Economic Community (EEC) must be fitted with catalytic converters and the series of controls introduced in California to try to solve their chronic air pollution problems. Such policies are at best partial, and as yet the fine words on environmentalism have not been translated into deeds. There has been a series of international conventions at which our leaders vie with each other to sound the most concerned and outraged. These global talk-shops have led to a series of calls for international consensus, funds for further research and, at best, the setting of international standards for pollutants such as carbon emissions, with little thought on how these standards can be achieved. All laudable goals, but in many cases they are a means of appearing to do something without taking the hard decisions needed to affect material change now.

There are many things which can be done *now* to start the process of change needed to build a sustainable future. For example, emissions of greenhouse gases in the UK could be drastically reduced *if* aggressive policies to control power station emissions, to encourage public transport and limit private car use and to improve energy efficiency throughout the economy were introduced. The trouble is that these steps would upset powerful vested interests, such as car manufacturers, the road freight lobby, housing developers and power generation companies. They would also require the government to interfere more vigorously in many areas of life *and* to spend a great deal more in direct investments (thereby upsetting the most powerful lobby of all—the Treasury). So far, there is little evidence that many of our political leaders have the courage to challenge these powerful interests. As with many journeys, the first steps are the most difficult.

It is clear that existing economic systems cannot survive unchanged into the future. This is true for the developed world, where the present high consumption, high waste lifestyle is depleting the planet's productive capacity and ability to support life. This is particularly true for the global environmental sinks—the atmosphere, oceans and soils—which are being rapidly degraded by pollution from the world's industries and consumers. The developed world's lifestyle is, in addition, based on an economic system which denies the billions of the world's poor a way out of their poverty; for them there is little prospect of real development, sustainable or otherwise, under the current system. This can be seen in the developing world, where both poverty and environmental degradation are the result of the history of exploitation and the patterns of economic development which now characterize them. For rich and poor, the logic of a sustainable development approach argues that the future must be based not on more of the same, but on a qualitatively different form of development. This idea of mutual dependence is central to the argument put forward in the Brundtland Report, and is one of the most important strands of the sustainable development approach.

Part of the changes in the approach to planning which must be sought is the need to build a long-term time horizon into the way needs are assessed, resources used and economies developed. This will often entail a substantial devolution of power over political *and* economic decision making. This question of control and accountability is about more than better democratic processes in the nation state (although these are important). It includes better mechanisms of control over industrial and financial systems. It is necessary to consider not only the immediate costs and benefits of investment decisions, resource developments and so on, but also their impact on future generations. Nothing illustrates this more forcefully than the spectre of global warming.

To do this, it is important to develop a vision of a future pattern of growth and change, and consider individual developments and changes introduced now in relation to this strategic goal. This extension of the time horizon for economic decisions will often depend on an ability to make political decisions which are equally far-sighted. Politicians tend to be even more near-sighted and reactive to immediate pressures than economic managers. It will be difficult to make the decisions needed for long-term planning while political systems persist where power depends on the ability to manipulate or respond to events which have a time horizon of weeks or even days. In this context, politicians cannot be blamed, as they are simply responding to the environment in which they exist. The people as a whole need to exert pressures based on an understanding of their long-term needs and interests. To do this, they will need a better understanding of these long-term issues, a greater sense of control over the process of development and change *and* a greater degree of security in their current standard of life.

As explained in Section 2.3, one element of sustainable development is to ensure that the current generation pass on to the next at least as much net 'capital' wealth as was inherited from past generations. This in turn requires people to accept a very different view of what is meant by 'capital', which is conventionally seen as simply the physical products of our economic system. It includes resource and environmental wealth, and should be viewed in qualitative terms as much as in traditional monetary terms. It also includes knowledge, social and physical infrastructure and a set of institutions which provide security for all and give people a genuine voice in the decisions which affect their lives. As such, sustainable development requires policies which produce a process of change which is sustainable in environmental, economic and institutional terms.

Sustainable development recognizes the inevitability, and the desirability, of resource exploitation. Nothing could be produced without humankind's ability to harness the Earth's resource base. In contrast to existing patterns of exploitation, sustainable economic systems will exploit resources in ways which are as efficient and sustainable as possible. For example, as Chapter 5 on forests shows, the exploitation of forest resources in the developing world can be both highly productive *and* environmentally sustainable if it is done in a way which works within the limits of the system's capacity to reproduce itself. It can produce a wide range of goods for the needs and development of forest peoples in these regions, and do so in a way which does not jeopardize the integrity of a complex and sensitive ecosystem.

An important point to emphasize is that sustainable development policies should improve environmental quality and, in many cases, could *increase* the value of the environmental stock. This is particularly true for resources such as land, forests, water and the built environment (as Chapter 17 on sustainable urbanization illustrates, urbanization can be compatible with a high quality of life). Old ideas of a trade-off between environmental maintenance and economic growth are explicitly rejected in sustainable development. Instead, the importance of economy–environment interactions as complementary features of the same process are taken as a starting point.

An essential component of this process is a need to value resources correctly. This does not necessarily mean finding a correct monetary value, but involves having a better sense of the worth of the environment. Of course, there are many problems in actually doing this if it is to be more than an empty process, but this is a challenge to which policy makers must rise.

The concept of equity, and of meeting the needs of the last first, is central to sustainable development. This entails providing for the basic survival needs of the world's poor as a first goal, but goes beyond this to strive for a standard of living which provides security and dignity for all.

These factors form the heart of a sustainable development approach. Taken together, they entail new forms of decision making and control that seek development paths which are sustainable over the long term and which are capable at any one time of making the decisions needed to create and sustain this process of change. There is no grand panacea for the mess the world is in. Sustainable development entails changes in all facets of the processes of production and consumption. These changes cannot happen overnight (which is why the idea of a process has been continually stressed), and will often entail sacrificing short-term profits in favour of long-term benefits. Implementing such policies will often involve hard decisions which challenge the interests of powerful groups or require a different form of consciousness on the part of society as a whole. It is difficult to say which of these is hardest to achieve.

A move to a sustainable future will only be possible if humankind begins to think differently about many aspects of the way goods are produced and used. Change cannot occur overnight; a point illustrated by Chapter 16, in which it is argued that the transport systems of cities must evolve in a way which brings them back into harmony with the development of the urban form. Many of the changes will only come about gradually as the effects of new policies and new resource costs work their way through into investment and consumption patterns. Such changes need time to mature if they are to be effective without seriously disrupting economic systems. The world can ill afford more of the devastating dislocation that occurred when oil prices quadrupled almost overnight in 1973. The long-term effect has been that energy is now used more efficiently, particularly in the economies of the developed world (see Chapter 8), but these beneficial effects are far outweighed by the adverse economic impact the oil shock had on the economy of the developed and developing worlds alike. In particular, many of the world's poorest nations suffered from disastrous recessions and foreign exchange problems from which they have yet to fully recover. Going for quick, easy solutions is a seductive proposition, but rarely works in practice. Change will be complex, difficult and inevitably disruptive to someone. Having said this, the non-sustainability of the present course means that the effects of not changing will be even more destructive as the world is faced by widespread environmental and, from it, economic collapse. These points are illustrated in Section 2.5, which looks at a key sphere of human activity: agriculture.

2.5   SUSTAINABLE DEVELOPMENT: THE ENERGY EFFICIENCY OF AGRICULTURE

No industry is more vital to our present and future well being than agriculture, and few show as clearly the non-sustainability of existing patterns of what is conventionally regarded as development. Agriculture provides food and a range of other vital inputs into industrial processes. It is based on a closer relationship with the resource base and is more varied and sensitive to local environmental conditions than any other industry. Humankind has been farming for thousands of years, and has shown an amazing ability to adapt and control land, water and the biosphere to increase dramatically their productive potential. Agriculture is truly one of the triumphs of human civilization. And yet throughout the modern world there is soil erosion, water pollution, increasing pests and diseases, falling yields and collapsing agricultural economies.

In the developed world agriculture has become increasingly industrialized. Modern farming systems depend as much on the chemists and geneticists of agribusinesses as on the farmer's knowledge of his or her environment. In the most intensive agriculture of the developed world, soils have become little more than a sterile medium through which to pass chemicals, and irrigation systems make rainfall an irrelevance. Many farm animals never see the light of day, spending their brief existences in buildings more like those of an industrial estate than a farm. They are force-fed processed foods mixed with cocktails of chemicals. Fruit and vegetables are loaded with other chemicals intended to improve their appearance irrespective of their nutritional content. In these and other ways, producing food in the developed world has as little to do with traditional notions of 'nature' as producing motor cars.

And for what? Throughout the developed world the farming industry has become bloated and distorted by subsidies and protectionist policies. True, more food than ever before is being grown, but it is the wrong food in the wrong places while much of the world still goes hungry. This bizarre combination of over-production and malnutrition does not even reflect efficiency, however measured. It is clearly not economically efficient, as the trend towards paying farmers in the developed world *not* to grow vividly illustrates. Nor is it an efficient stewardship of the land; modern farming is causing increasing environmental problems such as deteriorating soils and pollution of waterways. Nor is it efficient in other ways. One aspect of a sustainable development approach is to look at different types of production from a fresh angle. This is done here, through the consideration of the energy efficiency of different agricultural systems from around the world. Readers should consult Conway and Barbier (1990) for a more detailed consideration of sustainable agricultural development than is presented here.

As a general rule, the availability of energy is a key input into the ability of an economic system to transform its environment. This is illustrated by comparing different agricultural systems, an exercise worth doing if it is remembered that food is no more than a 'fuel' for the human race, and as such it is possible to argue that the efficiency of agriculture should be measured by its efficiency in converting energy.

The intensive, high input agriculture of the modern world is more productive (whether measured in terms of yields per unit of land or per unit of labour time) than would have been dreamt possible in the past. Throughout the developed world governments are faced with massive surpluses and are increasingly looking for ways to deal with the economic and political consequences of past agricultural policies. All this has been made possible by huge increases in inputs of energy to the agricultural system.

Whether this intensive agriculture is more efficient is another question. In economic terms it is clearly not; farming in the USA, the EEC and elsewhere in the developed world is more heavily subsidized than any other industry. Its energy efficiency is also questionable, with increasing amounts of artificial energy (and other inputs) needed to produce the high yields of modern farming. These problems reflect the way efficiency in agriculture has been measured, which is in yield per land unit. If other measures are used, such as the energy output per unit of energy input, a very different story unfolds.

The data in Table 2.1 compare a number of farming systems in terms of the energy inputs and the energy content of the crops grown. The inputs include human and animal labour, machinery, fertilizers and pesticides, seeds, fuel for transport, irrigation and so on. They show that, in energetic terms, non-mechanized, peasant farming systems are the most efficient, performing better than either the hunter–gatherer economies which predate them or the intensive mechanized farming which is increasingly replacing them.

Crop production which relies solely on human labour is particularly efficient, typically producing over ten times more energy per unit of energy put in. This is true both within the developing world and when peasant production is compared with developed world agriculture; indeed, estate crop production in Mexico and India is among the worst performers. Within the agriculture of the USA, the production of grains such as rice, wheat and maize is far more efficient than the production of fruit and vegetables. The production of crops such as spinach and apples is remarkably inefficient, requiring two or more units of energy in for each unit of energy produced. These crops are indeed 'luxury' items, and the way they are produced is clearly not sustainable nor, indeed, desirable.

The production techniques of modern agriculture were largely developed in the

Table 2.1 Energy efficiency of agricultural production systems. Units are kilocalories per hectare per annum. Adapted from Pimentel (1984)

Agricultural systems	Total energy input (A)	Energy production (B)	Energy efficiency (B/A)
Hunter/gatherer	2 685	10 500	3.9
Pastoralism (Africa)	5 150	49 500	9.6
Peasant farming (Mexico)	675 700	6 843 000	10.1
Estate crop production (Mexico)	979 400	3 331 230	3.4
Estate crop production (India)	2 837 760	2 709 300	0.9
Maize (USA)	1 173 204	3 306 744	2.8
Wheat (USA)	4 796 481	8 428 200	1.7
Rice (USA)	14 586 315	21 039 480	1.4
Apples (USA)	18 000 000	9 600 000	0.5
Spinach (USA)	12 800 000	2 900 000	0.23
Tomatoes (USA)	16 000 000	9 900 000	0.61

post-war era of cheap and abundant energy. During this period many developed world governments were worried about levels of food production, and put great efforts into increasing outputs regardless of the levels of inputs needed. Agriculture went through a technological revolution, as farming methods based on greatly increased levels of external inputs were developed. Later on, through the so-called 'Green Revolution', similar techniques were transferred to many parts of the developing world, where regions such as the Punjab in India saw equally dramatic increases in crop yields.

In both the developed and developing worlds the new energy intensive techniques have been successful in terms of increased yield per worker or hectare. Where they are patently unsuccessful is in terms of yields per unit of energy. In an energy-rich but land- and labour-short world this does not matter, but it is becoming increasingly apparent that many places are not like this.

The more realistic energy prices of the post-1973 era have been a key factor in undermining the economic viability of developed world agriculture; the cost of inputs of fuels and chemicals are more than the market price of the crops produced. Developed world consumers pay for this, through higher food prices (protected by trade barriers) and taxes to subsidize their farmers.

The energy efficiency of modern food production is even poorer if transport and processing costs which occur after the crops leave the farm are included. For example, the energy production:input ratio for corn in the USA drops from 3.5:1 at the farm gate to 0.53:1 for canned corn and 0.44:1 for frozen corn. Once the corn is cooked, the energy ratio of the food on the plate is as low as 0.11:1, or in other words every unit of energy a person eating corn receives requires ten units of energy to grow, process, transport and cook.

In extreme examples, the energy content of food has been lost sight of altogether. For example, a 12 oz can of a diet soft drink (with one kilocalorie of energy) takes 2200 kcal to produce. The rationality of the energy inputs to modern food production is likely to be increasingly questioned as energy costs increase and the environmental impacts of energy use in the developed world's high consumption society leads to calls for reduced energy use.

In many parts of the developing world cash to pay for the energy inputs (including chemicals) and machinery is a scarcer commodity than labour or land. For these areas, the energy efficiency of traditional agricultural systems is an important pointer to the future, as it shows that agriculture does not need to be modern to be efficient. This adds power to calls for locally based development strategies, and in particular for greater support for agricultural techniques with low external inputs. Many parts of the developing world experience severe problems in acquiring fuels, chemicals and other inputs in contemporary times. These problems are likely to be compounded as resources become scarcer globally.

2.6 CONCLUSIONS

This chapter has set out the main themes of the sustainable development approach. It represents a radically new way of approaching resource exploitation and economic development, and is seen by many as the way out of the increasingly difficult resource problems facing the modern world. To date, these fine principles have not been developed into a coherent programme of action, and it is clear that there are many formidable barriers to achieving this goal. Overcoming these barriers is the challenge facing the human race in the 1990s and through to the next century. This will need some, but not many, technological innovations and a lot of new economic tools. Above all it depends on a new way of thinking about economic change and society's relationship to the environment. This new psychology is perhaps the greatest challenge of all: those in the position to make the key changes needed are unlikely to do so until the need for a different pathway to development is accepted. The rise of a wider environmental consciousness in recent years gives cause for hope, but little will change unless these hopes are translated into action.

2.7 REFERENCES

Conway, G. and Barbier, E. (1990) *After the Green Revolution*. London: Earthscan.
International Union for the Conservation of Nature (IUCN)/WWF/UNEP (1980) *World Conservation Strategy*. Gland, Switzerland: IUCN.
Meadows, D. H., Meadows, D. L., Randers, J. and Behrens III, W. W. (1972) *The Limits to Growth: A Report to the Club of Rome's Project on the Predicament of Mankind*. New York: Potomac Associates.
Pearce, D., Markandya, A. and Barbier, E. (1989) *Blueprint for a Green Economy*. London: Earthscan.
Pimentel, D. (1984) Energy flows in food systems. In: Pimentel, D. and Hall, C. (Eds). *Food and Energy Systems*. New York: Academic Press.
Schumacher, E. (1973) *Small is Beautiful*. London: Blond and Briggs.
World Commission on Environment and Development (1987) *Our Common Future*. Oxford: Oxford University Press.
World Resources Institute and International Institute for Environment and Development (1988) *World Resources 1988–89,* published in collaboration with the UN Environmental Programme. New York: Basic Books.
World Resources Institute (1990) *World Resources 1990–91*, published in collaboration with the UN Environment Programme and UN Development Programme. Oxford: OUP.

2.8   FURTHER READING

Barde, J-P. and Pearce, D. (1991) *Valuing the Environment*. London: Earthscan.

Holmberg, J., Bass, S and Timberlake, L. (1991) *Defending the Future*. London: Earthscan.

Pearce, D., Markandya, A. and Barbier, E. (1989) *Blueprint for a Green Economy*. London: Earthscan.

Redclift, M. (1987) *Sustainable Development*. London: Methuen.

World Commission on Environment and Development (1987) *Our Common Future*. Oxford: Oxford University Press.

SECTION II

GLOBAL ISSUES
(a) Change in the Physical Environment

CHAPTER 3

Environmental Change: Lessons From the Past

A. M. Mannion

3.1 INTRODUCTION

At a time when the media are publicizing the 1990s and beyond as decades of unprecedented environmental change it is easy to forget that, throughout the earth's existence, its surface has been moulded by both gradual and catastrophic events. Among the most important processes that bring about environmental change are the biogeochemical cycles which effect the transfer of materials between the atmosphere, hydrosphere, lithosphere and biosphere (Section 1.2.3). These exchanges could not, however, occur without inputs of solar energy which also fuel the atmospheric system and ecosystem (Section 1.2.2). Variations in the amount of solar energy reaching the earth's surface are also involved in climatic change and for the most recent geological period there is evidence that they are controlled by the earth's orbital features. This period, the Quaternary, is characterized by the oscillation of cold glacial and warm interglacial stages. At least 20 such cycles have been recognized, culminating in the present interglacial, the Holocene, which began about 10 000 years BP (before present). On this basis, the earth should now be moving towards another glacial stage.

The climatic changes involved in the glacial–interglacial cycles are also related to changes in the global carbon cycle. As this biogeochemical cycle is linked with biotic processes, the existence of a close relationship between biotic activity and climate is suggested. This lends support to the controversial Gaia hypothesis (Section 1.2.3), which proposes that life plays a significant role in regulating conditions at the earth's surface, which, in turn, influence the nature of the biota. Moreover, the link between past climatic change and atmospheric carbon dioxide concentrations implies that the substantial increases in the latter since the Industrial Revolution will induce global climatic change. What the new global environment could be like and how it may impinge on the world's agricultural systems, which in turn will affect social and political configurations, is a matter for speculation.

Environmental Issues in the 1990s. Edited by A. M. Mannion and S. R. Bowlby
© 1992 John Wiley & Sons Ltd

Human communities have, however, been able to bring about such overt environmental change for only a relatively short time in the history of human evolution. Even before the establishment of glacial–interglacial cycles, bipedal (two-legged) hominids had evolved. Modern humans (*Homo sapiens sapiens*) evolved some 200 000 years BP in Africa, from where they colonized all the continents except Antarctica by 12 000 years BP. These early hunter–gatherers had no more than an ephemeral impact on their environment. This situation changed 10 000 years BP with the initial domestication of plants and animals and the inception of agricultural systems.

These developments marked the emergence of human communities as controllers of, rather than integral components of, ecosystems. As agriculture spread, more of the earth's surface was modified. The food surpluses produced enabled permanent settlements to flourish. Metal-working ensued and provided effective tools for more intense environmental manipulation. Against this backdrop of resource exploitation, civilizations rose and fell, each leaving its mark on the earth's surface.

Throughout history resource use has intensified. Agriculture developed at the expense of the natural interglacial forests, which also provided a vital source of charcoal for metal smelting. Moreover, the 15th and 16th centures AD witnessed a surge in exploration as European nations annexed the lands of the Americas, Africa and Asia for their resources. Coal began to replace wood as a fuel and the foundations of the Industrial Revolution became established as cottage industries in north-west Europe. With the rise of industry and the rapid growth of fossil fuel consumption came new agents of environmental change that profoundly affected biogeochemical cycles. The resulting legacy is manifest as problems such as acid pollution (Chapter 11) and stratospheric ozone depletion (Chapter 4). Agriculture and industry, while they have been creators of food security and wealth for some nations, have also become threats to the earth's life-support systems.

What follows is a synopsis of natural and cultural environmental history during the last three million years to provide a temporal perspective for those issues that occupy the remainder of this text, and to examine what lessons can be learnt that may contribute to reliable predictions about the climatic future and its impact on civilization.

3.2   ENVIRONMENTAL CHANGE DURING THE QUATERNARY

The acceptance of the glacial theory in the 1860s was a turning point in scientific thinking because it acknowledged the importance of glacial processes as agents of environmental change. There are numerous sources of information on Quaternary environmental change, notably terrestrial deposits (such as glacial sediments, peats and lake sediments), ocean sediment cores and polar ice cores. These archives of environmental history provide data on which many theories about climatic change have been formulated. These in turn provide precedents for conjecture about future climatic change.

3.2.1   Evidence from terrestrial deposits, ocean sediment cores and ice cores

Defining and dating the Pliocene–Pleistocene (Tertiary–Quaternary) boundary is itself problematic. The International Commission on Stratigraphy has formally adopted a stratigraphic horizon in a section at Vrica in Calabria, Italy, where a cold-tolerant foraminifer (formanifera are single-celled marine organisms with calcareous shells) appears and which is dated to about 1.8×10^6 years BP. There is, however, evidence indicating that the onset of glaciation occurred much earlier. In the north Atlantic region, for example, glaciation occurred about 2.4×10^6 years BP, as the amplitude between the warm and cold phases of each climatic cycle gradually increased. If a date of 1.8×10^6 years BP is accepted as the age of the base of the Quaternary, then the onset of glaciation actually occurred during the Tertiary period.

Interpreting the record of Quaternary environmental change is also problematic because of difficulties in correlating terrestrial deposits over large areas. This is due to the fragmentary nature of the deposits and the inadequacy of dating controls. There have been many improvements in methods of correlation since 1950, but much remains to be achieved before a comprehensive scheme for the sequence of glaciations can be devised. The International·Geological Correlation Programme (Šibrava et al., 1986) has correlated data from the northern hemisphere (Fig. 3.1). This is a preliminary scheme which will require revision as new evidence is collected, but it illustrates the general synchronism of the major glacial advances of the last 1×10^6 years.

Figure 3.1 gives the terminologies in use within each region and relates each stage to its equivalent marine stage. The record from ocean sediment cores is now considered to be the most appropriate means of establishing a globally applicable model of Quaternary climatic change. In particular, oxygen isotope ratios ($^{18}O/^{16}O$) present in the calcareous tests (shells) of foraminifera have been determined from a range of cores. The resulting data show parallel changes that are believed to represent changes in continental ice volume. Variations in oxygen isotope ratios result from the natural fractionation of the isotopes, involving the preferential evaporation of water containing the lighter isotope, ^{16}O, which becomes entrained in the atmosphere. As ice sheets accumulate they become repositories for increased ^{16}O deposited in precipitation, while ocean waters become relatively enriched in the heavier isotope, ^{18}O, as do the foraminiferal tests. The reverse situation obtains during interglacial stages as the incarcerated ^{16}O is released from melting ice. The isotopic signal thus presents a record of glacial and interglacial stages as shown in Fig. 3.2. Dates have been ascribed using a variety of methods (Imbrie et al., 1984); odd numbered stages represent interglacials and even numbered stages represent glacials.

Several important features are apparent in this record (Fig. 3.2). Firstly, there have been many more glacial–interglacial cycles than is reflected in most continental records and it is now accepted that there have been at least twenty such cycles. Moreover, the glacial stages last about five times longer than the intervening interglacial stages. Thus, cold conditions appear to be the norm for the earth, with each glacial stage lasting about 120 000 years in contrast to interglacial durations of 10 000 to 20 000 years. The oxygen isotope record also shows that the glacial stages were not uniform, as is also suggested by terrestrial evidence. During the last cold stage, for example, there were at least two short-lived warmer periods known as interstadials. Similarly, (see Section 3.2.2) the interglacials were not climatically uniform. The ocean sediment record also shows that the transitional stages from glacial to interglacial conditions, (the 'terminations' given in Fig. 3.2) occurred relatively rapidly, as did the transitions in the reverse direction.

	NORTH EUROPEAN STAGES	U.K.	NETHERLANDS	NORTH GERMANY	POLAND	U.S.S.R.	ALPINE REGION (AUSTRIA & GERMANY)	NORTH AMERICA	MARINE STAGES	DATES IN 10³ YEARS
IG	HOLOCENE	FLANDRIAN					HOLOCENE	HOLOCENE	1	13
G1	WEICHSEL	LATE DEVENSIAN	WEICHSELIAN	WEICHSEL GLACIATION	MAIN STADIAL / PREGRUNDZIAD STADIAL (VISTULIAN)	LATE VALDAI G.	MAX WÜRM G. STILLFRIED B.	LATE WISCONSIN MIDDLE WISCONSIN EARLY WISCONSIN	2 / 3	32-35 / 64-65
		MIDDLE DEVENSIAN			KASZUWSTADIAL	LOESS FLUVIAL DEPOSITS	FIRST WÜRM G.	EOWISCONSIN	4	75-79
		EARLY DEVENSIAN		LOW TERRACE	KASZULY STADIAL		STILLFRIED A		5 a / b / c / d	122
IG1	EEM	IPSWICHIAN	EEMIAN	EEMIAN	EEMIAN	MIKULINO	MONDSEE & SOMBERG	SANGAMOAN	e	128-132
G2	WARTHE	Lower organic deposits at Marsworth?		SAALE 3 G. RUGEN I.G. SAALE 2 G.	WARTA GLACIATION	MOSCOW	LATE RISS LATE RISS RISS? LATE RISS TERRACE	LATE ILLINOIAN	6	195-198
IG2	SAALE / DRENTHE (TREENE?)	Lower organic deposits at Marsworth?	SAALIAN COMPLEX	TREENE WENNINGSTEDTER & KITTMITZER PALAEOSOLS	LUBLIN I.G. POLICHNA I.S. (CENTRAL POLISH GLACIATION)	ODINTSOVU I.S. (DNEIPER STAGE)	PARABRAUN-EARTH	ILLINOIAN	7	251-262
G3	DRENTHE	WOLSTONIAN COMPLEX	DRENTHE GLACIATION	DRENTHE G. MAIN TERRACE	ODRA G. PODWINEK I.S. PRE-MAXIMUM STADIAL	DNEIPER GLACIATION	ANTEPENULTIMATE G. EARLY RISS MINDEL HIGH TERRACE	EARLY ILLINOIAN	8	297-302
IG3	DOMNITZ (WACKEN)		HOOGEVEEN I.S.	FREYBURGER BODEN	MAZOVIAN	ROMNY	REDDISH PARABRAUN-EARTH		9	338-347
G4	FUHNE (MEHLECK)	HOXNIAN		ELDERITZER TERRACE ERKNER ORGANIC SEDIMENTS	PRE-MAXIMUM STADIAL? WILGA G.?	ORCHIK STAGE (PRONYA GLACIATION)	GL4 TERRACE (PRE-RISS TERRACE)		10	367-352

Stage	No.	Age (ka)	(Muldsberg) / German	Anglian (Britain)	Peelo Fm. / Holstein (Netherlands)	Holstein	Ferdynandow (Poland)	Lichvin (Russia)	Alps
IG4	11	367-352	HOLSTEIN (MULDSBERG)		HOLSTEIN				REDDISH PARABRAUN-EARTH
G5	12	440-428	ELSTER 2	LOWESTOFT TILL		ELSTER 2 G.	SAN G.	OKA GLACIATION	GL5 GLACIATION OF ALPS / LATE MINDEL / GUNZ / MINDEL / GL5 TERRACE
						FLUVIAL GRAVELS	KOCK STADIAL?		
IG5	13	472-480	ELSTER 1/2	CORTON SAND		VOIGSTEDT	LUSZAVA I.G.?		RIESENBADEN PALAEOSOL
G6	14	502-?	ELSTER 1	CROMER TILL		ELSTER 1 G.	NALECZOIRG / SERNIKI STADIAL?		LOWER INTERTERRACE GRAVEL / EARLY MINDEL DONAU
						FLUVIAL GRAVELS			
IG6	15	542-562	CROWER (INTERGLACIAL 4)	CROMERIAN	INTERGLACIAL 4	VOIGSTEDT	PODLASIE I.G.		RIESENBODEN PALAEOSOL
G7	16	592-630	GLACIAL C	BEESTONIAN	GLACIAL C		PRE-CROMERIAN GLACIATIONS		MIDDLE INTERTERRACE GRAVEL
1G7	17	627-687	INTERGLACIAL 3	PASTONIAN	INTERGLACIAL 3				RIESENBODEN PALAEOSOL
G8	18	647-718	GLACIAL B		GLACIAL B				UPPER TERRACE GRAVEL / GL8 GLACIATION
IG8	19	688-782	INTERGLACIAL 2		INTERGLACIAL 2				RIESENBODEN PALAEOSOL
G9	20	706-790	GLACIAL A		GLACIAL A	HELME FLUVIAL GRAVELS			EARLY GUNZ?
IG9	21	729-812	ARTEN (INTERGLACIAL 1)		INTERGLACIAL 1	UPPER MUSCHELTONE			INTERGLACIAL

Spanning labels: ANGLIAN; PEELO FORMATION; ELSTERIAN; PRE CROMERIAN GLACIATION PLATEAU DRIFT; SOUTH POLISH GLACIATION; OKA GLACIATION; FLUVIAL GRAVELS; PRE-ILLINOIAN.

Lower age limit: 782-?

Figure 3.1 Correlations between the various Quaternary sequences in the northern hemisphere. Adapted from Šibrava *et al.* (1986) by permission of Pergamon Press PLC.

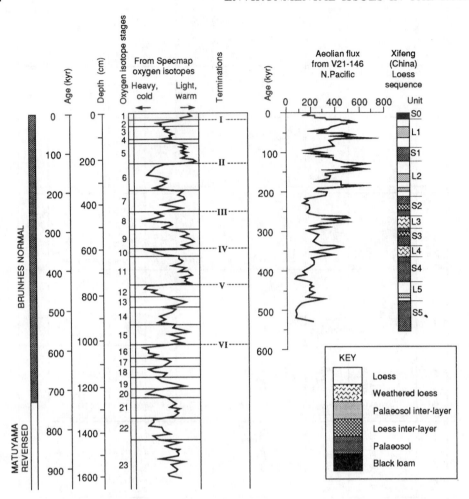

Figure 3.2   Oxygen isotope stages, terminations and loess stratigraphy during the middle and
late Quaternary period. Compiled from sources mentioned in the text.

Whereas the oxygen isotope signal provides a proxy record of global ice volume, the terrigenous material in ocean sediment cores provides information on wind direction and intensity and on the degree of continental aridity. This has been highlighted by Hovan *et al.* (1989), who have examined a sediment core from the north-west Pacific at a site 3500 km downwind from central China. As well as deriving an isotope stratigraphy, Hovan *et al.* (1989) measured the flux rates of aeolian (wind blown) material. As Fig. 3.2 shows, flux rates were much higher during glacial stages than during interglacial stages, indicating that mid- to high-latitude continental regions were subject to increased wind erosion. When deposited on land this aeolian material occurs as loess, which is particularly extensive in central and northern China. Here, sequences of silty, unweathered loess, which represent cold, arid conditions, are interrupted by palaeosols or weathered loess that form during humid interglacial conditions. The unweathered loess sequences correlate with the periods of highest

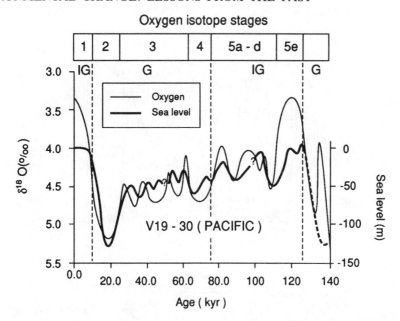

Figure 3.3   Sea-level curve for the Huon Peninsula, New Guinea, in relation to the oxygen isotope stages from the Pacific deep sea core V19–30. Adapted from Chappell and Shackleton (1986) and Shackleton *et al.* (1983).

influx of aeolian material in the ocean sediment cores. This establishes a direct link between the continental stratigraphy and the marine oxygen isotope record.

Changes in global ice volume also cause changes in sea level as both are related to the global reservoirs of water. The relationship between the two is complex because variations in sea level can occur due to changes in the total amount of water in the oceans and/or as a response to the earth's crustal readjustment to ice accumulation or loss. The former are known as eustatic changes, and, where glaciations are the prime cause of such changes, they are termed glacio-eustatic changes. Crustal readjustment gives rise to isostatic movements. Distinguishing between the two is difficult, especially in high latitudes which were directly affected by ice accumulation and were thus subject to both eustasy and isostasy. Chappell and Shackleton (1986) have, however, indicated that the low latitude reefs of the Huon peninsula, New Guinea, provide a record that is representative of global sea-level changes during the last 250 000 years. Here a flight of raised coral terraces have been dated and correlated with the oxygen isotope record of an equatorial Pacific core (V19–30). As Fig. 3.3 shows, the sea level was approximately 6 m above the present level during the last interglacial (oxygen isotope stage 5e). In contrast, the sea level was about 130 m lower during the last glacial maximum 18 000 years BP.

Ocean sediment records are also being corroborated and supplemented by data from polar ice cores. The longest ice core so far available is that from the Soviet Antarctic station, Vostok. The oxygen isotope record (Lorius *et al.*, 1989) reflects the depletion of ^{18}O in the atmosphere during glacial stages and its enrichment during interglacial stages. As Fig. 3.4 shows, the resulting stratigraphy can be used to divide

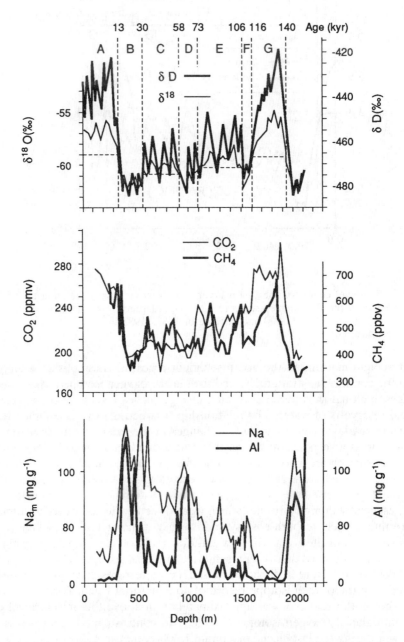

Figure 3.4   Palaeoenvironmental data from the Vostok ice core, Antarctica. Adapted from Lorius *et al.* (1989) and Chappellaz *et al.* (1990).

the 150 000 year period into a series of zones. Zone A is the Holocene and zone G is the last interglacial, whereas stages B to F represent the last glacial stage in which zones E and C are interstadials. Lorius *et al.* (1989) have also suggested that the change from the penultimate glacial to the last interglacial (stages H to G) involved a 10°C temperature difference, with an 8°C difference between the last glacial (zone B) and the present Holocene (zone A). This is also reflected in the deuterium profile (Fig. 3.4) because, as a heavy isotope of hydrogen, it is affected by temperature variations in a similar way to oxygen isotope ratios.

Analyses of the air encapsulated in the pores of the ice have also revealed changes in the atmospheric concentrations of carbon dioxide and methane. Figure 3.4 shows a clear relationship between these concentrations and glacial–interglacial conditions. For example, there is approximately 25% more carbon dioxide in the atmosphere in interglacial than in glacial stages and approximately 100% more methane. This indicates a link between climatic change and the global carbon cycle, but how the two interact is yet to be discerned. If living systems are involved it could mean that global organic productivity was higher during glacial stages, providing a sink for carbon either in the oceans and/or in mid and low latitude land areas which were more extensive due to lower sea levels. This interplay between climatic change and the global carbon cycle has important implications for understanding the possible impact of an enhanced greenhouse effect (see Section 3.2.4).

Analyses of a range of ions in the Vostok core also reveal the importance of marine and aeolian inputs. Aluminium concentrations, for example, are surrogate measures for terrestrial dust and as Fig. 3.4 shows, inputs of aluminium were significantly greater during glacial stages than during interglacial stages. Again, these data mirror those from ocean sediment cores (Fig. 3.2), indicating that glacial stages were characterized by more arid conditions in continental regions. Similarly, the concentrations of sodium, a measure of marine salts such as sodium chloride and sodium sulphate, were higher during glacial stages. This reflects the increased presence in the atmosphere of these aerosols that may also have contributed to climatic change by producing cloud condensation nuclei. An enhanced cloud cover may have accentuated the cooling trend instigated by changes in the earth's orbital characteristics (Section 3.2.3).

3.2.2   Environmental change during the post-glacial period

Much evidence for environmental change since the last major ice advance derives from peats and lake sediments. The transitional period between the last glacial stage and the Holocene was complex. The fossil record in north-west Europe, for example, indicates that just before 13 000 years BP rapid warming occurred as average summer temperatures increased to levels similar to those of today. This was a short-lived amelioration and by 12 000 years BP average temperatures were declining. The depression was sufficient to allow the recrudescence of local glaciers in many upland regions. By about 10 500 years BP, however, renewed climatic amelioration marked the opening of the Holocene. Atkinson *et al.* (1987) have calculated, using fossil beetle evidence, that temperature increases of about 2.6°C per century were occurring at 13 000 and 10 500 years BP in the UK, which indicates just how quickly climatic regimes can change.

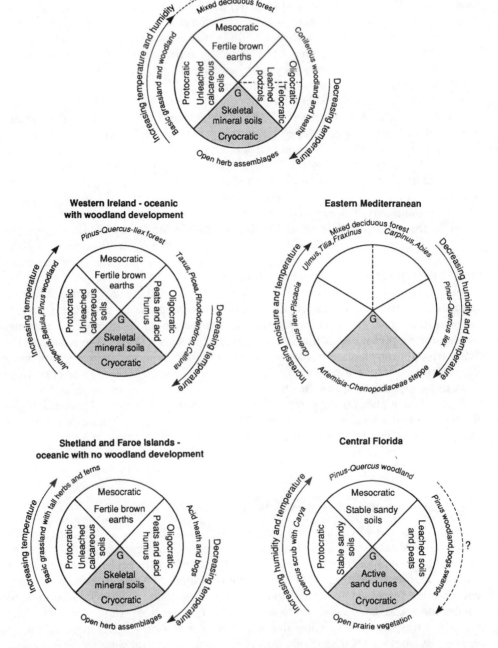

Figure 3.5 Ecological changes occurring in various regions during the interglacial cycle. Reprinted by permission from Birks (1986).

It is also of interest to compare the environmental changes that have occurred in the current interglacial with those of past interglacials, not least because the latter proceeded unaffected by human activity. The comparison is effected by reference to a model of the ecological processes that are operative during interglacial periods. This was first established in the late 1950s, a modified version of which is given in Fig. 3.5. The mesocratic stage represents the culmination of protocratic successional processes and soil maturation to allow the maximum expression of forest cover. This is the so-called climax vegetation. Thereafter soil acidification, due to progressive leaching and then climatic deterioration in the telocratic phase, rendered the milieu unsuitable for broadleaved species which require warmth and nutrients. Open coniferous woodlands and heathlands then developed prior to the re-establishment of glacial conditions.

The model is also useful for examining ecological changes outside north-west Europe because it portrays the dynamic continuum of environmental change. In detail it is apposite to mid-latitude biomes whose climax vegetation is forest and in which a cryocratic phase requiring the presence of ice or snow is significant. This is obviously not the case in lower latitudes where glacial and periglacial conditions were confined to high altitude zones. Nevertheless, fossil pollen assemblages (which are indicative of vegetation change) from low latitudes show that they too were characterized by large-scale latitudinal and altitudinal shifts in vegetation belts. In many areas forest cover was replaced by sparse woodland or grassland as the earth cooled. The present tropical rainforest formations, for example, were just as transient as the deciduous broadleaved forests of north-west Europe. Overall, it appears that the glacial periods of the last three million years were characterized by a relatively poor global forest cover. Despite this there was much less carbon dioxide present in the atmosphere (Section 3.2.1). This appears anomalous, although it is feasible that depressed sea levels exposed extensive coastal areas in lower latitudes on which a dense forest cover developed.

This information is also relevant to the debate surrounding tropical rain-forest diversity. The premise that diversity is a function of hundreds of thousands, if not millions, of years of environmental stability cannot now be accepted. The idea that pockets, so-called refugia, of rain-forest persisted in specific areas during colder periods and then extended to their present limits is also questionable. Research from areas postulated to have been refugia contain evidence for vegetation change rather than continuity. Either refugia existed elsewhere, possibly in coastal regions that disappeared beneath post-glacial marine transgressions, or species, which had survived near to their environmental thresholds, reassembled as climate ameliorated. Why tropical forests are home to the most diverse vegetation communities on the earth's surface remains a matter for conjecture. It is most likely to be the result of a combination of factors such as the advantages of diversity in combating the spread of pathogens and diseases, and/or the plentiful solar radiation and rainfall that are often limiting factors to growth in other biomes.

However, the model is not sufficiently refined to facilitate prediction of when the next ice age will begin. In general, interglacials last between 10 000 and 20 000 years and as the Holocene began some 10 000 years BP a cooling trend might be presently expected to mark the transition towards the next glacial stage. In addition, evidence from European Holocene deposits indicates that vegetation characteristic of the mesocratic phase of the interglacial cycle was well established by about 5000 years BP. Although the record of the later Holocene is difficult to interpret because of the complicating factor of human activity (Section 3.3) there is evidence to show that about 3000 years BP climatic deterioration had set in. Did this mark the onset of a cooling trend and has this since been halted? There is also the possibility that post-Holocene glaciation actually began, as there is evidence for the occurrence of a particularly cold period between 800 and 300 years BP, known as the 'Little Ice Age'. By this time deforestation, in Europe at least, was widespread and atmospheric carbon dioxide concentrations were increasing. Even if the 'Little Ice Age' was not the start of the next glacial stage, can it be assumed that the next ice age has not been indefinitely postponed? And is this good or bad for the earth and its inhabitants?

3.2.3 The causes of climatic changes

Climatic change has always been one of the most debated issues of environmental science and will remain so in the 1990s (Chapter 4). Sun-spot cycles have achieved notoriety, volcanic eruptions are known to cause short-term declines in average temperatures and the vagaries of ocean currents such as the El Nino southern oscillation have been associated with drought and floods (Chapter 4). All of these factors do indeed contribute to climatic fluctuations. However, the primary cause of the climatic change that has resulted in glacial–interglacial cycles is now considered to be changes in the earth's orbital characteristics.

This astronomical theory was refined by the Yugoslavian mathematician Milutin Milankovitch in the 1930s. At that time it was untestable, but the advent of ocean sediment research in the 1950s brought it once again into the climate change debate. There is now much evidence (e.g. Fig. 3.2) to corroborate the predictions of this theory for the timing of major climatic excursions. There are three orbital parameters that influence the position of the earth in relation to the sun, as shown in Fig. 3.6. The longest of these is the degree of circularity of the earth's orbit around the sun, with a periodicity of 100 000 years. This is thought to drive the glacial–interglacial

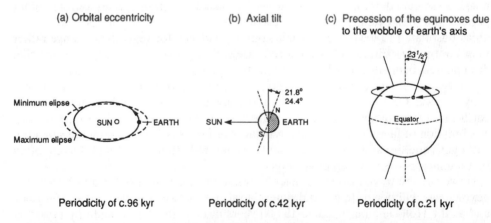

Figure 3.6   Orbital and axial variations of the earth which affect the receipt of solar radiation at the earth's surface.

cycle. The periodicities of axial tilt and the precession of the equinoxes, 42 000 and 21 000 years respectively, influence the pattern of stadials (phases of maximum ice advance) and interstadials (phases of short-lived warming) that occur during glacial stages.

These cycles affect the amount of solar radiation received at the earth's surface, especially in high northern latitudes, but of the three cycles, orbital eccentricity has the least effect on insolation. It is thus surprising that it triggers the most pronounced climatic switch from glacial to interglacial conditions. Nor does it account for the synchronism of events between the northern and southern hemispheres. Rapidly increasing atmospheric concentrations of carbon dioxide (Section 3.2.1) also closely follow initial interglacial warming.   Broecker and Denton (1990) suggest that these factors are interrelated via a climate–ocean link. The mechanism they propose is complex. In general terms it involves warming in the northern hemisphere, which somehow prompts changes in the so-called biological pump of carbon dioxide from the oceans to the atmosphere. At the same time changes in oceanic circulation occur which translate climatic change into a global phenomenon. This involves the formation of the north Atlantic deep water (NADW), which is characteristic of the current interglacial but which did not operate during glacial times. Modern obser-vations show that NADW formation involves north-flowing water of high salinity which is brought to the surface by wind action. As it travels north, it loses heat which, together with its high salinity, makes it dense so that it sinks to the ocean deep at the latitude of Iceland. It then contributes to the world's deep ocean currents, driving into the waters that surround Antarctica and into the Pacific. As NADW formation occurred at the end of an ice age, it created a different pattern of global oceanic circulation to that which characterized the glacial stages. Given the close coupling that occurs between the atmosphere and the oceans, this change in circulation pattern would undoubtedly influence global climatic regimes, although the precise mechanism has not yet been established.

There also remains the problem of what precipitated the onset of warm–cold oscillations in the Tertiary period. Again suggestions abound; to give one example, it is possible that mountain building in active uplift zones such as Tibet, the Himalaya and the American south-west may have reached such a height that atmospheric circulation was sufficiently altered to enhance the impact of orbital variations. There are also the questions as to if and why the glacial–interglacial cycles of the Quaternary period will end as those of earlier geological periods did.

3.2.4   Precedents for the future

There are a number of important points that arise from the research detailed in the preceding sections. Firstly, the 'normal' state of the earth is cold. This implies that the life-support systems, namely energy flows and biogeochemical cycles (see Section 1.2), probably achieve an equilibrium state when the average temperature of the earth is at least 5°C less than at present. Moreover, the glacial atmosphere is characterized by 25% less carbon dioxide than is the interglacial atmosphere. One implication of this, assuming that the additional carbon dioxide is interred in organic material, is that global productivity was greater during glacials than during intergla-cials. Did this increased productivity occur on exposed continental shelf regions in the tropics and/or in the oceans? In the oceans, productivity is mainly controlled by light and nutrient availability; with sea levels lower by 100–150 m, and nutrients such as nitrates and phosphates more evenly distributed, the open ocean may have been more productive and was thus a larger sink for carbon dioxide than it is now. In addition, the shallow coastal region, which is the most productive oceanic zone, was more extensive.

Data from terrestrial environments also imply that glacial stages were periods of reduced forest cover. If this was so, then high biological productivity is not necessarily synonymous with an extensive forest cover. This appears anomalous if the present day productivity patterns of the earth's major biomes are compared. Furthermore, if ice age earth is the equilibrium state, the more pertinent question is: what causes interglacials? Will the earth return to its equilibrium state within the next few millennia, assuming that anthropogenic activity has not already set in motion a climatic change of sufficient magnitude that the natural trends will be overridden?

This introduces a second precedent derived from the palaeoenvironmental record. The higher carbon dioxide concentrations of interglacials are therefore associated with a state of disequilibrium and any increases are likely to upset the balance further, creating positive feedback in the climate system. This could be the situation that obtains at present. Lorius *et al.* (1990) have estimated that the increase in concentrations of greenhouse gases in interglacial periods contributed about a 2°C warming. As the magnitude of carbon dioxide and methane increase is similar to that which has occurred since the Industrial Revolution, it is unlikely that it will have no repercussions. The fact that a distinctive warming trend is not obvious in modern temperature measurements (Chapter 4) may be due to a cooling trend, created by orbital forcing as the earth moves towards another glacial phase, which is counteracting global warming. Perhaps, as suggested in Section 3.2.2, increased greenhouse gas concentrations have already delayed or even prevented the onset of the next cold stage.

The lesson that the palaeoenvironmental record presents is one of an unequivocal, although complicated, relationship between atmospheric greenhouse gas concentrations and climatic change. Thus, society cannot afford to ignore the potential impact of the enhanced greenhouse effect. A most important question is whether such changes could alter the earth's life-support systems to such an extent that they will no longer support the human organisms that created the disturbance (see discussion on the Gaia hypothesis, Section 1.2.3).

3.3 THE EMERGENCE OF HUMAN COMMUNITIES AS AGENTS OF ENVIRONMENTAL CHANGE

Even before the onset of glacial conditions in the later part of the Tertiary period, a new agent of environmental change was emerging. Nearly 4×10^6 years BP bipedal hominids evolved from ape-like ancestors. Through a series of intermediate hominids, modern humans, *Homo sapiens sapiens*, evolved some 200 000 years BP, probably in Africa, and subsequently radiated into other parts of the globe. This migration began against the backdrop of the penultimate glacial stage (see Fig. 3.1). Prior to about 10 000 years BP *H. sapiens sapiens* was an integral component of ecosystems, first subsisting like other animals and then manipulating biotic resources in a way that did not radically alter ecosystem structure and function.

After 10 000 years BP, a major change occurred: permanent agriculture replaced hunting–gathering as the major food procurement strategy. This began in the Near East and resulted in social and environmental changes that subsequently transformed the earth's surface. Soon after this metal technology began, again in the Near East. By the time the Romans had become a world power in the first century BC agricultural systems were well established in many parts of the world, especially in Europe, where human communities had already wrought considerable environmental change. Throughout the historic period Europe was a centre of agricultural and technological innovation. Eventually, in the mid-1700s, the Industrial Revolution not only paved the way for the development of modern society, but also brought with it new and potent agents of environmental change.

3.3.1 Human evolution

Palaeoanthropology, the study of human evolution and behaviour, is fraught with controversy. This is especially so since the application, in the 1980s, of molecular biology techniques. DNA (deoxyribonucleic acid) annealing has shown how closely related humans are to the great apes. This technique involves the separation of the strands that comprise the double helix of DNA, the mixing of single strands with those from other animals to form a hybrid helix, and then the application of heat to prise them apart. The amount of heat required is a measure of how closely related the two species are, i.e. the more heat that is necessary, the tighter the bonds within the hybrid helix and thus the closer the species are related. The controversial aspect of this technique is the view that changes (mutations) in the nucleotide sequences, the fundamental building blocks of DNA, occur on a regular basis. Thus the difference that exists between two species is a measure of the time that has elapsed since the species diverged from a common ancestor. Despite the controversy, the dates for hominid evolution derived from biomolecular evidence and fossil finds are similar. This divergence is now widely accepted as occurring some 5×10^6 years BP in Africa.

As Fig. 3.7 shows, bipedal hominids had evolved from a quadripedal (four-legged) ancestor by about 3.75×10^6 years BP. Bipedalism may have conferred an evolutionary advantage, possibly because it freed the hands for tool-making which may, in turn, have stimulated brain development. These bipedal hominids were species of *Australopithecus afarensis*. The most famous fossil discovery of this species is the skeleton 'Lucy' that was unearthed in Ethiopia in the mid-1970s. From *A. afarensis* the genus *Homo* evolved about 2×10^6 years ago, starting with *H. habilis*, which developed the ability to make crude tools. *H. habilis* was probably a meat-eater, in contrast to its ancestor, which was mainly a fruit-eater. Within another 500 000 years another hominid species had evolved. This was *H. erectus*, a species that probably lived in organized groups, practised hunting and gathering, made tools and used fire.

H. erectus was the ancestor of modern humans which evolved about 200 000 years ago, but there is much debate as to where *H. sapiens* evolved. One hypothesis, the multiregional hypothesis, suggests that the species evolved from *H. erectus* populations which had already radiated out from their African centre of origin. The single region hypothesis, which is also favoured by molecular biological evidence, indicates that modern humans evolved in Africa and then spread into Europe and Asia.

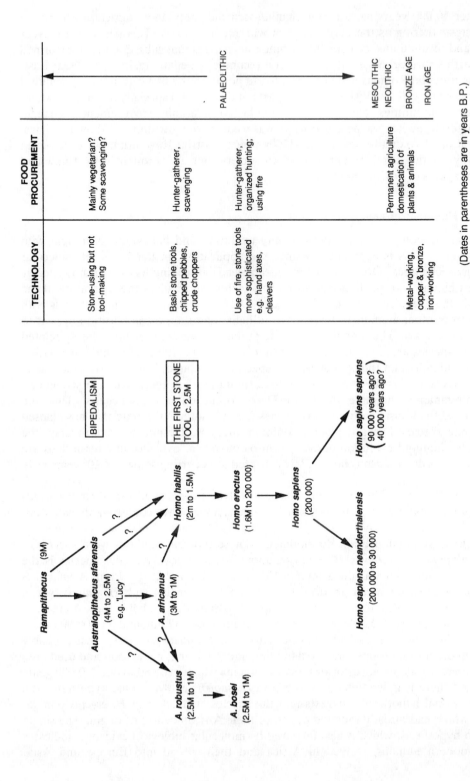

Figure 3.7   Human evolution in relation to technological and food procurement strategies (M denotes millions of years ago). Reprinted from Mannion (1991) by permission of Longman Group UK Ltd.

Neanderthals (*H. sapiens neanderthalensis*) are also thought to derive from the same ancestor as modern humans, but became extinct 30 000 years ago.

By at least 50 000 years BP modern humans had arrived in Australia. In the Americas, the earliest unequivocal dates for human settlement fall within the range 14 000 to 12 000 years BP, but there are several sites in South America which contain disputed evidence for a human presence at least 20 000 years earlier. However, by the time the current interglacial opened, all the continents, except Antarctica, had been colonized by an overwhelmingly successful species that was soon to alter radically the very environment that produced and sustained it.

3.3.2 Human communities as hunter–gatherers

For most of their history human communities have procured food through hunting and gathering (Fig. 3.7). How this was organized, if it was organized, is a matter for speculation. Plant resources were probably the only source of food for these hominid ancestors, but precisely what was exploited is unknown. Indeed, because plant remains disintegrate rapidly, there is little evidence for the use of plant resources throughout the early history of human communities. It is also difficult to establish with certainty when hominids turned to omnivory. This, however, may have conferred an advantage for the survival of hominid groups in the predominantly savanna environment of East Africa, where biomass is considerably reduced during the dry season. Nevertheless, mammal bones with cut marks found in many East African archaeological sites indicate that *H. habilis* was omnivorous. Whether the animals were procured through hunting or scavenging is also debatable. Both can be opportunistic, requiring little organization and pre-planning, but hunting can also be undertaken on a pre-meditated basis that requires a knowledge of prey movements. It is reasonable to suppose that scavenging not only encouraged omnivory but also the practice of sharing meat that, in turn, led to hunting on a community basis. Certainly, there is evidence that *H. erectus* was an active hunter.

As the last glacial advance was drawing to a close, about 15 000 years ago, hunting had developed into selective herding. This was apparent by the later part of the palaeolithic period, or old stone age. It marks the emergence of human groups as controllers of their environment. By the time the last ice sheet finally disappeared, human communities were well organized and skilled hunter–gatherers with specific tools for butchering and the ability to control fire. Moreover, the infrastructure they created provided the opportunity for even more sophisticated activities that began about 10 000 years ago.

3.3.3 Development of agricultural systems

Organized herding is itself an agricultural system because it represents a manipulation of biomass that would not occur under natural circumstances. It is also likely that such organization paved the way for the development of permanent agriculture. By about 10 000 years BP the scene was set for what is often referred to as the first green revolution. It was not, however, inherently revolutionary as it was the culmination of a long history of plant and animal exploitation, but it was certainly radical in terms of environmental and social repercussions.

The impetus for the transition from food procuring to food producing strategies remains enigmatic. Perhaps necessity (due to increasing population numbers) and greed (as food surpluses provide bargaining power for the procurement of other important resources) both played pivotal roles in instigating one of the most important stages in human and environmental history. Whatever the motives, there is evidence, mainly from the Near East, for the domestication of plants and animals dating to about 10 000 years BP. It is not fully appreciated how the domestication of plants and animals began. Was it deliberate or accidental? It is possible that encampments of hunter–gatherers, which were often in unafforested river basins and on lake edges, attracted open habitat species such as grasses, which were also the most successful in these environments. In turn, these species may have been favoured by soils enriched with the wastes of human subsistence. The higher than average food resources of many grasses may have been recognized as a useful and proximal food supply so that it became worthwhile to reserve some of the seed for future planting. Whatever the motives and the mechanisms, domestication provided the means for ecosystems to be transposed into agro-ecosystems within which human communities could control energy flows.

Excavations at sites such as Jericho, one of the earliest permanent settlements, indicate that by 9000 years BP cereal crops of emmer and einkorn wheat and barley were being grown, in addition to pulses such as lentils and peas. These were domesticated from wild species endemic to the Near East where an organized hunter–gatherer society was well established. All of these crops ultimately became components of the agricultural systems that developed in prehistoric Europe and Asia. Numerous other species were domesticated later. Oats and rye, for example, originated as weeds of wheat and barley and were eventually recognized as useful crops in their own right. Animal species were being domesticated at the same time, with the earliest domestications again occurring in the Near East. Most of the animals which underpin modern pastoral and mixed farming economies were domesticated between 8000 and 10 000 years BP. The dog, however, was domesticated some two millennia earlier, in the Near East, and was used for herding and hunting. In the ensuing 5000 years domesticated species spread throughout Europe and Asia. In Britain, for example, the first domesticated species were apparent about 5000 years BP.

Not all domestications occurred in the Near East. The llama and alpaca were domesticated about 6000 years ago in the Andes. The chicken originated even earlier around 8000 years BP from red jungle fowl in South-east Asia, whereas the horse was domesticated 4000 years ago, in what is now the Ukraine. Maize has its origins in wild grasses of the Tehuacán region of Mexico and first appears in archaeological sites that are dated about 7500 years BP. Rice also derives from wild grasses and was domesticated by at least 5000 years BP in South-east Asia.

As agricultural systems developed and spread they did not immediately replace hunting and gathering, which continued to supplement the produce of permanent agriculture. In Europe, for example, hunting continued as a means of food procurement well into the historic period. The establishment of agricultural systems did, however, initiate environmental change of a hitherto unprecedented nature and scale. Biotic changes occurred as domesticated plants and animals replaced naturally occurring species, often resulting in a reduction of species diversity and a reorganization of energy flows and biogeochemical cycles. Geomorphological, pedological and hydrological systems were altered when agro-ecosystems, especially arable ecosystems, replaced natural ecosystems. Both biotic and abiotic changes are particularly acute when fire is used as a clearance and management technique. Deforestation, soil erosion and soil degradation were just as important, although more localized, during the early days of agriculture as they are now. The major difference between early agriculture and that of the present day is a matter of degree. Currently a wider battery of mechanical and chemical tools exists, mostly produced by fossil fuel exploitation, with which society can manipulate the energy flows and biogeochemical cycles that underpin all food production.

3.3.4 Emergence and spread of metal technology

Metal technology, like agriculture, has been a characteristic of human communities for only a comparatively short time. Exactly when and how metal-working originated is obscure, but it post-dates the onset of settled agriculture. There is evidence for the widespread smelting of copper oxide ores in the Near East by 7000 years BP. The copper age was short-lived, giving way initially to arsenic bronze and then to tin bronze by 5000 years BP. Bronze production may have been the unintentional result of mixed-ore smelting which produced a more durable product than the intended copper. Nevertheless, copper and bronze implements did not replace stone tools at the workface but supplemented them and expanded the range of weaponry and ceremonial accoutrements. By about 4000 years BP the use of bronze implements had spread as far west as Britain and as far east as China and probably encouraged trade and contact as the demand for both raw materials and finished products increased. The advent of metal smelting probably brought about social changes in human groups, notably greater division of labour and craft specialization. Inevitably, this would only have been possible if food surpluses were produced. Inevitably, metal technology also encouraged deforestation as large amounts of charcoal were essential to smelt the metal ores and the implements produced were more efficient tools for clearing woodland for agriculture.

Throughout Europe and the Near East there is evidence for accelerated deforestation as metal technology became established. This, along with the construction of irrigation systems in the Near East where agriculture flourished and settlements expanded, helped to provide food security. It is probable that the rise and fall of many ancient civilizations related to food production and food security. Metal technology provided more efficacious tools with which to manipulate the environment and as such put a price on metal ores and tools, which became trading commodities. Interest also focused on precious metals such as gold and silver, which were worked to produce ornaments and jewellery. At the same time, 4000–5000 years BP, much effort was also being expended in the construction of large monuments. The first pyramids were erected in Egypt about 4500 years BP; the Minoans constructed the palaces of Crete at Knossos, Phaistos and Malia about 4000 years BP; Stonehenge in Britain was being constructed and the burial chamber of New Grange in Ireland's Boyne Valley had been built. Food production must thus have been of sufficient magnitude over large parts of Europe and elsewhere to sustain a significant proportion of the populace in non-agricultural pursuits.

As bronze-working began in Britain about 4000 years BP, iron-working was beginning in the Near East. Iron smelting technology may have developed by chance. Smelting of the ore with charcoal may have accidentally produced more durable goods and focused attention on a cheaper and more widespread raw material. The more general use of iron occurred about 3000 years BP, beginning in the Near East and North Africa and then spread into the rest of Europe, although there were also independent centres of origin in India and China. The changes in agriculture and society which the advent of iron technology brought about probably contributed to the emergence of the Greek and Roman empires that replaced the Assyrians and Egyptians in the Eurocentric world.

There is evidence from many parts of Europe that the development of bronze and, later, iron implements contributed to environmental change. Soil erosion events are apparent at many archaeological sites; salinization problems occurred in arid and semi-arid regions where irrigation was established and was widespread. Thus, in the space of 4000 years or so, the vestiges of modern agricultural landscapes had become apparent, as had many of the problems associated with agricultural practices that are even more significant today.

3.4 CONCLUSIONS

The palaeoenvironmental record provides convincing evidence for a link between the global biogeochemical cycle of carbon and climatic change. There is also evidence that the climate can change rapidly, at a rate that is discernible in a human lifetime. However, there are many questions surrounding climatic change and the rate at which such change provokes responses in ecosystems, earth surface processes and societies. Changes in the earth's orbital characteristics are clearly involved in climatic change, but what the links are between this, the oceans and concentrations of greenhouse gases in the atmosphere remain to be determined. Thus there is considerable scope for further work on the palaeoenvironmental record. This needs to proceed in tandem with research designed to elucidate the nature of flux rates that underpin biogeochemical cycles, especially in relation to the carbon cycle. What is also needed is a reliable index of global organic productivity and its changes over time.

The significance of biogeochemical cycles is also pertinent to the role of agriculture as an agent of environmental change. Since the inception of settled agriculture people have learnt to manipulate energy flows and biogeochemical cycles. This ability has increased as the provision of food surpluses has facilitated the growth of settlements and divisions of labour, in addition to prompting the exploitation of other natural resources. These activities have not only influenced energy flows but also affected hydrological, geomorphological and pedological processes. Removal of the natural vegetation cover has altered these earth surface processes to create environmental problems such as soil erosion (Chapter 14), salinization and desertification (Chapter 15). All that has happened with the passage of time is that the array of tools with which society manipulates the environment has increased. Prehistory and history reveal the poignant fact that the rise and fall of political power is intimately related to the ability to provide food surpluses.

The wave of unrest in the USSR in 1990 was, at least in part, due to limited food availability. What would the global situation be, if, within a decade, global warming disrupted agricultural systems to such an extent that global food shortages ensued? This might be the worst possible case that could occur, but it is the responsibility of the scientists of the 1990s to address such questions. The answers will need to take account of the lessons from the past that are presented in this chapter, as will the response of the politicians and policy makers of the 1990s and beyond.

3.5 REFERENCES

Atkinson, T. C., Briffa, K. R. and Coope, G. R. (1987) Seasonal temperatures in Britain during the past 22 000 years, reconstructed using beetle remains. *Nature (London)*, **325**, 587–592.

Birks, H. J. B. (1986) Late-Quaternary biotic changes in terrestrial and lacustrine environments, with particular reference to north-west Europe. In: Berglund, B. E. (Ed.). *Handbook of Holocene Palaeoecology and Palaeohydrology*. Chichester: Wiley, pp. 3–65.

Broecker, W. S. and Denton, G. H. (1990) What drives glacial cycles? *Scientific American*, **262**, 42–50.

Chappell, J. and Shackleton, N. J. (1986) Oxygen isotopes and sea level. *Nature (London)*, **324**, 137–140.

Chappellaz, J., Barnola, J. M., Raynaud, D., Korotkevitch, Y. S. and Lorius, C. (1990) Ice core record of atmospheric methane over the past 160 000 years. *Nature (London)*, **345**, 127–131.

Hovan, S. A., Rea, D. K., Pisias, N. G. and Shackleton, N. J. (1989) A direct link between the China loess and marine records: aeolian flux to the north Pacific. *Nature (London)*, **340**, 296–298.

Imbrie, J. *et al.* (1984) The orbital theory of Pleistocene climate: support from a revised chronology of the marine ^{18}O record. In: Berger, A., Imbrie, J., Hays, J., Kukla, E. and Saltzman, B. (Eds). *Milankovitch and Climate Part I*. Dordrecht: D. Reidel, pp. 269–305.

Lorius, C., Jouzel, J., Raynaud, D., Hansen, J. and Le Treut, H. (1990) The ice-core record: climate sensitivity and future greenhouse warming. *Nature (London)*, **347**, 139–145.

Lorius, C., Raisbeck, G., Jouzel, J. and Raynaud, D. (1989) Long-term environmental records from Antarctic ice cores. In: Oeschger, J. and Langway, C. C. Jr (Eds). *The Environmental Record in Glaciers and Ice Sheets*. Chichester: Wiley, pp. 343–361.

Shackleton, N. J., Imbrie, J. and Hall, M. A. (1983) Oxygen and carbon isotope record of East Pacific core V19–30: Implications for the formation of deep water in the late Pleistocene North Atlantic. *Earth and Planetary Science Letters*, **65**, 233–244.

Šibrava, V., Bowen, D. Q. and Richmond, G. M. (1986) Quaternary glaciations in the northern hemisphere. *Quaternary Science Reviews*, **5**, 1–510.

3.6 FURTHER READING

Mannion, A. M. (1991) *Global Environmental Change: A Natural and Cultural Environmental History*. Harlow: Longman.

Roberts, N. (1989) *The Holocene: An Environmental History*. Oxford: Blackwell.

The Times (1986) *The World: An Illustrated History*. London: Times Books.

Zohary, D. and Hopf, M. (1988) *Domestication of Plants in the Old World*. Oxford: Clarendon Press.

CHAPTER 4

The Changing Atmosphere and its Impact on Planet Earth

R. D. Thompson

4.1 INTRODUCTION

There is little doubt that one of the most interesting, and perhaps most threatening, environmental issues of the 1990s relates to the impact of human activities on the atmosphere. The debates focus on two main issues. Firstly, there is the possibility of global warming due to the release of heat-trapping gases. Secondly, depletion of stratospheric ozone and increasing ultraviolet solar radiation, associated with the use of chlorofluorocarbons (CFCs), has implications for human health. Deep concerns about these issues have initiated intergovernmental debates. For CFCs, the outcome was the Montreal Protocol of 1987, an international agreement to curtail their use. The meeting of the Intergovernmental Panel on Climatic Change (IPCC) in November 1990 was also a significant event in the global warming debate.

However, the direct evidence for global warming and increases in ultraviolet radiation is not convincing. For example, over the last century, the overall rise of global temperatures is probably only 0.5°C. As most stations recording this rise are located in areas of expanding urbanization, then this overall rise may reflect an accelerating urban heat island phenomenon and not the enhanced greenhouse effect. In terms of ultraviolet radiation, increases have not been recorded over Antarctica, which would provide evidence for the implications of the infamous ozone 'hole'. Conversely, measurements over the USA between 1974 and 1985 revealed that UVB levels actually decreased at five out of eight centres (with no change at the other three), despite the 5% ozone depletion recorded globally in the early 1980s. The current dilemma facing scientists is to reconcile the degree of measured atmospheric change with the actual and predicted increase in greenhouse gases and ozone destruction. There is currently much controversy about the burning of fossil fuels, increased carbon dioxide levels and global warming. However, it is important to remember that throughout the geological time-scale (and long before *Homo sapiens sapiens* arrived on the scene), there have been major changes of climate (Fig. 3.3)

Environmental Issues in the 1990s. Edited by A. M. Mannion and S. R. Bowlby
© 1992 John Wiley & Sons Ltd

associated with naturally changing carbon dioxide levels and a host of other causative factors. For example, studies of polar ice cores and deep-sea sediments show that atmospheric carbon dioxide concentrations have varied with the ice ages and interglacials (Section 3.2.1). The causes of climatic change are complicated and involve wide ranging natural and human mechanisms. This chapter examines the evidence for climate change provided by atmospheric observations and environmental indicators, particularly over the last century or so. It also accounts for this change in terms of the main causative factors, especially the variation in solar radiation intensity and atmospheric constituents which control the global heat balance and world climatic regimes.

4.2   EVIDENCE OF CLIMATIC CHANGE

Over the geological time-scale, fossil evidence and the stratigraphic record indicate the mode of sedimentation and the prevailing climate. For example, the discovery of subtropical flora and fauna fossil remains in the sedimentary rocks of Antarctica is evidence of a dramatic change of climate over the past 180 million years. More precise evidence for the last two million years is available from a range of palaeoclimatological techniques (Chapter 3). Since the mid-nineteenth century, specific climatic data have been provided by actual measurements of the weather elements. The use of such records to interpret climatic change is, however, hampered by the lack of standardization of instruments, sites and units of measurements in the early days. Nevertheless, the most compelling evidence of climatic change is provided by the measurement of surface air temperatures over global landmasses since 1880, as shown in Fig. 4.1. The overall trend has been upwards, with an average rise of 0.5°C, although half of this increase has been recorded since the mid-1960s. During this time there have been four distinctive phases of marked temperature oscillations, namely: cooling up to the late 1880s; warming from the 1890s to the 1940s; cooling to the mid-1960s and a warming since the late 1960s (Fig. 4.1). The current global warming phase is of particular interest as seven of the 11 warmest years on record occurred between 1978 and 1988, with 1983, 1987 and 1988 clearly the three warmest years since observations began.

These data indicate that the current climatic trend is complex, being characterized by a series of wide ranging extremes in weather. At any time, and at any place, it is possible to experience droughts, floods, heatwaves, severe cold spells and devastating storms, which can completely mask the underlying long-term climatic trend. Furthermore, lasting climate change may mimic short-term fluctuations so that it becomes difficult to recognize the pattern of change until it has been in progress for decades. For example, the recent increased frequency of freak weather has been supported by observations in the south of England (Thompson, 1989). Out of the eight heatwave summers recorded since 1948 (i.e. with temperatures more than 3°C above the mean and at least 15 days of drought), six have occurred between 1975 and 1990. Interestingly, the increasing frequency of summer heatwaves has also been associated with a record number of severe winters in the British Isles. Again, south of England observations confirm that the six heatwaves since 1975 have been accompanied by seven severe winter cold spells or 'big freezes' (defined as monthly temperatures at least 3°C below the mean), compared with only two in the previous 27 years. Three

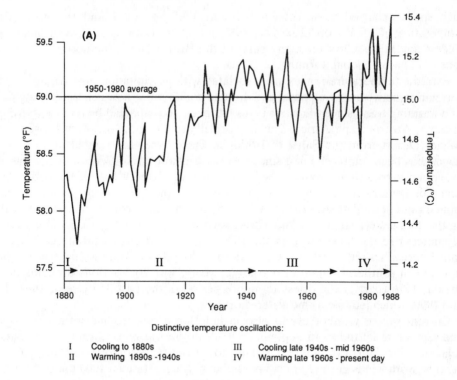

Distinctive temperature oscillations:

I	Cooling to 1880s	III	Cooling late 1940s - mid 1960s
II	Warming 1890s -1940s	IV	Warming late 1960s - present day

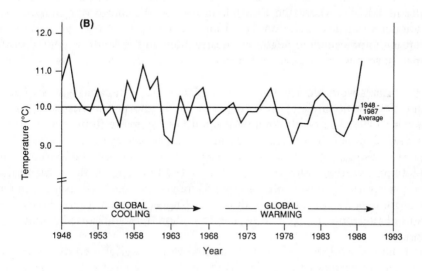

Figure 4.1 (A) Annual average global temperatures, 1880–1988. Adapted from sources quoted in Thompson (1989) by permission of Hodder & Stoughton. This first appeared in *New Scientist*, London, the weekly review of science and technology. (B) Mean temperatures at Reading, UK, 1948–89.

such spells occurred in succession between 1985 and 1987 and the mean daily temperature of −9.1°C on 12 January 1987 in the south of England was probably the coldest day this century for many parts of the British Isles (Thompson, 1989), at a time of apparent global warming.

Extreme temperature ranges are associated with precipitation fluctuations and the correlation between warm phases and drought conditions is now generally recognized. The warming trend since the early 1970s has been accompanied by recurrent drought in India, Australia and Africa. This is especially so along the southern fringes of the Sahara Desert from the Sahel to Ethiopia. Since 1968, the rainfall decline in the region has been unprecedented since records began in 1900. Central Sudan has also experienced persistent drought conditions since the late 1960s. 1984 was the driest year this century in the region and 1978 was the only year in the last 20 with an annual rainfall total greater than the 1951–80 mean. The record temperatures of 1988 in the midwestern states of the USA were accompanied by a crippling drought reminiscent of the 'dust bowl' in the 1930s, with disastrous crop failures in the so-called 'breadbasket' of the USA. Extreme drought conditions have been more frequent in southern Britain in the last 15 years, with four serious droughts in this period: 1976 (part of the driest 12-month period in the past 350 years), 1983, 1989 and 1990, which led to chronic water shortages.

Current severe weather events also include more catastrophic storms in tropical and temperate latitudes. In September 1988, in the Caribbean, after seven years of relative calm, Hurricane Gilbert proved to be the fiercest storm in the region this century, with winds gusting up to 349 km/hour. It devastated a 4000 km swath across Jamaica, the Cayman Islands and Mexico, killing 300 people and making 750 000 homeless, mostly in Jamaica. Gilbert caused damage estimated at £6 billion, especially in Jamaica where the £4.8 billion estimated damage was four times the gross national product. In addition, Hurricane Joan devastated Nicaragua in October 1988, causing unprecedented death and destruction. In the South Pacific, disastrous hurricanes have become a regular feature of the last decade and are not isolated occurrences as they are in the Caribbean. For example, between 1941 and 1980, five severe hurricanes were recorded in the Fiji Islands area compared with six between 1981 and 1985, with three of these experienced in the period March–April 1985 (Thompson, 1989). The north Atlantic has not been immune from severe storms in the last few years and in Britain, three major hurricane-type depressions between October 1987 and January 1990 have emphasized the disturbed and extreme nature of mid-latitude weather. The winds gusted to 190 km/hour on the night of 15–16 October 1987, with the lowest pressure of 955 mbar. The gusts represented the most violent storm experienced in the south-east of England for 300 years. It resulted in the deaths of 19 people, toppled more than 15 million trees and caused damage worth £1 billion.

Apart from the evidence provided by annual climatological records, other useful environmental indicators of climatic change include sea-level fluctuations and the regimes of glaciers and ice sheets. Average sea level has been rising since 1880 at a rate equivalent to 10–15 centimetres per century. The average global rate of sea-level rise has been estimated to be 1.51 mm per year between 1900 and 1975, with an accelerating rise since 1944 (Gribbin, 1988). Glacial melt due to global warming may account for about half of this rise in sea level and thermal expansion of the oceans for the remainder. This is, however, conjectural and the recognition of sea-level changes is complicated by land uplift or subsidence due to isostatic and eustatic processes (Section 3.2.1). The great ice sheets of Greenland and Antarctica have remained at equilibrium over the last century although, at present, their net mass balances appear to be close to zero, which makes the ice masses vulnerable to further climatic change. However, global warming will increase the moisture capacity and precipitation potential over these ice sheets, which would lead to strong positive mass balances and increased glaciation. In contrast, temperate glaciers have undergone significant wastage and retreat for most of this century, although many of them have advanced over the last decade due to the warming trend and increased precipitation rates.

At present, there is no overwhelming climatic or environmental evidence to support long-term global warming. Despite the 1989–90 consecutive heatwaves and droughts in the UK, it is not firmly established. Instead, periods of wildly erratic and freak weather have been experienced which have been reliably observed as indicators of a distinctive, if uncertain, climatic change. For example, during the 'Little Ice Age', which occurred in Britain in the 17–18th centuries, extreme weather events were experienced before the start of and after the end of this infamous cold period. Consequently, it appears that the current world-wide weather extremes are likely to be associated with a dramatic, if unpredictable, long-term climatic trend. It could lead to major global warming or, indeed, another ice age, although the former trend is the favoured scenario at present.

4.3 CAUSES OF CLIMATIC CHANGE

The evidence for climatic change is complex, and sometimes conflicting and wide ranging mechanisms are involved which are largely natural in origin. This is particularly so with changes over the long term (e.g. the geological time-scale discussed earlier), when dramatic fluctuations from tropical to polar climates can be related to major changes in the earth's orbital characteristics and terrestrial geography. The former changes are mainly associated with the astronomical or Milankovitch cycles (Chapter 3). Changes in terrestrial geography are mainly associated with the theory of continental drift. This once controversial mechanism is now supported by plate tectonics theory where, for example, the break-up and drift of Gondwanaland is assumed to account for the deterioration of climate in Antarctica since subtropical Permian times.

Current climatic change, and that over the next few decades (including the proposed progressive warming in the 1990s) will involve more subtle mechanisms than those mentioned above, which will include changes in solar radiation intensity and atmospheric composition. However, even though these changes are less dramatic than astronomical variations and drifting continents, their impact on climate and the resulting environmental systems could be considerable, especially if they are superimposed on reinforcing changes of a different nature. This forms the basis of the global warming or returning ice age theories. Consequently, this section will concentrate on these more immediately pertinent mechanisms which can modify the global heat balance and the resultant general circulation of pressure and wind systems.

4.3.1 Sun-spot cycles

Conspicuous variations in solar energy output are referred to as sun-spot cycles, which have been identified at regular intervals, particularly every 11 and 100 years. During these cycles, solar temperatures during dark cool spots (minima) and bright hot flares (maxima) can vary by up to 1100°C. This should influence the intensity of insolation, radiation balance and heat transfers at the earth's surface, with pronounced climatic deterioration and amelioration. However, the correlation between sun-spot cycles and temperature is uncertain and conflicting; for example, tropical temperatures showed a positive correlation with sun-spots between 1930 and 1950, but an inverse relationship from 1875 to 1920. The correlation is masked by variations in stratospheric ozone concentrations. Ozone is more abundant about two years before a sun-spot minima and the associated stratospheric warming weakens the subtropical anticyclones and the mid-latitude westerlies. This leads to cool and dry weather slightly out of phase with the sun-spot cycle.

Current research by the George C. Marshall Institute in the USA has correlated the worst winters of the Little Ice Age with 100-year sun-spot cycles. These winters appear to relate to the so-called 'quiet sun', or sun-spot minima, with correspondingly low temperatures (namely the 17th and early 19th centuries). Conversely, the 20th century has been dominated by sun-spot maxima and high temperatures, which correspond with the recognized overall global warming. The Institute forecasts that the 21st century will be a time of enhanced sun-spot minima and the return of Little Ice Age conditions. In time, this enhancement could balance the predicted global warming from an accelerating greenhouse effect and will considerably reduce the proposed environmental impact. It is obvious that the sun-spot cycles do not provide a reliable and conclusive explanation for climatic variations. Current climatic change is probably best explained in terms of fluctuations of the constituent gases and particulate matter within the atmosphere as they control the radiation balance and heat transfers at the earth's surface (see following sections).

4.3.2 The dust veil effect

The role of dust in climatic change is generally acknowledged and has been cited as the reason for global cooling in the Little Ice Age and the more recent 1940–1960s cool period. Dust is particulate matter temporarily suspended in the atmosphere which has been released from natural sources (especially volcanic activity) and human activities, such as the combustion of fossil fuels. Its concentration reduces the transparency of the air (i.e. increases the turbidity) and acts as a 'shield' or 'veil' in the troposphere. It is effective in scattering solar radiation back to space at the expense of insolation, which lowers surface temperatures. For example, it has been estimated that global mean temperatures will fall by 0.8°C for every 1% decrease in solar radiation (Thompson, 1989).

Natural dust-loading variations over time are associated with the changing frequency of volcanic activity. The increased vulcanicity between the late 1940s and mid-1960s (e.g. Agung in Indonesia, 1963, and Awu in the Celebes, 1966) could account for the global cooling experienced at that time. The climate–volcanic eruption association is very complex and sometimes confused as the current global warming trend has continued for a decade or so during which there have been two major eruptions at Mount St Helens in the USA (June 1980) and El Chichón in Mexico (April 1982). The prevailing dust-loading effects were evidently short-lived as the years following both eruptions (i.e. 1981 and 1983) were very warm world-wide, particularly 1983, which was one of the warmest on record. It must be emphasized that these natural emissions of dust have been aggravated by human activities over the last century or so, particularly with the combustion of fossil fuels, deforestation and overgrazing. Indeed, such anthropogenic supplementations mimic the natural dust-veil effect and associated global cooling. For example, global cooling due to increased vulcanicity between the late 1940s and mid-1960s was accentuated by increased combustion–dust loading (Fig. 4.2A), mainly due to the increasing motor vehicle exhaust emissions. This dust supply must have overwhelmed the increasing greenhouse effect because global warming was negated, despite a considerable increase in carbon dioxide (Fig. 4.2B) and the appearance of CFCs (Fig. 4.2C). At this time, dust loading and global cooling were linked with theories heralding a returning ice age in the 21st century.

Figure 4.2A shows the increased turbidity at Hawaii resulting from accelerating dust loading during the late 1950s and early 1960s. Even excluding the aerosol emission from the Mt. Agung volcanic eruption (i.e. the broken line in Fig. 4.2A), it is apparent that the turbidity increase over the period was about 30%. As Mauna Loa in Hawaii is remote from pollution sources, this increase represents an important change in the so-called background turbidity. Data from the more polluted Washington DC, USA, reveal a 57% increase in turbidity over the first 60 years of this century. Furthermore, about 70% of this increase was attributed to local pollution from coal and oil combustion (Thompson, 1989).

These measured turbidity, changes reveal significant (if localized) increases of between 30 and 57% which, on a global scale, would lead to a serious decline in temperatures. For example, Bryson (1968) estimated that a turbidity increase of about 3.5% would lower surface temperatures by 0.4°C. Consequently, a world-wide increase in the range 30–57% would be associated with global cooling of between 3.5 and 6.5°C, which would herald a super ice age. Ironically, the gain in heat from an increased greenhouse effect (of about the same order of magnitude as the global cooling envisaged from increasing dust veils) would be a welcome event as it would just be sufficient to counteract this dust-veil cooling. However, it could not match the cooling from a return to enhanced sun-spot minima as well (Section 4.3.1), if this trend becomes a reality.

4.3.3 The enhanced greenhouse effect

The current debate about increasing carbon dioxide concentrations and global warming has made a considerable impact on society. This results from constant media publicity and has been exacerbated by irresponsible and scaremongering predictions. There is no doubt that greenhouse gases have a vital role to play in the atmospheric heat balance as their absence would lower the global surface temperatures by 30–40°C, rendering the earth lifeless. Although these gases are mainly transparent to incoming short-wave solar radiation, they are able to absorb some of the outgoing long-wave terrestrial radiation (in the infra-red part of the spectrum in the wavelength

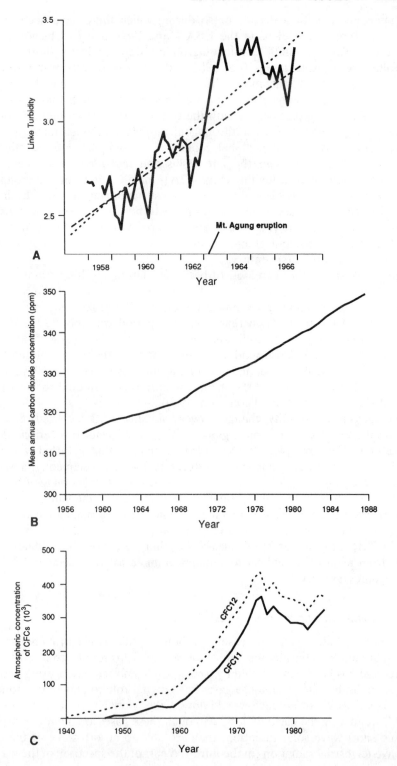

range 4–100 μm). This absorbed radiation is then re-radiated back to the earth's surface producing a warming effect. Greenhouse gases occur naturally in the atmosphere and include carbon dioxide (produced by volcanic activity and respiration), methane (from bacteria in swamps and marshes) and nitrous oxide (from denitrifying bacteria from nitrate use). These trace gases occur in minute proportions and even though carbon dioxide is the major trace constituent, it only represents some 0.03% by volume of dry air (equalling 345 ppm).

As was discussed earlier, trace gas concentrations are not constant and the variation of carbon dioxide in particular has been correlated with interglacials and ice ages (Fig. 3.3). Over the last 150 years, inadvertent human activities (particularly the burning of fossil fuels) have increased the concentration of carbon dioxide from 300 ppm in 1900 (the first actual measurements), to 316 ppm by 1958 and 350 ppm by 1990. Carbon dioxide in the atmosphere has increased by 25% in the last century (adding some 300×10^9 tonnes of this gas to the atmosphere since the Industrial Revolution began) and about half of this increase has occurred in the last 25 years. Figure 4.2B illustrates the recorded carbon dioxide increase at Mauna Loa, Hawaii, between 1956 and 1988. Carbon dioxide is currently increasing by 1.5% per year (equalling 1.5 ppm). At this rate, carbon dioxide concentrations will increase by 150% over the next century compared with a mere 14% recorded over the last 100 years. Furthermore, predictive modelling suggests that carbon dioxide levels will reach 540 ppm earlier, by the year 2050 (i.e. doubling the pre-industrial level). This is the basis of the concern about global warming which is discussed in the following.

There is no doubt that artificial carbon dioxide concentrations have increased the absorptive strengths for terrestrial infra-red radiation in the band between 13 and 100 μm. At the same time, inadvertent human activities have increased other important greenhouse trace gases. For example, methane is produced by rice paddies, rubbish tips and livestock populations and has increased from a level of 700×10^9 ppb in the 18th century to about 1669 ppb today. The atmospheric concentration of methane today is 11% higher than 10 years ago and the current annual rate of increase is assumed to be 17 ppb or 1% of the current concentration (Rowntree, 1990b). Nitrous oxides (produced artificially from chemical fertilisers, oil and coal combustion and as a spray-can propellant) show a more modest increase of 0.7 ppb per annum (or 0.25%). However, as they can remain in the atmosphere for 150 years or more, their cumulative impact could accentuate global warming for some considerable time.

Methane and nitrous oxides contribute about 30% of the greenhouse effect and are effective absorbers of infra-red radiation between 7 and 13 μm. This range is the so-called atmospheric 'window', where infra-red radiation can escape into space, as both water vapour and carbon dioxide are ineffective absorbers at these wavelengths. A new group of greenhouse gases appeared in the 1940s (Fig. 4.2C). These are CFCs, which are also notorious for their role in ozone depletion (Section 4.3.4). It is now

---

Figure 4.2. (A) turbidity levels and (B) mean annual carbon dioxide levels at Mauna Loa, Hawaii. Adapted from sources quoted in Thompson (1989) by permission of Hodder & Stoughton. This first appeared in *New Scientist*, London, the weekly review of science and technology. (C) Concentration of CFCs in the atmosphere 1940–85. Adapted from Gribbin (1988) by permission of the Controller of Her Majesty's Stationery Office. (A) (—) Measured turbidity; (- - -) linear trend; (– – –) linear trend excluding Mt Agung eruption.

accepted that CFCs contribute 15% of the greenhouse effect, and are produced from a wide range of processed compounds used for spray-can propellants, air conditioners, foam-blowing agents, refrigerators and foam hard plastics. CFCs are particularly strong absorbers in the atmospheric 'window' where carbon dioxide is ineffective. They are also 20 000 times more effective in greenhouse activity than carbon dioxide.

Figure 4.2C shows that there has been a dramatic rise in the two most commonly used CFCs, namely CFC11 (trichlorofluoromethane) and CFC12 (dichloroflurome-thane). Atmospheric concentrations have risen from zero in 1940 to 320 000 tonnes in 1985. Interestingly, the peak concentrations occurred in the early 1970s before the US Government partially banned CFC based sprays in 1978 (discussed later), which caused a decline in the concentration of these compounds by about 25% between 1974 and 1982. Until the Montreal Protocol in 1987, CFCs increased by 5% per year. Despite their current post-Protocol reductions (and even the revised zero production by the year 2000), the main concern associated with CFCs is their stability over time. They can accumulate at a rate five times faster than they are destroyed by ultraviolet radiation in the stratosphere. Consequently, the estimated 'life spans' of CFC11 and CFC12 are 75 and 110 years, respectively. Whatever controls are introduced in the future, the impact of CFCs already in the atmosphere will continue until the end of the 21st century, with implications for global warming and ozone depletion (Section 4.3.4). Nitrous oxides have an even longer life span and will aggravate the impact of CFCs well into the 22nd century. Although these trace gases contribute less than half of the current combined greenhouse effect, their cumulative effect on climate over the next 50 years will probably exceed that of carbon dioxide (Schneider, 1987).

The combined build-up of all trace gases over the last century or so has contributed to the enhanced greenhouse effect and (probably) the recognized overall 0.5°C temperature rise. The marked temperature oscillations observed during this period (Fig. 4.1) are not easily explained in terms of variations in concentrations of trace gases, but could represent the temporary or periodic overwhelming of the greenhouse effect by sun-spot minima and increasing dust veils. Future levels of trace gases and their possible impact on global warming have been simulated by climate models. These are mostly based on the projected carbon dioxide doubling (from the 1850 level of 270 ppm) and incorporate equilibrium response and feedbacks, including water vapour increases, decreases in snow and ice albedo (reflectance) and cloud changes (Rowntree, 1990a). Table 4.1 shows the results for four such models published around 1986 (Rowntree, 1990b), between which there is good agreement. The data confirm an annual average global warming of between 3.5 and 5.2°C and increased global precipitation of between 7 and 15%. The limitations of such models are associated with the simple representation of the ocean, with constant oceanic heat advection, and basic cloud characteristics, with fixed reflectance and radiative properties (Rowntree, 1990a; 1990b).

These limitations highlight the inadequacies and subjectivity of predictive modelling. This uncertainty is emphasized by a recent experiment re-using the Meteorological Office model (Mitchell et al., 1989), with more realistic cloud microphysics and radiative properties than those used in the Table 4.1 models. In this experiment, the global temperature rise associated with a doubling of carbon dioxide was reduced from 5.2°C (Table 4.1) to 2.7°C (using cloud microphysics properties) and to only 1.9°C using variable cloud radiative properties (Rowntree, 1990b). It is also necessary to include increases in all trace gases and the dust veil to accurately predict future

Table 4.1   Modelled equilibrium responses for a doubling of atmospheric carbon dioxide levels. Global mean and eastern UK values are given. Adapted from Rowntree (1990b) by permission of The Royal Meteorologial Society

Equilibrium response	Model*			
	GFDL	GISS	MET.O	NCAR
Warming (°C):				
Global mean	4.0	4.2	5.2	3.5
Eastern UK JJA+	5.0	4.0	5.0	4.0
Eastern UK DJF+	5.5	4.0	6.0	6.0
Precipitation:				
Global mean (%)	+9	+11	+15	+7
Eastern UK JJA+(mm/day)	−(0–1)	−(0–1)	+(0–1)	+(0–1)
Eastern UK DJF+(mm/day)	+(0–1)	+(0–1)	+(0–1)	+(0–1)

* Models are low resolution versions from: Geophysical Fluid Dynamics Laboratory, Princeton, NJ, USA (GFDL); Goddard Institute for Space Studies, New York, USA (GISS); Meteorological Office, Bracknell, UK (MET.O); National Center for Atmospheric Research, Boulder, CO, USA (NCAR).
+JJA = June, July, August; DJF = December, January, February.

climatic change. Hansen *et al.* (1988) used data on the three main greenhouse gases (with different emission rates) in a model incorporating three carbon dioxide scenarios (Fig. 4.3). The results emphasize that the only acceptable alternative is scenario C, which demands drastic cuts immediately. This conclusion was confirmed by delegates at the first IPCC meeting in Windsor, UK, in 1990. After this meeting, even the British Government became committed to stabilizing carbon dioxide emission at their 1990 level by the year 2005, which in effect amounts to a 30% cut based on current projections. However, Hansen *et al.* (1988) stressed that the model's results could be askew by as much as 50%. Thus, planned cuts in emissions must be constantly reviewed to allow for the limitations of simulated modelling.

Table 4.2   Probable effects of increased carbon dioxide levels in the atmosphere. Adapted from Perry and Perry (1986) by permission of Bell & Hyman (now Unwin Hyman of HarperCollins Publishers Ltd)

Latitude	Average annual change in surface temperature (°C)*	Change in precipitation (%)
60°N	7.5	+18
50°N	6.0	+ 4
40°N	6.0	−14
30°N	4.5	0
20°N	2.5	+20
10°N	1.5	+20
Equator	3.0	0
10°S	4.0	−20
20°S	4.5	− 5
30°S	4.0	+ 5
40°S	4.0	+12
50°S	3.0	+13
60°S	2.5	+12

* All positive changes

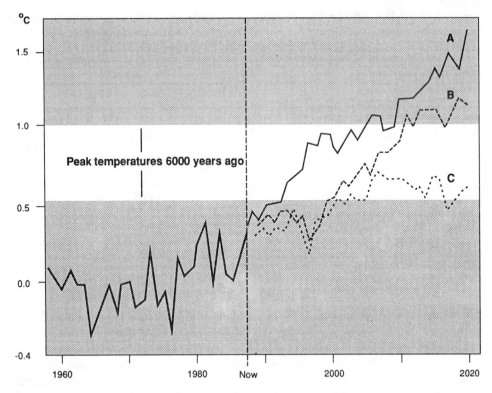

Figure 4.3 Climate model predictions of annual mean global temperature change. (A), (B), (C) represent scenarios for possible future carbon dioxide emissions: (A) Trace gas emissions continue to grow at current rates of about 1.5% per year, producing an escalating 'forcing' effect for global warming; (B) growth rate slows so that the 'forcing' effect remains constant; (C) drastic cuts in emissions over the next 10 years cause the 'forcing' effect to cease growing by the year 2000. Adapted from Hansen *et al.* (1988). Copyright © 1988 by the American Geophysical Union.

To conclude, it must be noted that the global climate change proposed by these models (Table 4.1) would not be uniform over the earth's surface. Table 4.2 shows the considerable variability in predicted changes in average surface temperatures and precipitation at different latitudes over the next 60–80 years, following a doubling of the carbon dioxide concentration in the atmosphere. Temperature changes would be larger at high latitudes (e.g. 60–80°N), and in winter, thus reducing the intensity of seasonal fluctuations and the thermal contrasts between high and low latitudes. Rainfall changes are more difficult to predict, but it appears that the increase will be mainly in subtropical regions between 10 and 20°N (including the Sahel and Ethiopia), following an intensification of the intertropical convergence zone. Conversely, the USA, the Mediterranean lands, Brazil, Peru, Zaïre and Angola will receive much less rainfall and more regular severe droughts. The resultant average global climatic conditions will be well beyond the range of climates experienced in the last two to three million years. For example, a global mean temperature rise of 3°C (which the warming consensus agrees is likely to happen during the next century) would represent even warmer conditions than those in the previous interglacial 120 000 years ago (Section 3.2.1). However, sceptics argue that the impact of feedback mechanisms, sun-spot cycles and dust veils should not be ignored and that only 20 years ago, fears were being expressed that another ice age was on the way!

4.3.4 Ozone depletion

In 1984–85, another major environmental issue was highlighted with the discovery of the so-called ozone 'holes' over Antarctica. The international outcry that followed resulted in the Montreal Protocol in 1987, which called for cuts in CFC production within five years.

Ozone (O_3) is concentrated in the stratosphere at elevations between 20 and 30 km and is formed by the reaction of ultraviolet solar radiation (UV), less than 190 nm wavelength, with oxygen (O_2). This process is speeded up by a neutral molecule (M), normally nitrogen, which acts as a catalyst for the reaction. The formation is represented by the following equations:

$$O_2 + UV \rightarrow O + O \tag{4.1}$$
$$O + O_2 + M \rightarrow O_3 + M \tag{4.2}$$

Ozone is depleted naturally as it reacts in a number of ways: firstly (equation 4.3), with UV radiation at wavelengths between 230 and 290 nm; secondly (equation 4.4), with nitric oxide (NO) produced by the oxidation of nitrous oxide (Section 4.3.3); and finally, (equation 4.5) with chlorine (Cl) released during volcanic eruptions.

These depletions are represented by the following equations:

$$O_3 + UV \rightarrow O + O_2 \tag{4.3}$$
$$NO + O_3 \rightarrow NO_2 + O_2 \tag{4.4}$$
$$Cl + O_3 \rightarrow ClO + O_2 \tag{4.5}$$

At present, natural depletion is accelerated by human activities, especially the production of CFCs, which are decomposed by UV radiation and release free chlorine atoms. These react with ozone (equation 4.5) to produce chlorine monoxide (ClO) and oxygen in a catalytic cycle, which effectively removes two molecules of ozone without consuming chlorine. Continued ozone depletion will lead to a pronounced cooling of the stratosphere (between 10 and 20°C) and a bombardment of the earth's surface by UV radiation. It has been estimated that every 1% decrease in ozone concentration would produce a 2% increase in UV radiation. The increased solar radiation intensity would also accentuate global warming from increased carbon dioxide emissions. Furthermore, this magnitude of UV radiation increase could result in an 8% increase in non-melanoma skin cancer in white-skinned people. However, it must be remembered that the interaction between human activities and ozone concentrations is complex. The ozone balance between depletion from chemical pollution and production due to oxygen photodissociation involves over 50 chemical compounds and processes and 150 chains of chemical reactions (Gribbin, 1988).

Predictive models, involving both stratospheric chemical and energy budgets, have in the past produced exaggerated forecasts of the depletion of ozone and its environmental impact. The early models of the 1970s associated ozone depletion with the emission of oxides of nitrogen (equation 4.4) from supersonic transport (SST) aircraft. They suggested that such an emission could deplete ozone by as much as

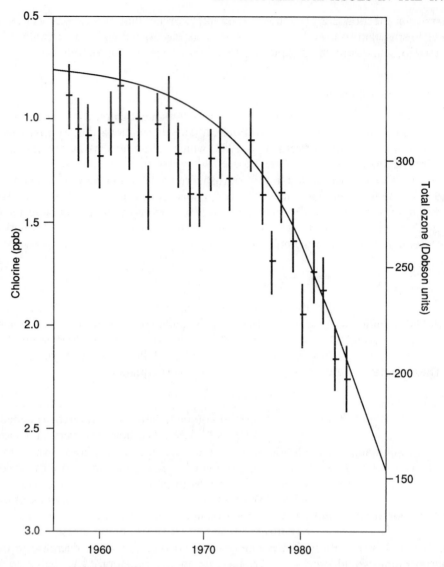

Figure 4.4   Mean monthly values of total ozone (length of vertical bar indicates data range) and concentrations of CFCs, Halley Bay, Antarctica 1957–84. Adapted from sources quoted in Gribbin (1988) by permission. One Dobson Unit = one thousandth of a centimetre of ozone.

50% and increase the surface UV radiation by about 100%, with a massive increase in the incidence of skin cancer. Later models were more comprehensive and conservative and ozone depletion by the exhausts of SST aircraft was reduced to a meagre 1%. Actual measurements showed that ozone increased by 5% following the release of oxides of nitrogen from atmospheric nuclear bomb tests in 1961–62 (in which the release equalled that expected from 600–1000 fully operational SST aircraft).

The concern with SST aircraft therefore disappeared in the USA (along with the

anti-Concorde lobby based on the tenuous exhaust causing increased cancer argument) and attention was switched in the mid-1970s to the possible destruction of
ozone by CFCs, used extensively as spray-can propellants. Without actual proof of
this destruction, the US Government accepted this new 'environmental scare theory'
(as it was reported in the British press), and in 1978 removed 1000 million aerosols
from stores in the USA. Evidence to support this theory was slow to appear because
even though 20% ozone depletion was first noticed in Antarctica in October 1982, it
was considered to be anomalous and the result of outdated instrumentation. Two
years later, improved British Antarctic Survey instruments confirmed the ozone
depletion. Earlier data were then re-checked and showed that the decline had been
accelerating since the late 1970s (Fig. 4.4). In addition, measurements of CFCs
showed a significant increase during this period of ozone depletion.

The Antarctic springtime ozone hole is now an observed phenomenon and became
almost complete over Halley Bay at 16.5 km on 7 October 1987, following a 97.5%
destruction of the amount of ozone recorded on 15 August 1987 (Farman, 1987). The
Antarctic stratosphere has stable, very cold air below $-85°C$, where methane is an
ineffective chlorine scavenger. It is a perfect environment for the operation of the
chlorine catalytic cycle when the sun returns to the Antarctic stratosphere in spring,
with a continuous supply of UV radiation. Similar holes outside Antarctica have not
been discovered, although Arctic depletion rates are causing concern. However,
global ozone depletion between 1979 and 1986 was 5% (as large as the ozone increase
recorded in the 1960s). By the mid-1980s, CFCs were increasing by about 5% per
year (Section 4.3.3.) and their long residence time in the atmosphere gave rise to
great concern.

Consequently, the Montreal Protocol of September 1987 (i.e. formulated a few
weeks before the discovery of the almost complete hole over Antarctica) aimed at a
50% cut-back in the production of CFCs by 1999. The hole discovery of 7 October
1987 suggested that this reduction was totally inadequate and a Protocol review in
1990 imposed a total ban by the year 2000 and promised developing countries
financial aid to produce alternatives to CFCs, especially for refrigeration. Despite the
success of the Protocol review, it is not clear that CFC reduction will halt ozone
depletion, nor is it certain that the existing ozone depletion has caused substantially
increased UV radiation and skin cancer. Firstly, measurements of the 5% ozone
decrease between 1979 and 1986 were not matched by the expected 10% increase in
UV radiation and 40% increase in skin cancers. Secondly, ozone appears to be at risk
from bromine, used in industrial compounds and fire extinguishers. Indeed, British
Antarctic Survey scientists believe that the main anthropogenic destroyer of ozone is
bromine, especially two widely used compounds (1211 and 1301) which have a
particularly long residence time in the stratosphere.

4.4   CONCLUSIONS

In this chapter the evidence and causes of climate change have been examined,
particularly the contributions of dust veils, ozone depletion and enhanced greenhouse
gases as on-going mechanisms which could well be influential in the 1990s and
beyond. In the last three decades, environmental bandwagons have been evident for
all three contributions, namely the returning ice age fear of the late 1960s to early

1970s, the skin cancer scare of the late 1970s to early 1980s and the current global warming controversy. In all three instances, the actual evidence for the associated climate change is often contradictory. Furthermore, the projected changes over the next century or so are based on climate modelling, which has serious limitations and is mostly too simple and generalized to represent all possible feedback mechanisms. Despite these problems and reservations, it is important for society to control the output of all types of pollutants that may seriously harm the earth and its atmosphere. Also, all pollution controls need to be integrated components of sustainable development plans to preserve future environmental quality.

If CFCs and greenhouse gases are allowed to increase at an accelerating rate, then the environment would be at considerable risk from climatic change, and the predicted impacts could well reach disaster levels within the next century. These consequences were emphasized in Toronto, Canada, in 1988 at a meeting of the World Commission on Environment and Development, which concluded that the 'time has come to develop an action plan for protecting the atmosphere . . . we have come to a threshold. If we cross the threshold, we may not be able to return'. Details of the environmental consequences of the discussed scenarios have been outlined by Thompson (1989) and it must be stressed that the repercussions could be catastrophic for all societies.

This is exemplified by the rise of sea level that would follow prolonged warming, causing sea water to expand and polar ice sheets to melt. Projections of such rises range from 8 mm per year in future decades to a maximum of 2 m by the end of the next century. Sensitive areas such as river deltas and low-lying coastal areas, such as The Netherlands, Bangladesh and coral islands in the Indian and Pacific Oceans, are at risk of flooding. The mass exodus of people from such areas would create a new class of environmental refugees on a massive scale. Even Britain would suffer from a 1 m rise of sea level, and it has been estimated that this would make 15 million people homeless (especially in East Anglia and the Thames estuary). US estimates of the cost of efforts to protect populated coastlands from a similar rise exceed 100×10^9. Terrestrial and aquatic ecosystems, agricultural production and human health would be at risk. Thus, the immediate control of global warming and ozone depletion is vital for the future of the earth.

The world's leading chemical manufacturers (especially ICI in the UK and Du Pont in the USA) are already producing alternatives to CFCs. Bromine 1211 and 1301 should also be banned, with substitutes made available. Atmospheric 'engineering' has been suggested in the media as a solution, which has been described by the scientific community as impractical, if fun. Techniques include the release of ozone pellets, destroying CFCs with lasers and reducing global warming by covering the world's oceans with white polystyrene chips to increase the global albedo. The so-called carbon tax has also been considered as a solution to reduce carbon dioxide emissions. Perhaps more traditional conservation methods are the best way forward, particularly with current energy consumption and the exploitation of the wide ranging alternative, renewable (largely atmospheric) energy resources. Even afforestation on a gigantic scale would help. For example, the US Department of Energy has estimated that new forests covering 7×10^6 km^2 (and the associated increased photosynthetic uptake) could absorb all the predicted releases of carbon dioxide from fossil fuels. However, this would mean covering an area the size of the USA (excluding Alaska) with trees, unless future genetic engineers can increase their growth rate and hence their ability to absorb carbon dioxide.

Society faces an uncertain future, which arises from the lack of substantial, direct evidence for the climatic change scenarios presented and from the inherent problems of predictive modelling. Thus, there is an urgent need to begin immediately a global network of surface baseline measurements of the atmospheric and environmental factors discussed in this chapter. For example, if serious ozone depletion is taking place, then there must be evidence of greatly increased UVC and UVB radiation in Antarctica and adjacent middle latitudes. Similarly, if the enhanced greenhouse effect is a reality, then there must be global evidence of greatly increased infra-red counter-radiation at the earth's surface and cooling in the lower stratosphere. A network of surface and upper air observations is therefore necessary over the next decade or so to test model predictions. Supporting environmental evidence will be available for ozone depletion from the incidence rates of eye cancer and pink eye in the penguin population of Antarctica, signs of increased tissue damage and modified plant growth on sub-Antarctic islands and increased skin cancer and eye cataracts in long-term residents in the Arctic. Similar supporting evidence for global warming would include sea-level changes, glacial mass balance deficits, permafrost degradation and latitudinal and altitudinal shifts of plant species.

These patterns will only become evident over the ensuing decades, although they will most certainly occur if predictions from current climatic modelling are correct and if the promised cut-backs are ineffective or inadequate. The implementation of such an ambitious, world-wide monitoring project over the next decade requires considerable funding and a high degree of international co-operation. Such a project is urgently needed to end wide ranging speculations about the future of the planet earth and to confirm or deny the apocalyptic role of current atmospheric changes.

4.5 REFERENCES

Bryson, R. A. (1968) All other factors being constant . . . a reconciliation of several theories of climatic change. *Weatherwise*, **21**, 56–61.

Farman, J. (1987) What hope for the ozone layer now? *New Scientist*, **116** (1586), 50–54.

Gribbin, J. (1988) *The Ozone Hole*. London: Corgi Books.

Hansen, J., Fung, I., Lacis, A., Rind, D., Lebedeff, S., Ruedy, R., Russell, G. and Stone, P. (1988) Global climatic changes as forecast by Goddard Institute for Space Studies three-dimensional model. *Journal of Geophysical Research*, **93**, 9341–9364.

Mitchell, J. F. B., Senior, C. A. and Ingram, W. J. (1989) CO_2 and climate: a missing feedback. *Nature (London)*, **341**, 132–134.

Perry, A. and Perry, V. (1986) *Climate and Society*. London: Bell and Hyman, pp. 78–79.

Rowntree, P. R. (1990a) Estimates of future climatic change over Britain. Part 1: Mechanisms and models. *Weather*, **45**(2), 38–42.

Rowntree, P. R. (1990b) Estimates of future climatic change over Britain. Part 2: Results. *Weather*, **45**(3), 79–89.

Schneider, S. H. (1987) Climate modelling. *Scientific American*, **256**, 72–81.

Thompson, R. D. (1989) Short-term climatic change: evidence, causes, environmental consequences and strategies for action. *Progress in Physical Geography*, **13**, 315–347.

4.6 FURTHER READING

Bolin, B., Döös, B. R., Jäger, J. and Warrwick, R. A. (1989) *The Greenhouse Effect, Climatic Change and Ecosystems*, SCOPE 29. Chichester: Wiley.

Houghton, J. T., Jenkins, G. J. and Ephraums, J. J. (1990) *Climate Change*. Cambridge: Cambridge University Press.
Kemp, D. D. (1990) *Global Environmental Issues*. London: Routledge.
National Research Council, USA (1991) *Confronting Climate Change: Strategies for Energy Research and Development*. Washington, DC: National Academy Press.
Parry, M. (1990) *Climate Change and World Agriculture*. London: Earthscan.

CHAPTER 5

Forests, Woodlands and Deforestation

J. G. Soussan and A. C. Millington

5.1 INTRODUCTION

Forests and related land cover types (woodlands, shrublands, degraded forests and forest fallows) are important terrestrial ecosystems. They account for over 52×10^6 km^2 of the earth's land area, occupy three times the area of croplands and 75% more area than grasslands. They are vital to the ecological functioning of the planet, producing 60% of the net primary productivity of all terrestrial ecosystems, of which tropical forests account for approximately two-thirds. They are also the habitat of a large proportion of the earth's plant and animal species, providing the basis for the biodiversity which is essential for the biosphere's future.

The world's forests serve a series of vital environmental services, without which the functioning of the biosphere would be endangered. Woodlands and forests regulate water regimes by intercepting rainfall and regulating their flow. This is particularly important in areas with highly seasonal rainfall, a characteristic of many tropical regions. Trees are vital to the maintenance of soil quality, providing organic matter through leaf fall, limiting soil erosion through the binding effects of root systems and protecting soil from the direct impact of rainfall. On a regional and global scale forests play a part in modulating climates and are the lungs of the planet; deforestation is believed to contribute as much as 25% of the increased carbon dioxide which is the principal cause of the enhanced greenhouse effect (Chapter 4). This occurs because heat-trapping gases are released by the burning of felled woodlands and because the absorption of carbon dioxide from the atmosphere via photosynthesis is reduced.

Tree resources are also of major economic importance. They form the basis of a range of industries, notably timber, processed wood and paper, but including products such as rubber, fruits and coffee. The amounts of wood involved are immense: in 1988 total world roundwood production was estimated at 3.431×10^{12} m^3 (a figure which is probably an underestimate, as non-commercial production in developing countries is not included). Similarly, world paper production was an estimated 1.7×10^{11} tonnes in 1988. Most roundwood and paper is used within the country of

Environmental Issues in the 1990s. Edited by A. M. Mannion and S. R. Bowlby
© 1992 John Wiley & Sons Ltd

production, but wood products are still major exports. Over 2.5×10^{11} m^3 of wood and 5.1×10^{10} tonnes of paper were traded internationally in 1988. The value of this trade was more than US$$9 \times 10^{11}$ in 1988, 3% of total world trade. Many of these products come from plantations and other heavily managed areas, but natural forests are still the source of important industries, especially in tropical areas.

The world's forests are also the home of millions of people, many of whom live pre-industrial lives and have unique cultural heritages. These ways of life depend totally on the survival of the forests, and are under threat from the spectre of deforestation. Forests and woodlands (including patches of woodland in agricultural areas) are also the source of many products which are vital to the viability of agricultural communities throughout the developing world, e.g. fuel, fruits, building materials, medicines and herbs. These forest products provide diversity to the rural economy, and security when times are difficult. The loss of tree resources undermines the viability of agricultural economies, and in particular makes the poor more vulnerable to environmental or economic disruption. Deforestation, particularly in developing countries, is consequently not only an environmental problem; but also a development disaster.

These statements illustrate the main theme of this chapter: that forests and woodlands are one of the world's most important resources. Their true significance can only be evaluated from a perspective which looks at economic and human management issues alongside environmental issues. It is only from this basis that it will be possible to move to a truly sustainable stewardship of this great resource. This stewardship must recognize that the provision of such vital products and processes will continue only if the way they are exploited changes from the current rapacious approach to one which is more sustainable. It is thus essential to maintain the integrity of forest ecosystems.

5.2 CURRENT STATUS OF THE WORLD'S FORESTS AND WOODLANDS

Discussion of forestry issues is dominated by anxieties over deforestation, but these fears are not always based on incontrovertible scientific evidence. There is considerable uncertainty about the health and extent of the world's forests. Accurate assessments of the current level of global forest cover and rates of deforestation have proved elusive for four main reasons:

(a) variations in the definition of forest and forest land between countries
(b) variations in the techniques used to provide an inventory of forest cover and types, often because they are specific to the purpose of the inventory
(c) inventories concentrate on commercial forests and woodlands and neglect trees in other areas; particularly important in this respect are trees on farms, which in some countries are an important source of wood
(d) data may be withheld by governments and industry for strategic reasons.

Many estimates of forest areas and rates of deforestation are consequently little more than a compilation of already existing statistical information, which vary considerably in accuracy.

Despite the uncertainties, a number of important trends can be identified in Fig. 5.1 and Table 5.1. Conifers dominate large parts of the forest landscape of the Soviet

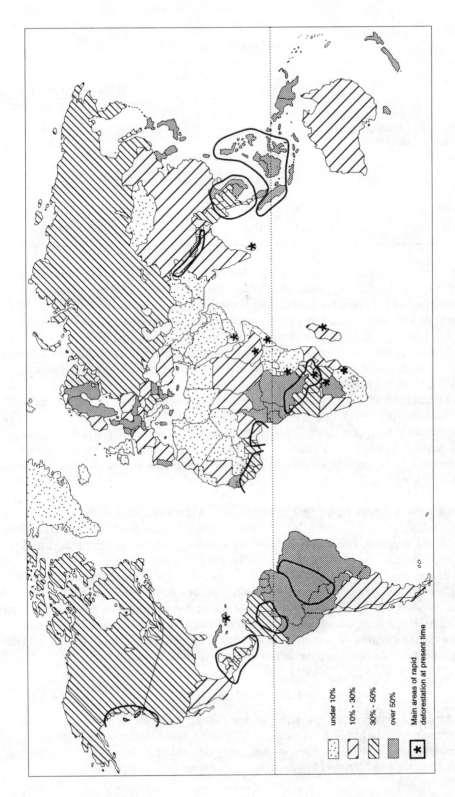

Figure 5.1 Percentage of land area of different countries occupied by forest and woodland.

Table 5.1   Estimate of land cover type as a proportion (%) of land area in the mid-1980s. Adapted from Westoby (1989)

Region	Closed forest	Other types of wooded land	Fallow land and shrubland
North and Central America (including the Caribbean)	24.1	12.5	4.3
South America	37.5	13.9	12.3
Europe	26.3	4.2	0
Asia (excluding the USSR)	14.3	4.2	2.5
Australasia (including insular South-east Asia)	20.8	7.1	4.3
USSR	35.4	6.1	0
World total	22.0	10.3	7.7

Union and, to a lesser extent, Europe and North America. In Latin America, Africa and the remainder of Asia, broadleaf evergreen and deciduous tropical forests dominate. This basic division has important implications for the types and rates of forest exploitation. Forest cover is highest in Latin America (46% of the land area) and the Soviet Union (41%), but is as low as 15% in the rest of Asia. If shrubland and forest fallows are included in these estimates, the proportion of wooded land in all of the developing world is increased. Latin America still has the greatest cover (61%), but the increase in Africa (from 24 to 45%) reflects the large areas of semi-arid shrubland and woody agro-ecosystems that dominate much of that continent. It is also interesting to compare the forest cover of developed and developing countries. The developed world generally has a lower proportion of its land forested than developing countries, but has greater levels of closed forest and woodland. The dichotomy between developed and developing countries will be discussed in Section 5.3.

The regional estimates of forest cover in Table 5.1 mask significant interregional variations in forest cover and composition. Although Latin America is the most wooded region in the world (Fig. 5.1), the forested areas are concentrated in Brazil, Bolivia, Paraguay, Peru, Guyana, French Guiana, Surinam and Uruguay, and there is little forest in Argentina and Chile. In Asia, the least forested region, there are relatively high proportions of forest in the Koreas, Japan, Taiwan, Malaysia, Vietnam, Cambodia, Nepal and Sri Lanka. Recent, more detailed estimates of forest and woodland cover for Sub-Saharan Africa at the subregional level (Millington *et al.*, 1992) also indicate significant regional variations in forest and woodland cover. These occur because of climatic and ecological variability, although further variations within ecological zones are a function of the extent of deforestation.

5.3   DEFORESTATION

Despite receiving much attention in the last decade, deforestation is not a new phenomenon; neither is it restricted to humid tropical forests, as some of the popular literature might suggest. This section provides insights into the global extent of deforestation, and Section 5.4 illustrates this problem with case studies.

5.3.1 Rates and causes of deforestation

Just as the accurate assessment of forest cover is difficult, so too is the determination of deforestation rates. Again the problems hindering such assessments are mainly methodological. Estimates made in the early 1980s for tropical forests (Table 5.2) suggest that about 0.6% of tropical forests are lost annually. This equates to about 11.3×10^6 ha worldwide, or an area about half the size of the United Kingdom. Deforestation rates are slightly higher in tropical America than in tropical Asia, but those in tropical Africa are significantly lower. Figure 5.2 shows that the rates of deforestation in each of these regions varies with forest type. In tropical Asia the highest rates of deforestation are found in closed forests which are exploited for timber, whereas in Africa deforestation is most common in the open woodland ecosystems.

Table 5.2 Mean annual rates of deforestation of tropical forests, 1981–85. All values are percentages as a proportion of the forest area in 1980. Adapted from FAO (1982) and Westoby (1989)

Region	Undisturbed productive closed forest	Logged productive closed forest	All closed forest (productive and unproductive)	Open forest	All forests
Tropical America	0.29	2.80	0.64	0.59	0.63
Tropical Africa	0.19	2.41	0.61	0.48	0.52
Tropical Asia	0.39	2.14	0.60	0.61	0.60
Total	0.28	1.98	0.62	0.52	0.58

In Africa the amount of open woodland lost due to land clearance between 1980 and 1985 was approximately 11.5×10^6 ha. In addition to this outright destruction of woodland, degradation of the remaining woodlands occurred through fuelwood and timber harvesting and tree-cutting for fodder and grazing. The main focus of this activity has been in the semi-arid parts of East, West and Southern Africa. The loss of African closed forests has mainly occurred in the west African countries of Ghana, Guinea, Ivory Coast, Liberia and Nigeria, due to the steady expansion of hardwood logging. The rates of forest destruction in these countries are approximately seven times the global average (Fig. 5.2). Significant closed forest destruction has also occurred in Ethiopia and Madagascar.

In tropical Asia closed forest destruction has been due to land clearance for agriculture, planned migration and resettlement projects (e.g. the Indonesian trans-migration project), and timber extraction. The rates are highest in Burma, India, Indonesia, Laos, Malaysia, Nepal, the Philippines and Thailand. The situation in China is difficult to assess because, despite numerous reafforestation projects, including the ambitious 'Green Belt' project to arrest desertification, some areas, for example Yunnan Province, have very high rates of deforestation. In addition, deforestation rates are low in large parts of China, India and elsewhere in Asia because there is very little forest cover left. This reflects the historical deforestation of these regions.

84

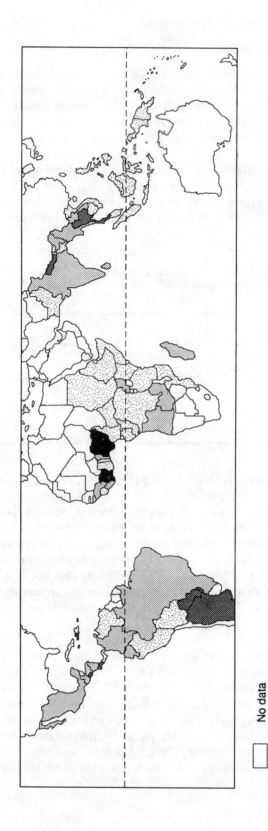

Figure 5.2 Deforestation rates as a percentage of closed forest area (1980–90). Sources: (1) Argentina, Brazil, Colombia, Equador, Mexico, Peru, Ivory Coast, Nigeria, Zaïre, India, Indonesia, Malaysia, Myanmar, Thailand from Wood (1990); (2) FAO (1981a, b), FAO/UNECE (1985).

5.3.2 Measuring and monitoring deforestation

Much research on the measurement of the rates of, and processes leading to, deforestation has been carried out in Latin America. The problem is acute in parts of Bolivia, Colombia, Ecuador, Mexico, Peru and Venezuela, but exceeding all of these in importance is Brazil, which accounts for about 35% of all deforestation in Latin America and 20% of all tropical forested land lost globally each year (Table 5.3).

Table 5.3   Countries with the highest annual losses of closed tropical forest, 1981–85. Adapted from FAO (1981a); FAO (1981b); FAO (1981c); FAO/UNECE (1985)

Country	Area of closed tropical forest lost each year (100 ha)	Area lost annually in country as a proportion of closed tropical forest land lost annually worldwide (%)
Brazil	1480	19.8
Colombia	820	10.9
Indonesia	600	8.0
Mexico	595	7.9
Ecuador	340	4.5
Nigeria	300	4.0
Ivory Coast	290	3.9
Peru	270	3.6
Malaysia	255	3.4
Thailand	252	3.4

A major advance in monitoring rates of forest destruction has taken place in the Amazon Basin over the last decade. Scientists from Brazil, the USA and the United Nations Environmental Program have been monitoring deforestation using satellite images (Malingreau and Tucker, 1988). Research carried out by the Brazilian Space Research Institute using such data has brought into question estimated deforestation rates in the Amazon of 50 000 km^2 per year, revising the estimates for 1979–89 down to 21 000 km^2 per year. Such large revisions indicate, once again, the variability of forest cover and deforestation estimates and the problems inherent in basing estimates on outdated statistics. Plans to continue such monitoring with the launch of further satellites are well underway in Brazil and attention is now being turned to monitoring deforestation using similar techniques in West and Central Africa and South-east Asia.

Set against this spectre of deforestation, at least in a global context, is the fact that many developed (and some developing) countries underwent extensive afforestation in the 1980s. Heading the list was China, with 4.67×10^6 ha afforested each year, closely followed by the Soviet Union with 4.54×10^6 ha. Other countries with high rates of afforestation were the USA (1.78×10^6 ha), Canada (0.72×10^6 ha), Brazil (0.35×10^6 ha), Japan (0.24×10^6 ha) and Sweden (0.21×10^6 ha). The inescapable fact is that while, with the exception of China and the Koreas, developing world forests are declining rapidly, deforestation in the developed world generally has been halted and in many countries reversed (Table 5.4).

Table 5.4   Regional rates of forest loss and renewal, *circa* 1980. Adapted from FAO (1985)

Region	Renewal (1000 ha/year)	Loss (1000 ha/year)	Net gain or loss
North and Central America (including the Caribbean)	2528	930	+1598
Europe	978	0	+ 978
USSR	4540	0	+4540
Africa	108	1333	−1225
South America	470	3191	−2721
Asia	5545	1037	−4508
Oceania	342	679	− 337
World total	14 511	7266	

5.4   CASE STUDIES OF DEFORESTATION

The importance of a historical perspective on current patterns of forest cover and deforestation in the tropics is illustrated in the following three case studies from Sierra Leone, the Amazon Basin and the Himalayas.

5.4.1   Sierra Leone

In the 16th century most of Sierra Leone was covered by extensive tracts of humid tropical rain forest with wet (Guinea) savanna woodlands in the north. Archaeological and anthropological evidence suggest that native population densities were very low, and that most people practised slash-and-burn agriculture which had only localized and minimal impacts on forest cover. Early European colonization, in the mid-16th century, was restricted to missionary settlements and trading posts on the coast and consequently these early settlers had little effect on the forests and woodlands.

Large-scale deforestation first occurred in the 17th century with the increased demand for ships' timbers by the British Navy, who were based in Sierra Leone to combat the slave trade. Forest destruction extended inland along lines of penetration provided by rivers. This period of extensive deforestation occurred at the same time as European demand developed for kernel oil from the oil-palm (*Elaeis guineensis*), a native tree of African humid tropical forests. Consequently the land left after logging was rapidly colonized by farmers, who found that the oil-palm, a fire-resistant tree, thrived in both fields and tree fallow regrowth and provided a ready cash crop at the European trading posts.

Sierra Leone's potential to grow other cash crops, particularly cocoa, coffee and wild rubber, was recognized by the colonial administration and a railway line was built into the interior at the turn of the century to exploit this agricultural potential. This promoted further forest destruction along the railway line as forest was converted to agricultural land (Millington, 1988). Land conversion was so rapid that during the first two decades of the 20th century reservation of the remaining forests became an important issue, and the first reserves were established in 1913. Further exploitation of the forests of Sierra Leone occurred throughout the 20th century in response to various agricultural policies. These include the clearance of mangroves to provide land for swamp rice cultivation, and savanna woodland clearance to provide land for nomadic Fulani herders during the Cattle Owner's Settlement Scheme of the late 1950s and early 1960s.

A number of important points can therefore be drawn from the study of deforestation in Sierra Leone:

(a) Deforestation has a long history, which is in part related to European colonization.
(b) Deforestation has occurred in phases during which the rates of forest clearance were very rapid, interspersed by periods when deforestation rates were low.
(c) Land, originally cleared of trees to provide timber, was rapidly colonized by farmers, thereby restricting natural forest regeneration.
(d) Later phases of forest and woodland destruction in the 20th century have been stimulated by government agricultural policies.

5.4.2 The Brazilian Amazon

Forest destruction in the Brazilian Amazon differs from that in Sierra Leone in two ways. First, the provision of land for agriculture has been the main cause of forest destruction and, secondly, it is a comparatively recent phenomenon. The Amazonian forests have long been used by Amerindians, who have exploited them intensively along the rivers and wetlands, and used them more extensively on the *terra firma*. These patterns of exploitation have generally reflected patterns of natural resource availability and impacts upon the forest environment have been localized (Eden, 1989). Early European colonization generally followed these subsistence and extractive patterns, although early attempts were made at commercial agriculture.

The earliest phases of commercial activity in the Amazon, between 1839 and 1912, expanded the rates of wild rubber tapping already found in the forest and did not promote deforestation. The first major phase of deforestation in the Amazonian forest came with the establishment of 20 agricultural colonies between Belem and Bragança (the Bragantina Zone), in northern Maranhao State, to produce food for Belem (Fig. 5.3). Settlers from France and Spain and *nordestinos*, migrants from drought-prone north-east Brazil, were given plots of land in this area from the late 19th century onwards. The agricultural system was based on simple forest clearance and burning, but the crop yields soon plummeted due to a rapid decline in soil fertility and weed invasion. The settlers, unable to eke out a living on their allocated plots, cleared more forest with the same results—rapidly declining yields. This caused an unanticipated expansion in land clearance which, exacerbated by forest exploitation for timber and fuel for the railways, reduced the entire area to a mosaic of cultivation and forest regrowth by the 1920s. From the late 1920s until the mid-1950s there was limited forest destruction in the Amazon. The main activity was the attempt to establish rubber plantations at Fordlandia (in 1927) and Belterra (in 1934) by the Ford Motor Company, and later by Goodyear and Pirelli in the early 1950s. These schemes involved complete deforestation followed by reafforestation with rubber trees (*Hevea* spp.), but more often than not the success of these plantations was thwarted by severe soil erosion and weed infestation.

Thus, most forest destruction and associated environmental degradation in Amazonia is a recent phenomenon (Fig. 5.3) which is continuing due to:

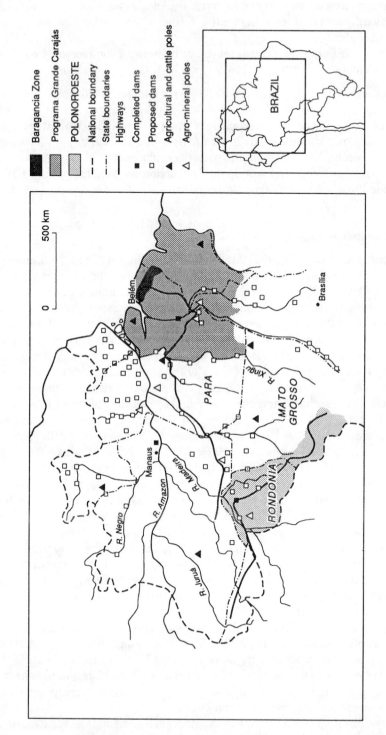

Figure 5.3   Forest utilization in the Brazilian Amazon.

(a) agricultural colonization along highways such as the Trans-Amazonian Highway and around planned growth poles
(b) cattle ranching, again along highways and more recently around growth poles in eastern Para and northern Mato Grosso Provinces and along the Trans-Amazonian Highway
(c) mining around various growth poles
(d) large hydroelectric power schemes
(e) forestry, although this is mainly restricted to Para and, to a lesser extent, Amazonas and northern Mato Grosso Provinces.

These schemes have been actively encouraged through a series of government projects and initiatives aimed at the economic development of the Amazon since 1956. Such initiatives have actively encouraged population migration and colonization, which is described by Westoby (1989) as 'the absurd Amazon dream of marrying the "men with no land" (from the north-east) to the "land with no men".' Such policies have also promoted massive forest destruction and land clearance by peasant farmers to the ultimate benefit of big business. Rondonia is a classic example of what is termed explosive deforestation (Fearnside, 1984). Two decades ago its forests were virtually untouched, but the development of roads in the 1970s was followed by surges of migration. Deforestation has accelerated dramatically in this time (Malingreau and Tucker, 1988) (Fig. 5.4), leaving some 35% of the state deforested by the late 1980s. The demand for land is so great that forest, biological and Indian reserves are coming under intense pressure to be exploited.

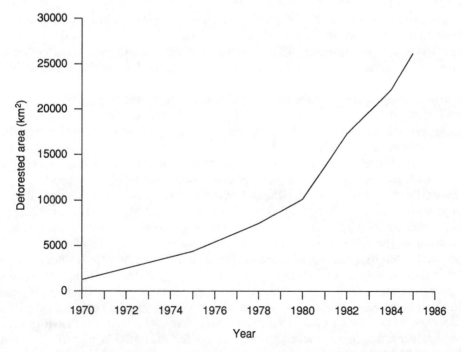

Figure 5.4   Deforestation in Rondonia (Brazil). Adapted from Malingreau and Tucker (1988) by permission of the Royal Swedish Academy of Sciences.

5.4.3 The Himalayas

Deforestation in the developing world is not confined to tropical moist forests, and indeed many commentators would argue that the loss of forest and woodland cover in more ecologically fragile and less productive environments such as semi-arid regions of Africa and the mountain regions of the Himalayas and the Andes is as serious a problem as the destruction of tropical moist forests. This is illustrated with reference to deforestation in the Himalayas. A case study of deforestation in a semi-arid region (North Yemen) can be found in Chapter 15.

The southern flanks of the Himalayas consist of a series of hill ranges stretching through Pakistan, India, Nepal and Bhutan. This area has been settled for many centuries, and an agricultural system based on the management of both private farmland (in valley bottoms and the lower slopes, often with extensive terracing) and communal forests and grazing lands (on the upper slopes and ridge crests) has evolved. This system can be both productive and stable, but the agro-ecological potential of this region places clear limits upon it. Trees are an integral part of this agricultural system in settled areas, providing food crops, fodder, fuel and construction materials, as well as reducing erosion hazards and contributing to water control. For example, Robinson and Thompson (1989) cite forests, trees and livestock as the cornerstones of hill farming systems in Nepal, providing essential alternative sources of diet and income to those provided by the main crops. Himalayan hill people cannot survive without the many essential services the forests provide, and the threat of deforestation is as much an economic as an environmental hazard.

This system is, however, vulnerable to resource pressures and labour shortages and in many parts of the Himalayan region it is breaking down. This is causing environmental deterioration, ultimately leading to severe hardship for the local population. The reasons for this are complex. In some areas too much forest has been converted to farmland. This results in farming on steeper, unstable slopes, which places excessive pressure on the remaining areas for fuel, fodder and other products, and disrupts the hydrological cycle leading to more rapid runoff and increased erosion, and ultimately downstream flooding. Terrace agriculture can be an effective method of environmental protection, even on the steepest slopes, but requires high inputs of labour for maintenance, which is hampered by out-migration causing labour shortages.

Commercial timber extraction is widespread in more accessible regions, with the wood from the Himalayas being exported to the populous areas to the south (e.g. the Ganges Plain in India and the Punjab and Peshawar Plain in Pakistan). It is no coincidence that the famous Chipko Movement started in a Himalayan village in India in response to forest clearance by outside logging companies. Shortages of fuelwood in this environment can lead to the increased pruning of fruit trees for fuelwood, thereby reducing fruit yields, the increased use of animal dung as a fuel rather than as a manure, which reduces soil fertility, and the use of crop residues (e.g. wheat straw and maize stalks) as a fuel rather than as fodder. All these alternatives increase pressure on the agricultural system and the environment.

Development projects such as dams and roads have also had adverse environmental consequences, and have opened up the forests for exploitation by outside agents. There is no doubt that in some areas growing population densities have had an adverse impact on the forest resource, but this is mainly through land clearance for agriculture and greater demands for fodder from increasing livestock populations. Fuelwood collection for urban markets is a problem in some regions. The hills surrounding Katmandu are a classic example of this, with large areas formerly covered by forests now largely denuded at the hands of urban wood dealers. Finally, the nationalization of the forests in India and Nepal has led to their alienation from the local population and the erosion of traditional methods of forest management. People still use forest products, but feel no sense of responsibility for their maintenance.

Mountain environments such as the Himalayas are fragile and have a limited agro-ecological potential. They contain agricultural systems which can be sustainable, but which are tending to break down under external pressures. Where this happens the natural resource base can deteriorate rapidly in a downward spiral as deforestation leads to increased erosion and less soil moisture retention. This lowers land productivity, which in turn causes more soil erosion and further decreases land productivity, leading to increasing poverty and the likelihood of either out-migration or additional land clearance.

Deforestation is a major problem in the Himalayas, one of the world's most beautiful but poorest regions. The same is true for many other areas outside the moist forests (e.g. the Ethiopian Highlands, Southern Africa); an important point which is often lost in the discussion of deforestation in the developing world.

5.5   WHY DO FORESTS MATTER?

The preceding sections clearly show that a large, but diminishing, proportion of the earth's surface is covered by forests and woodlands. Deforestation and woodland destruction is widespread, particularly in the tropics, and it is acknowledged to be one of the great environmental challenges of the late 20th century. Why should this be the case? Much of the world's agricultural land was formerly covered by forest or woodland, so without deforestation the modern world would not exist and the global population would only be a fraction of what it is today. Indeed, most of Britain and Western Europe used to be covered by forests, but it would be foolish to advocate the wholesale depopulation of this region to allow forests to re-establish themselves! Forest and woodland clearance, therefore, must generally be tackled by reducing the rates of deforestation rather than drastic solutions such as population transfer and giant, futile reafforestation schemes.

In the tropics the exploitation of forests provides vital income to some of the world's poorest countries, both directly through commercial logging, and indirectly through the sale of crops grown on land formerly under forest. Moreover, in many regions woodland clearance represents the only hope for poor, land-hungry people to gain access to the farmland they need. Deforestation, from a purely utilitarian standpoint, could be considered as a rational use of vital resources for economic development. Many people in developing countries resent lessons in the morality of forest clearance from the developed world that has already reaped the benefits from the clearance of its own forests and are responsible for most of the modern world's resource consumption and pollution.

Nevertheless, forests *do* matter, and much of the deforestation currently taking place is undesirable for a number of reasons. Forests and woodlands perform a series of vital environmental functions at local, regional and global levels. Firstly, and at the local level, the forest and woodland canopy protects soil from rainfall, reducing erosion and regulating runoff. Forested areas produce more even water flows, as the soil acts as a major store in the hydrological cycle. This buffering role is especially important in regions with irregular or strongly seasonal rainfall, such as the monsoon areas of Asia and semi-arid regions.

Secondly, forests create local microclimates, regulating temperature and humidity in their immediate vicinity. Generally, temperatures are moderated (minimum temperatures are higher and maximum temperatures lower), humidity is increased, and wind speed reduced. Forest environments create an ideal microclimate for a number of crops, although solar radiation can be limited by canopy cover. Such environments also provide habitats for many types of flora and fauna which add to the variety and durability of local environments.

On the regional and global scales forest resources play a major role in environmental regulation and preservation. Some local functions, such as soil protection and regulation of the hydrological cycle, also operate at a regional scale, and deforestation is often cited as a major contributing factor to disasters such as downstream flooding of previously forested areas of (e.g. Bangladesh). It is also possible that the effects of droughts in arid and semi-arid areas are worsened by the loss of woodland cover.

Forests have two other major environmental functions at the global scale, both of which are seriously threatened by current patterns of deforestation. These are, firstly, their role as carbon sinks in the global carbon cycle (Fig. 1.2) and, secondly, as pools of biodiversity, i.e. as the home of a vast number of species of plants and animals.

Global warming caused by the enhanced greenhouse effect (Chapters 3 and 4) is considered by many to be the greatest environmental problem currently facing the world. Although there is still debate, the evidence to support the fears that human activities are leading to a significant increase in mean global temperatures, which will result in major disruptions to climate patterns and ecological distributions, is very strong. The world's biomass resources, of which forests and woodlands are the largest component, play a key role in global climatic change.

Their contribution to climatic changes takes place through two mechanisms. Firstly, plants absorb carbon dioxide and release oxygen during photosynthesis. The loss of a significant proportion of the world's biomass stocks through deforestation and woodland destruction will consequently lead to a reduced capacity to absorb carbon dioxide and could, in the long term, cause other forms of disruption to the composition of the atmosphere. Secondly, the process of deforestation, and in particular tree burning during clearance, produces carbon dioxide and other greenhouse gases, and thereby has a direct effect on global warming. Not surprisingly, the reversal of deforestation trends and a move to significant re-afforestation is seen by many as a key component in global efforts to mitigate the impact and speed of global warming.

The question of biodiversity is an issue which particularly relates to tropical moist forests, as these areas contain the largest numbers of plants and animal species of any ecosystems and are therefore significant biodiversity reserves. Although these forests cover only 6% (1.296×10^{15} ha) of the earth's land surface, they contain at least 50% of all plant and animal species. Indeed, the proportion may be greater than this, as only a small proportion of the millions of species that are thought to occur in moist humid tropical forests have been identified so far.

The diversity of species over small areas in these forests is amazing. For example, a detailed study of 10 ha of rain forest in Borneo found over 700 different species of tree, the equivalent of all the tree species found in North America. Similarly, a small (13.7 km²) nature reserve in Costa Rica contains as many plant and animal species as the whole of Great Britain, while one tree in the Peruvian part of the Amazon Basin was found to be host to 43 species of ant, more species than are found in the whole of the British Isles. Many of these species have very limited distributions, so deforestation of a relatively small area of forest can lead to the extinction of a large number of species. The area of continuous forest needed to make a viable habitat for many rain forest species, in particular the larger mammals, is often substantial, so that the fragmentation of forest reserves in some countries (e.g. Thailand and the Philippines) means that even where some forest is left it is not enough to prevent the extinction of certain species, thereby disrupting food chains that may cause further extinctions.

The implication of this loss of biodiversity cannot easily be assessed, but extinctions lead to the loss of potentially important genetic characteristics. For example, many of the world's medicines come from tropical plants, while much of the success achieved through breeding new strains of many important plants (e.g. cocoa, rubber, coffee, rice) has come about by incorporating genes from wild species with domesticated plants. Little is known about the genetic characteristics of most rain forest plants, but the extinction of species (not only in the tropics, but also globally) reduces gene pools and destroys potentially commercially (as pharmaceuticals or crop protection chemicals) useful organisms. This, in turn, reduces biotic capital for inter- and intra-generational benefit by reducing the resource base. It is, therefore, in direct conflict with sustainable development (Chapter 2).

5.6 ECONOMICS OF FOREST MANAGEMENT

The environmental functions of forests and woodlands make a strong case for more effective action to reduce the rates of tropical deforestation. If it was simply a matter of balancing these environmental benefits against the economic cost of reducing deforestation (from lost income for timber and reduced agricultural production), then it could be difficult to defend this case in regions where so many people are so poor and development opportunities are so few. However, such calculations are far from simple as the decision to cut down forests can also lead to economic costs. Forests have many economic values over and above that of the timber they contain, and if properly managed can produce far greater economic benefits in their present form than the returns gained by their clearance. It is possible to identify three types of economic benefits which can accrue from the sustainable management of forests.

5.6.1 Direct economic benefits

Forests have a number of direct economic benefits. Firstly, timber can be produced in a sustainable manner. Levels of production are lower than in non-sustainable timber exploitation (and should be closely related to natural rates of regeneration), but large areas could be so managed to provide the income needed from timber production. The dispersed, small-scale nature of the production means, moreover, that it could be more directly under the control of local people than are large-scale commercial logging operations. This provides both environmental and economic advantages.

Forests also produce a range of non-timber products, which can be an important source of revenue. Tropical forests already supply many products, e.g. many species of fruit, oils such as camphor, spices like nutmeg, medicines, fibre products such as bamboo and rattan, and latex. Often the potential of these products has never been fully realized, and they offer a sustainable development opportunity for many tropical countries. The economic returns from managing non-timber products could be greater than those from clear-felling for timber alone. A study of potential yield from an area of Peruvian rain forest suggested that income from fruit and latex could be six times as high as that from clear-felling the land for timber (Pearce, 1990). The numbers of people who could be employed in these industries if they were properly developed is also far greater than those working in timber production.

Forests also provide a number of services which are of direct economic benefit. Countries such as Belize and Costa Rica earn income from 'eco-tourism' in forest areas, and the prospects for growth from tourists from developed countries are considerable, reflecting the growing interest in these environments in the developed world (Chapter 19). This growth of tourism is likely to be concentrated in specific areas; for example, forest reserves in Rwanda are already used by tourists to view mountain gorillas. The growth of tourism in rain forest areas is likely to result in some reserves being closely protected, but the need for well developed tourist facilities and good transport infrastructure means that tourism is unlikely to spread beyond these relatively small enclaves. Nevertheless, the success of the 'safari' tourism of countries such as Kenya and Zimbabwe shows the potential of this form of economic utilization of forests in their present form.

5.6.2 Indirect values

A number of the environmental functions of forests (Section 5.5) also have economic values. For instance, the role of forests in protecting watersheds helps to maintain agricultural productivity, reduces the siltation of reservoirs, diminishes dangers from flooding and saves considerable amounts of capital in measures such as erosion protection, reservoir and canal dredging and flood protection (Chapter 14). These 'indirect values' of forests are hard to give a price to, but nevertheless need to be taken into account when comparing the costs and benefits of different forms of management of forest areas.

5.6.3 Non-use values

Forests have non-use values, reflecting the aesthetic and cultural significance of forests and trees. There is no doubt that many people in many countries place great value on tropical forests even if they are unlikely to ever visit them; the growth of environmental opposition to tropical forest clearance in Europe demonstrates this. These values are intangible but real, and should be taken into account when assessing the value of forests.

Forests and woodlands also have important social and development functions. In many developing countries rural people rely on woodlands for a range of essential goods such as fuelwood, fodder, local medicines, building materials and different sorts of food. These 'common property resources' are freely and reliably available, and in some regions provide essential security for poor rural populations when crops fail or crop prices fall. The deforestation of many developing world regions undermines these forest functions, and places unsustainable pressures on the remaining woodland areas. As such, retaining woodland and tree resources is an essential component of efforts to support and maintain rural production systems in many developing countries.

For all of the reasons cited above, it is clear that the maintenance of the world's forest resources is a challenge which is vital to the future of the planet. As the foregoing discussion shows, this is compatible with a form of resource management which maintains the integrity of these vital ecosystems. As such, the battle is not between protection or exploitation, it is between sustainable or rapacious development.

5.7 CONCLUSIONS

Despite the uncertainties surrounding estimates of forest and woodland cover, and rates of deforestation, it is clear that the loss of forest and woodland cover is a serious problem in the developing world. This is in direct contrast with the developed world where, generally, afforestation is the norm. The key issues do not relate to the exact extent of forest cover, but to the causes and rates of forest and woodland destruction and the environmental and economic imperatives which can be used to arrest rapid deforestation.

The environmental roles played by forests and woodlands are closely related to scale. At the local and regional scales environmental protection (soil protection, regulation of the hydrological cycle and microclimate amelioration) are important environmental functions of forests. At the global scale the role of forests, and tropical forests in particular, in the carbon cycle and the regulation of climate, especially in the context of the enhanced greenhouse effect, are key issues. So too is the importance of maintaining biodiversity as a present and future resource.

The reduction of deforestation rates, and the conservation of the world's remaining forests, cannot be achieved by simply stressing the undoubted environmental importance of forests and woodlands regardless of scale. The economic rewards of commercial timber extraction and the perceived vital need to clear forest and woodland for agricultural production are considerations which must enter the equation. If the environmental role of forests and woodlands is to be used to counter deforestation, economic values must be attributed to the protection function of forests, at the same time as economically quantifying other forest values. As shown in this chapter, economic analyses carried out for certain tropical forests show that they have greater overall benefits if left intact, rather than destroying them for timber. Such analyses have, as yet, barely proceeded beyond academic posturing. Will the real world of timber concerns, bankrupt developing world governments, and western and Japanese users of tropical hardwoods take heed of these economic calculations before it is too late?

5.8   REFERENCES

Eden, M. J. (1989) *Land Management in Amazonia*. London: Belhaven.

Fearnside, R. M. (1984) Roads in Rondonia: highway construction and the farce of unprotected reserves in Brazil's Amazonian Forest. *Environmental Conservation*, **11**, 358–360.

Food and Agriculture Organization (FAO) (1981a) *Forest Resources of Tropical Africa*. Rome: FAO.

Food and Agriculture Organization (FAO) (1981b) *Forest Resources of Tropical Asia*: Rome: FAO.

Food and Agriculture Organization (FAO) (1981c) *Los Recursos Forestales de la America Tropical*: Rome: FAO

Food and Agricultural Organization (FAO) (1982) *Tropical Forestry Resources*, Forestry Paper 30. Rome: FAO.

Food and Agriculture Organization/UNECE (1985) *The Forest Resources of the ECE Region*. Geneva: ECE.

Malingreau, J. P. and Tucker, C. J. (1988) Large-scale deforestation in the southeastern Amazon Basin of Brazil. *Ambio*, **17**, 49–55.

Millington, A. C. (1988) Environmental degradation, soil conservation and agricultural policies in Sierra Leone, 1895–1984. In: D. Anderson and R. Grove (Eds). *Conservation in Africa*. Cambridge: Cambridge University Press, pp. 229–248.

Millington, A. C., Critchley, R. W., Douglas, T. D. and Ryan. P. (1992) *Estimating Woody Biomass in Sub-Saharan Africa*. Washington DC: World Bank Publications.

Pearce, D. W. (1990) *An Economic Approach to Saving the Tropical Forests. LEEC Paper 90/6*. London: London Environmental Economics Centre.

Robinson, P. and Thompson, I. (1989) *Fodder Trees, Nurseries and Their Control in the Hill Farm Systems of Nepal*. ODI Social Forestry Network, Paper 9a. London: Overseas development Institute.

Westoby, J. (1989) *Introduction to World Forestry*. Oxford: Blackwell Scientific.

Wood, W. B. (1990) Tropical deforestation: balancing regional development demands and global environmental concerns. *Global Environmental Change*, **1**, 23–41.

5.9   FURTHER READING

Gradwohl, J. and Greenberg, R. (1988) *Saving Tropical Forests*. London: Earthscan.

Poore, D. (1989) *No Timber Without Trees*. London: Earthscan.

Westoby, J. (1989) *Introduction to World Forestry*. Oxford: Blackwell Scientific.

CHAPTER 6

Marine Pollution

J. M. R. Hughes and B. Goodall

6.1 THE MARINE ENVIRONMENT— A COMMON PROPERTY RESOURCE

Seventy-one per cent of the earth's surface is occupied by seas. Of this, 59% covers the ocean floors, at depths in excess of 200 m, while 12% covers submerged continental margins and is less than 200 m deep. Sixty-one per cent of the northern hemisphere and 80% of the southern hemisphere is covered by ocean (Fig. 6.1), while smaller, shallower bodies of water known as seas are found where water is

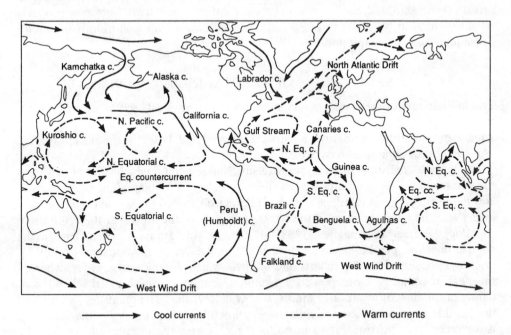

Figure 6.1 Oceans and (January) currents.

Environmental Issues in the 1990s. Edited by A. M. Mannion and S. R. Bowlby
© 1992 John Wiley & Sons Ltd

almost enclosed by land or island chains, e.g. the Mediterranean Sea, the Gulf of Mexico and the Baltic Sea.

Oceans and seas have always served as a primary source of sustenance for life, offering since the earliest times opportunities for discovery, trade and resources. Use of the oceans has increased significantly in the twentieth century as traditional land-based resources have come under increasing pressure in line with economic and population growth. Oceans are now viewed as a resource base to be exploited for food (mainly by hunting and gathering, rather than by mariculture), for metals and minerals (resource mining), for energy generation (from tidal and wave power), for water, for transportation, and for recreation and tourism. They also have a military and strategic significance. The potential for the sustainable utilization of the oceans and seas is enormous, especially as the resource (including both biotic and abiotic components) is, if properly managed, self-renewing (Chapter 2). Oceans and seas are also used as a resource in another sense—as an environmental sink or 'dustbin' for waste disposal (Chapter 1). Unfortunately, this has frequently ignored the absorptive capacity of the marine environment. The functioning of oceans and seas as an integral part of the global ecosystem is therefore threatened. Where absorptive capacity is exceeded, marine pollution results and waste disposal must be regarded as a destructive use of the marine environment.

Such resource exploitation, especially for waste disposal, hinges on the fact that oceans and seas are common property resources (Hinrichsen, 1990). They are considered to be 'owned in common by everybody', are not subject to individual or private ownership and can therefore be used without payment. Common property resources tend to be over-exploited and misused, and marine pollution is the major consequence of such exploitation.

6.2 CLASSIFICATION OF THE MARINE ENVIRONMENT

The marine environment may be classified on either physical, ecological or politico-economic grounds (Fig. 6.2). The former two are of importance when identifying the impact of pollution on marine environments and the latter is significant for the implementation of pollution control.

The marine environment forms an extensive saline habitat which is inhabited from the surface down to its greatest depth (deep trenches lie 10 000 m below the surface and the average ocean depth is 3700 m). There are two ways in which organisms live in the sea: they float or swim in the water (pelagic), or they dwell at the sea bottom (benthic) (Barnes and Hughes, 1988). There are two groups of organisms in the pelagic division, plankton and nekton, and these differ in their means of motion. Plankton consist of floating plants and animals which drift with the water, whereas the nekton comprise more powerful swimming animals which can travel independently of the flow of water, e.g. squids, sharks and whales. The plants and animals of the benthic division are known collectively as benthos, and they are either attached to or creep and burrow along the bottom, e.g. algae, sea urchins and crabs. The littoral zone is the most productive part of the ocean and within it may be found photosynthesizing benthos (salt-marsh plants, sea grasses, mangroves) and reef corals.

Several physical subdivisions of the marine environment can be made which are closely related to ecological and political classifications (Fig. 6.2):

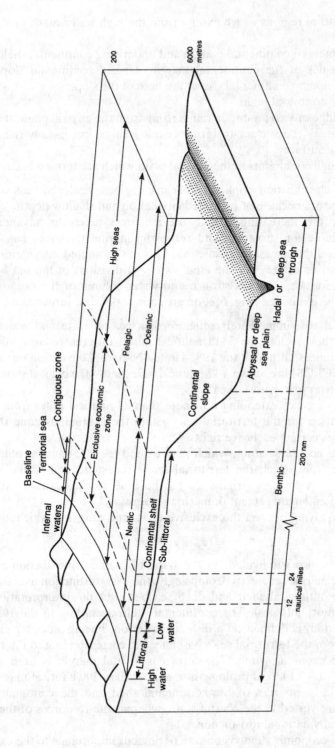

Figure 6.2 Classification of marine biome zones and political divisions.

(a) coastal or littoral regions which extend from the high water mark to about 30 m depth
(b) areas of submerged continental margin and underlying continental shelf (neritic)
(c) descending sides of the continental margin, termed continental slope, with a more gently sloping continental rise at the base of the slope
(d) oceanic floor or abyssal plain
(e) mid-ocean ridge or vast mountain range rising from the abyssal plain; these ridges sometimes break the surface to form oceanic islands, but usually rise to within 2000 m of the surface
(f) deep sea troughs or chasms in the abyssal plain which are termed hadal regions.

Such a physical classification applies to oceanic regions; enclosed seas often have unique characteristics because of restricted circulation and shallow depth.

Awareness of the increasing scarcity of resources allied to advances in the technology available for surveying and retrieving marine resources has led many countries to extend their coastal jurisdiction. This has prompted international efforts to delimit the marine estate, i.e. the areal extent of divisions of the sea bed, water column and sea surface. The sea surface and water column of the oceans, beyond internal waters, is subdivided (Fig. 6.2) on grounds of political jurisdiction, into:

(a) the territorial sea, immediately offshore of a coastal state and which extends seaward of the baseline for 3–12 nautical miles; different coastal states claim different distances, although the 1982 United Nations Convention on the Law of the Sea (UNCLOS) lays down a 12 nautical mile limit—the coastal state exercises sovereignty over its territorial sea
(b) the contiguous zone, extending not more than 24 nautical miles from the same baseline as used for the territorial sea; within the contiguous zone the coastal state exercises certain exclusive rights
(c) the exclusive economic zone, which may extend for 200 nautical miles beyond the baseline from which the territorial sea is measured and within which the coastal state has sovereign rights for the purpose of exploration, exploitation, conservation and management of marine resources
(d) the high seas, lying beyond the exclusive economic zone and which are open to all states.

The sea bed is also subdivided on the basis of political jurisdiction relating to ownership of the marine resources. Confusingly, the basic distinction uses the physical terminology of continental shelf and deep sea bed, but the interpretation is legal rather than geomorphological. The continental shelf, according to the 1982 United Nations Convention, is defined on a distance criterion, having an outer limit of 200 nautical miles from the territorial sea's baseline, i.e. corresponding to the outer limit of the exclusive economic zone. The term continental margin is used where the continental shelf has a natural prolongation greater than 200 nautical miles. The sea bed and substratal resources of the continental shelf and the continental margin where it exists are vested in the coastal state, whereas the resources of the deep sea bed are beyond any national jurisdiction.

These politico-economic subdivisions are of obvious importance to the exploitation of marine resources, including waste disposal, and must therefore also be relevant to attempts to control marine pollution. On the basis of such subdivisions it is only the high seas and deep sea bed which can be recognized as common property resources. However, considerable disagreement exists between many neighbouring coastal states as to the delimitation of zones, and some countries, including the UK, USA and Germany, are not yet signatories to the 1982 Convention.

6.3   SOURCES OF MARINE POLLUTION

Global concern for the marine environment stems from the lavish media coverage of supertanker oil spills and from proposals to dispose of radioactive wastes at sea. It is reinforced by ecological studies highlighting the concentration of pesticides and heavy metals in oceanic food chains, as well as concern about the unhygienic state of sea water and beaches at popular holiday destinations. Marine pollution is defined by the United Nations Group of Experts on the Scientific Aspects of Marine Pollution (GESAMP) as 'the introduction by (people), directly or indirectly, of substances or energy into the marine environment resulting in such deleterious effects as harm to living resources, hazards to human health, hindrance to marine activities, including fishing, impairment of quality of use of sea-water and reduction of amenities'. It is the result of the release of materials by human activities in the wrong places in amounts that are too large for the capacity of marine ecosystems either to neutralize or disperse to harmless levels.

However, identifying marine pollution in practice is not always clear-cut. Where the pollutants do not occur naturally, e.g. halogenated hydrocarbons and plastics, there is little problem, but most of the materials which are potential pollutants exist naturally in the oceans, e.g. organic material subject to bacterial degradation, particulate material from coastal erosion, metals in run-off from metalliferous deposits, seepages from oil-bearing strata, hot water from geothermal springs, and radioactivity. Where human activity causes increases in these materials but does not create the deleterious effects delimited by the GESAMP definition, the condition of the marine environment may be one of contamination rather than pollution. Even where a contaminant is toxic, the mortality caused in marine animals may be insignificant when set alongside the natural losses. Thus there is great difficulty in identifying the deleterious effects of many materials added to the marine environment; the picture is only clear in the most severe, localized cases of pollution.

If marine pollution is to be tackled successfully, basic scientific knowledge of the materials being discharged into the marine environment is needed, as well as knowledge of the sources of the pollutants and the effects that such additives may have on marine environments. It is useful to distinguish pollutants according to whether they remain unchanged when introduced to the marine environment. Non-conservative pollutants will eventually be assimilated into marine biological cycles via processes such as biodegradation and dissipation. Conservative pollutants, which are non-biodegradable and not readily dissipated, remain unchanged and therefore tend to accumulate in marine biological systems. This problem is exacerbated where they are toxic.

Materials disposed of in the oceans may be classified in a number of ways, but it is most useful to categorize them on the basis of the conservative/non-conservative distinction. Non-conservative additions take three forms: (a) degradable wastes (e.g.

Table 6.1 Types and sources of marine pollutants

Type of pollutant	Source of pollutant			
	Land-based		Sea-based	
	Operational	Accidental	Operational	Accidental
Organic or degradable wastes	Direct outfalls: sewage and industrial effluents; River inputs: sewage, agricultural wastes, industrial effluents from food, paper and chemical industries	Spillages: from coastal oil refineries and chemical works	Bulk dumping: land-derived sewage sludge; Mariculture: excess feedstuffs; Shipping: deballasting, tank-washing, bilge water, sewage	Spillages: shipping accidents, oil production, platform blow-outs, pipeline fractures
Fertilizers	River run-off: agricultural use of nitrates and phosphates	—	—	—
Dissipating wastes	Cooling water discharge: power stations (especially nuclear) and other industries; Industrial effluents: acids and alkalis	—	'Burnships': gases from waste incineration	—

Particulates	Mining wastes: from china clay, coal and gravel extraction Cellulose fibres: from pulp mills Coal-fired power stations: pulverized fuel ash Atmospheric fall-out: incomplete combustion of fossil fuels	Accidental leakage from coastal industrial plants	Dredging spoil Dumping: mining wastes Plastics: litter dumped overboard from ships Offshore mining; drilling muds	Plastics: nylon ropes, nets and other fishing gear
Persistent/conservative wastes				
(a) heavy metals	Sewage and cooling water Combustion of fossil fuels Industrial wastes	Accidental leakage from coastal industrial plants	Sewage sludge and dredging spoil Use of antifouling paints containing tributyl tin	Accidental losses of deck cargoes in stormy weather
(b) halogenated hydrocarbons	Agriculture: pesticides, herbicides Plastics: polychlorinated biphenols	Accidental leakage from coastal industrial plants	Mariculture: pesticides	Accidental losses of deck cargoes in stormy weather
(c) radioactivity	Routine discharges: nuclear power stations and reprocessing plants	Accidental leakage from coastal industrial plants	Dumping of low level waste (now discontinued)	Leakage from sunken nuclear powered submarines

sewage, slurry), wastes from the food processing, brewing, distilling, pulp and paper and chemical industries, and oil spillages; (b) fertilizers, principally from agricultural use; and (c) dissipating wastes, principally energy in the form of heat in cooling water discharge, but also acids and alkalis. There are two forms of conservative additions: (a) particulates, such as mining wastes (e.g. colliery spoil, pulverized fuel ash) and inert plastics; and (b) persistent wastes, which take three forms (i) heavy metals (e.g. mercury, lead, zinc), (ii) halogenated hydrocarbons (DDT and other chlorinated hydrocarbon pesticides and polychlorinated biphenyls) and (iii) radioactive materials (primarily from nuclear power stations and nuclear reprocessing plants).

Often a particular waste input to the ocean is complex, and may contain both conservative and non-conservative pollutants. For example, urban sewage consists mainly of organic waste but also contains metals, oils and greases, detergents, organochlorines and other industrial waste.

The various additions with the potential to become marine pollutants are listed in Table 6.1 (pp. 102–3). The addition of the pollutant may be accidental or intentional (a routine operation). Accidental pollution may involve large volumes of materials such as the oil supertanker disasters, e.g. the *Torrey Canyon* (Cornwall and Brittany, 1967, spilling 117 000 tonnes), the *Amoco Cadiz* (Brittany, 1978, spilling 233 000 tonnes), the *Exxon Valdez* (Alaska, 1989, leaking 11.2×10^6 tonnes along 3800 km of coastline) and *Mega Borg* (Texas, 1990, leaking 500 000 gallons). More important, however, are routine operations which regularly release potential pollutants into marine environments because this is the cheapest method of waste disposal for the polluter. Examples include the discharge of raw or partially treated sewage, liquid effluents and cooling water from industry, sewage sludge, colliery and dredging spoil, obsolescent munitions and low and intermediate level radioactive waste. Similarly, the deballasting and tank-washing operations of oil tankers and the discharge of oily bilge water by other ships are commonplace.

The sources of pollutants also vary and are due to either land- or sea-based activity. Some wastes are discharged directly into the sea from coastal locations or offshore activities, or indirectly as materials brought via river flows; other wastes are deliberately taken to offshore dumping grounds. The most important sources of land-based pollutants are the regular releases of industrial effluents and sewage from direct outfalls. For example, around the coast of Great Britain 450 outfalls pump out raw or partially treated sewage, three-quarters of which are from short sea outfalls exposed at low tide. Routine discharges of liquid radioactive effluent from the Sellafield nuclear reprocessing plant have made the Irish Sea the most radioactively contaminated sea in the world.

Sea-based pollution sources, especially on the high seas, are limited, although accidental and operational vessel discharges are major sources. Sea-bed mining disposes of its drilling wastes 'on-site' and specified dumping grounds have been designated for a range of substances. Mariculture, which requires clean water, may itself be highly polluting as excess food, falling from marine fish farming cages, encourages eutrophication (Chapter 11) and the pesticides used to keep the fish free from parasites may kill other invertebrates. In addition, discharges to the atmosphere are returned to the sea in rain or, if particulate, as fallout. For example, 'burnships' incinerate toxic liquid wastes such as chlorinated hydrocarbons and release hydrogen chloride near the sea surface where the humidity is high; this is rapidly transferred in water droplets to the sea where it is dissipated.

As most of the pollution sources are land or nearshore based, the greatest concentration of pollutants occurs at the land–sea interface. The coastal zone, which accounts for less than 1% of the volume of sea water, is the most used and abused part of the oceans. Estuaries receive a further input of pollutants from river discharge. Thus, the greater the level of economic development of a coastal state and the higher the proportion of its population and industry with a coastal location, the greater the concentration of pollutants in its coastal waters.

6.4   IMPACT OF MARINE POLLUTION

Once energy has been fixed within the primary producer component (i.e. plankton), it passes through the system by means of a food chain (Fig. 6.3), which describes the successive consumption of one type of organism by another (Meadows and Campbell, 1988). Among the longest food chains are those in the pelagic organisms, where five or more 'steps' may be involved from primary producer to higher order carnivores such as tunas, squids and dolphins, via intermediaries such as krill and crustaceans. Steps in the transfer of food within the chain or web are termed trophic levels, and in general the first trophic level is composed of producers, the second of herbivores, the third carnivores, and then successive levels of carnivores. An understanding of food chains is extremely important to the whole question of marine pollution, because pollutants can accumulate to lethal doses. This involves bioaccumulation, e.g. the accumulation of pesticide residues in plankton, oysters and fish. Moreover, the loss of organisms at one trophic level will affect the whole food web.

Disturbance, as a result of wave action, predator invasions and temperature fluctuations, is a natural process in the marine environment, especially in the littoral zone. The oceanic zone, in contrast, exhibits greater stability. Wide population fluctuations can also occur naturally. These factors make it difficult to distinguish between pollution induced changes and natural changes. Furthermore, some natural fluctuations are big enough to produce a major ecological change comparable to a pollution incident. For example, major faunal changes due to winter sea water temperature fluctuations have been recorded in the benthic fauna off the Northumberland coast (Clark, 1989).

The ecological impacts of pollution are thus dependent not only upon local physical and biological conditions, but also upon the type of pollution and whether it is continuous, episodic, deliberate or accidental. Figure 6.4 summarizes the possible distribution of a pollutant following dispersal into the marine environment. The pollutant will either find its way into the food chain, or it will accumulate in bottom sediments and affect invertebrate benthos. Transportation, accumulation and dispersal of the pollutant is very much dependent on local physical conditions.

Within the coastal zone, pollution is most severe in estuaries and semi-enclosed bays or harbours, or waters which have a slow rate of replacement in relation to their volume. In severe instances of marine pollution it is obvious where a polluting substance has occurred, e.g. sea birds coated in oil. Similarly, anoxic conditions (devoid of oxygen) can occur where there are inputs of organic effluents such as sewage (Chapter 11). In regions where there are high levels of organic matter and

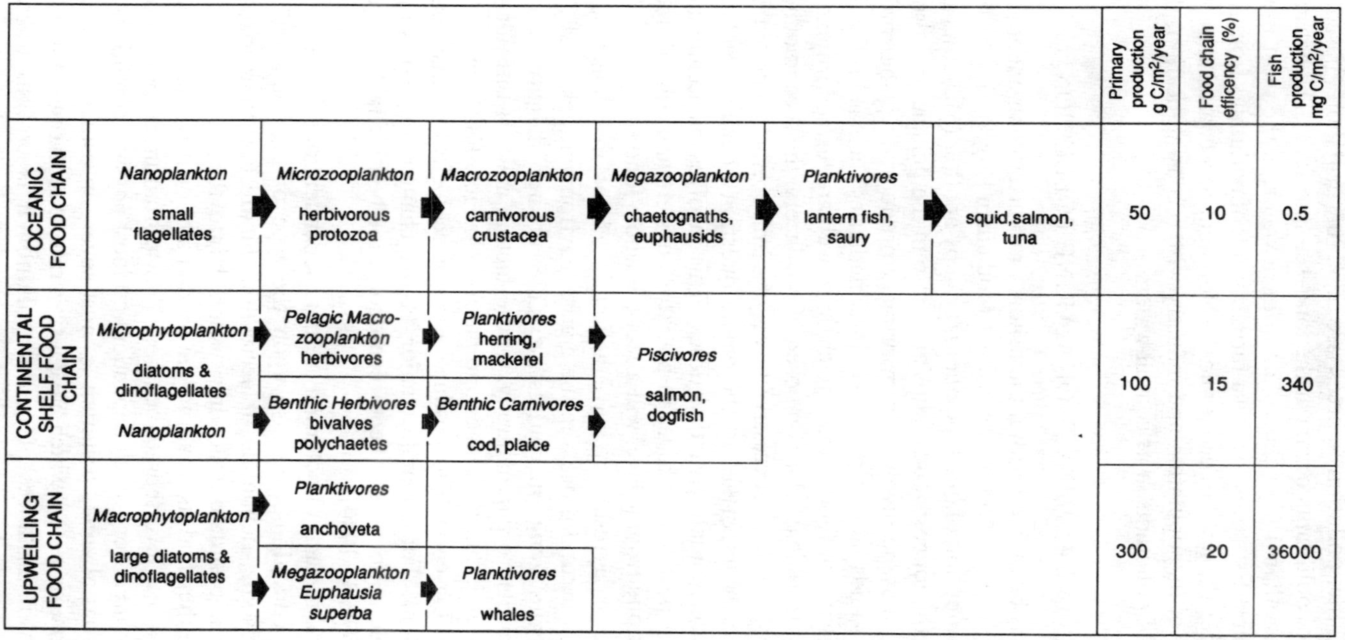

Primary production g C/m²/year	Food chain efficiency (%)	Fish production mg C/m²/year
50	10	0.5
100	15	340
300	20	36000

OCEANIC FOOD CHAIN

Nanoplankton — small flagellates → Microzooplankton — herbivorous protozoa → Macrozooplankton — carnivorous crustacea → Megazooplankton — chaetognaths, euphausids → Planktivores — lantern fish, saury → squid, salmon, tuna

CONTINENTAL SHELF FOOD CHAIN

Microphytoplankton — diatoms & dinoflagellates → Pelagic Macrozooplankton — herbivores → Planktivores — herring, mackerel

Nanoplankton → Benthic Herbivores — bivalves polychaetes → Benthic Carnivores — cod, place → Piscivores — salmon, dogfish

UPWELLING FOOD CHAIN

Macrophytoplankton — large diatoms & dinoflagellates → Planktivores — anchoveta

Megazooplankton — Euphausia superba → Planktivores — whales

Figure 6.3 Ocean, continental and upwelling food chains. Adapted from Meadows and Campbell (1988).

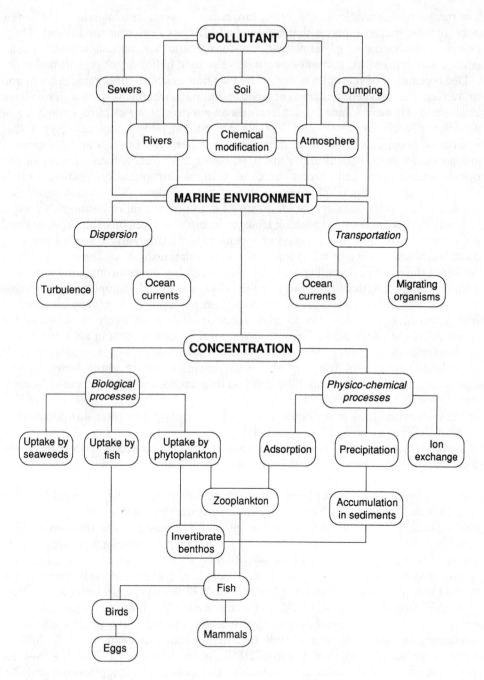

Figure 6.4   Distribution of a marine pollutant. Adapted from Ketchum (1967).

low oxygen concentrations, the biota (apart from bacteria) are reduced to a few opportunistic species (species that can take advantage of extreme conditions). These include polychaetes (e.g. fanworms, ragworms) and several oligochaete species (sludge worms). In fact, such species can serve as good indicators of organic pollution.

Deoxygenation of sea water is caused by bacterial oxidation of organic effluent and to reduce the oxygen demand of such pollution, treatment plants have been established. However, each human excretes an average of 9 g of nitrogen and 2 g of phosphorus each day (9000 and 2000 kg per million people, respectively). Other sources of phosphate come from detergents in sewers and nitrates and phosphorus are produced from agricultural run-off (Chapter 11). Nutrient enrichment of the marine system from such sources can cause cultural eutrophication, and this in turn will encourage blooms of phytoplankton and benthic algae. The overall result is a reduction in species diversity. Some forms of pollution are more insidious. Changes can only be detected by monitoring changes in biomass (weight of the species) and population in relation to the range of species present (the early stages of pollutive disturbance will cause a break in this straight line relationship).

One of the most important considerations in assessing the environmental impact of pollution is the length of time the damage lasts (as opposed to population perturbations and mortality). In temperate waters severe pollution damage may be reversed within two or three years, but in most situations it is more likely to be six to ten years. After the wreck of the *Torrey Canyon* off the Cornish coast in 1967, when tens of thousands of sea birds, local fisheries and other marine life were destroyed, a considerable degree of recovery was achieved after two or three years. Where dominant herbivores were killed by toxic oil dispersants and *Fucus* seaweed became established, it inhibited the re-establishment of limpets for several years. Restoration of the community took seven or eight years, but for some time after this impoverishment of the fauna could still be detected.

Certain organochlorine pesticides, notably DDT, have become widely dispersed throughout the oceans and are only very slowly degraded. For some marine organisms, especially crustacea, organochlorine and organophosphorus compounds need only occur in low doses to be lethal. These compounds are more soluble in fats than in water, and so they become increasingly concentrated at each stage of a food chain. Until 1971, the Los Angeles sewerage system deposited large amounts of DDT residues, and offshore sediments were heavily contaminated over a wide area (Clark, 1989). Despite the prohibition of the use of DDT in the USA in 1971, contamination of fish persisted for years. Dover sole, which feed on invertebrates living in the bottom sediment, had higher DDT residues in 1975 than 1971, and over 80% of fish contained more than 5 ppm of DDT (the recommended limit for human consumption). Kelp bass feed on fish and invertebrates in mid-water depths, and their contamination was less than bottom-feeding species, but even in 1977, 40% of samples contained more than 5 ppm of DDT. In the summer of 1976 at Los Angeles Zoo, almost all the collection of cormorants and gulls died. Autopsies revealed that the gulls had 300 ppm of DDT residues in their livers and the cormorants 750 ppm. The cormorants and the gulls had been in the zoo six to seven years, and the fish on which they had been fed came from a fishery in the Los Angeles area and contained 3.1–4.2 ppm of DDT residues.

Marine invertebrates display a high degree of genetic variability and fish are genetically the most varied of all vertebrates. There is evidence that for some organisms genetic variability is an adaptation to survival in contaminated environments. Selection may occur for whichever genotype (hereditary make-up of the individual) is adapted to the prevailing conditions. For example, there is rapid selection for resistance to the water-soluble components of crude oil in the copepod *Tisbe*. Similarly, organisms in Cornish estuaries contaminated with copper accumulate amounts of the metal which would be lethal to members of the same species living in uncontaminated habitats. This phenomenon may explain the surprising robustness of many marine ecosystems to contamination by pollutants.

6.5 CASE STUDIES OF MARINE POLLUTION

6.5.1 Operational pollution: radioactive waste disposal

The disposal of liquid and solid forms of radioactive waste into the marine enviroment, even though restricted to low and intermediate level wastes, is the cause of much controversy. It should be noted, however, that the average naturally occurring radioactivity of sea water is 12.6 Bq/l (the SI unit for measuring radioactivity is the becquerel, or one nuclear disintegration per second). This is due to the presence of potassium-40, decay products of uranium and thorium and a continuous input of tritium (hydrogen-3) from cosmic rays. Levels in marine sand (200–400 Bq/l) and mud (700–1000 Bq/l) are much higher due to the adsorbance of heavy radionuclides onto particulate matter and the large surface area of marine sediments (Clark, 1989).

6.5.1.1 *Liquid waste*

Cooling water and other liquid wastes from nuclear reactors and fuel reprocessing plants contain radioactive substances. In 1987 there were 433 nuclear reactors for electricity generation in operation or under construction in 30 countries. All depend on the fission of uranium to produce electricity. Large volumes of cooling water and some liquid wastes from nuclear reactions are discharged from coastal and estuarine nuclear power stations directly into the sea. In the UK there are regulations to control total radioactive discharge and the levels of certain radionuclides, but these regulations vary from site to site depending on local conditions. For example, at Hinckley Point in the Bristol Channel, levels are restricted to a total discharge of 7.4 TBq/yr and 74 TBq/yr of tritium with no restrictions on zinc-65. This contrasts with the nuclear power station at Bradwell in Essex, which is in the vicinity of commercial oyster beds and is permitted a total discharge of 7.4 TBq/yr, 55.5 TBq/yr of tritium and no discharge of zinc-65.

Nuclear reprocessing plants use spent fuel rods from nuclear reactors to recover uranium, plutonium and other radionuclides. Major reprocessing plants in Europe exist at Sellafield on the Cumbrian coast, UK, La Hague near Cherbourg in France and Karlsruhe in Germany. Reprocessing plants discharge waste water with a relatively low radioactive content, but because of the large amounts of water involved the radioactive output far exceeds that of power stations. Permitted levels of alpha- and beta-emitters have decreased from 222 and 11 100 TBq/yr in the 1970s to 14 and 950 TBq/yr in 1986, respectively. The behaviour of radionuclides in the sea depends on their chemical form and physico-chemical characteristics. For example, caesium-137 remains in solution and has a half-life of 30 years, whereas ruthenium and plutonium are adsorbed on to particles and carried to the sea-bed where they accumulate in the substratum. Sellafield discharges account for contamination of up to 3.9 Bq/g of plutonium-239 in sea-bed sediments in the area. Some of these radioactive particles are redeposited in estuarine muds and salt-marshes on the Cumbrian and Solway coasts, thus exposing fishermen in the area to radiation.

6.5.1.2   Solid waste

Marine dumping of radioactive solid wastes has been in operation since 1946, but the 1972 London Dumping Convention now regulates all dumping at sea and it prohibits the dumping of high level radioactive waste. High level radioactive waste is defined as that containing more than 37 000 TBq of tritium, 37 TBq of beta- and gamma-emitters, 3.7 TBq of strontium-90 and caesium-137 or 0.037 TBq of alpha-emitters with half-lives over 50 years (Clark, 1989). Solid wastes consist of contaminated concrete, piping and protective clothing, etc., and it is packed into concrete-lined steel drums, embedded in resin or bitumen, or installed with a pressure equalization device. Once dumped at great depth the containers slowly corrode, with the delay resulting in a loss of radioactivity as the short-lived radionuclides decay.

The fate of these radioactive residues (from liquid and solid wastes) on the sea-bed and their ecological impacts is still uncertain, but it is known that seaweeds and fish can rapidly absorb a variety of radionuclides, with implications for bioaccumulation in food chains. For example, near to Sellafield the alga *Porphyra umbilicalis* (processed into laver bread and consumed in South Wales) accumulates ten times the concentration of caesium-137 found in the water, 400 times the concentration of zirconium-95 and niobium-95, 100 times the concentration of cerium-144 and 1500 times the concentration of ruthenium-106 (Clark, 1989). Bivalves are very rapid accumulators of high concentrations of metals. Scallops (*Pecten maximus*) accumulate manganese, oysters (*Ostrea*) large amounts of zinc and mussels (*Mytilus*) accumulate iron. The accumulation of such radioactive residues in seafood has obvious implications for human health.

6.5.2   Regional pollution: the Mediterranean Sea

The Mediterranean Sea covers an area of 2 523 000 km^2 with a mean depth of 1470 m. The eastern and western Mediterranean is separated by the relatively shallow straits between Sicily and Tunisia. Water exchange between the land-locked sea and oceans is slow and limited (a 90 year turnover) and stratification of salinity and nutrients concentrates pollutants in specific areas. The main centres of coastal population are on the northern side of the western Mediterranean and on the north shore of the Adriatic. Here the coastal population is predicted to increase from 133 million in 1990 to 230 million in the next 40 years. These areas are also the most industrialized regions and receive most of the 100 million tourists who visit the Mediterranean each year.

All parts of the Mediterranean are severely polluted with tar balls due to deballasting, the tank-washing of oil tankers and the discharge of oily bilge water by other shipping. Some 250×10^6 tonne/year of oil are transported through the Mediterranean. Secondary sources of oil pollution come from oil refineries. A ban on the discharge of oily wastes was made at the 1976 Barcelona Convention, which has effectively reduced the amounts of floating tar in the eastern Mediterranean from 37 000 $\mu g/m^2$ in 1969 to 1175 $\mu g/m^2$ in 1987. Oil pollution has affected the Mediterranean fishing industry by tainting bivalves and fish and by damaging the spawning grounds of mackerel and bonito. The Bay of Muggia at Trieste, for example, is now almost a biological desert due to pollution from the petrochemical industry.

Domestic waste contributes considerably to the pollution of the Mediterranean. The worst affected areas are along the Spanish, French and Italian rivieras, which all have inadequate sewage treatment facilities. Untreated sewage is both aesthetically unpleasant and a health hazard. Eighty-five per cent of anthropogenic inputs to the Mediterranean are from the land (Clark, 1989), with river inputs being the most dominant. River discharge accounts for most of the phosphorus and nitrogen inputs as well as 90 tonne/yr of organochlorine pesticides into the Mediterranean. The highest input of heavy metals and polychlorinated biphenyls is from the rivers Rhone and Po and the heavily industrialized zones near their mouths. As discussed in Chapter 11, the Adriatic Sea suffers adversely from cultural eutrophication due to the effluent it receives being rich in detergents, agricultural fertilizers and sewage. Marine ecosystems, fishery industries and tourism have all been adversely affected.

The task of remedying the pollution problems of the Mediterranean Sea is herculean in practical terms as it requires the co-operation of 18 nations. One such attempt is the Mediterranean Action Plan formulated in 1975 by the signatories of the Barcelona Convention. It has been reinforced by many initiatives and working groups, including the 1990 Nicosia Charter, which defines the immediate action that needs to be taken.

6.5.3 Oil spills

6.5.3.1 Accidental pollution

Oil spills can lead to the destruction of much of the intertidal population as well as being a major threat to sea birds and mammals. Oil readily penetrates and mats the plummage of sea birds, making flight impossible, and leads to a loss of buoyancy and heat insulation. Attempts to preen cause the ingestion of oil and gut irritation, and at present, hundreds of thousands of sea birds are lost annually as a result. The species at greatest risk are those that live mainly on or in the water, e.g. puffins, divers and razorbills, but seals can also be affected as oiling of their fur reduces heat insulation.

The extent to which oil is directly toxic to the marine biome is related to its content of light aromatic hydrocarbons. These evaporate quickly so that the direct toxic effects are short-lived. The heavier fraction is not as poisonous, but organisms may die if heavily coated. The main toxicity associated with oil is the dispersant, which can be highly poisonous to marine life. If the oil is left untreated, it will gradually disperse, mainly through biodegradation by micro-organisms. The mechanical break-up of oil lumps through water movement and by grazing and passage through the gut of animals also acts as a dispersant.

6.5.3.2   *Operational pollution*

Oil refineries discharge waste water containing petroleum hydrocarbons, and the receiving waters are therefore subjected to low level pollution. For example, the oil refinery at Fawley in Southampton Water on the south coast of the UK discharged two effluent streams across a salt-marsh. Between 1953 and 1970, these discharges caused the loss of vegetation from a considerable area of the marsh, but in 1970 an effluent improvement scheme was established. By 1975 the oil content of the effluent had been reduced from 31 to 10 ppm and the amount of discharge reduced by a third. By 1980 the vegetation was re-established, but the sediments remain contaminated and the fauna of the area impoverished. Although remedial measures can be taken to combat oil pollution, the effects on the marine environment can be persistent.

6.6   PERSPECTIVE: THE IMPORTANCE OF THE OCEANS AND THE CONTROL OF MARINE POLLUTION

Given their resource potential, the oceans and seas, in particular the coastal zone, will come under increasing pressure during the next century, with continued population growth and economic development. The oceans will certainly continue to be exploited for the purposes of waste disposal, despite increasing awareness among governments and the general public that pollution poses risks to the marine environment itself and its use as a food source and recreational resource. The public's perception of the issue has been highlighted by the media, e.g. the public health risk associated with swimming from polluted beaches, and at a local and national level commercial damage to fisheries and coastal tourism is acknowledged. Moreover, the importance of the interplay between the oceans and the atmosphere in global environmental systems is now recognized. The coastal zone, as the most ecologically productive part of the oceans, absorbs huge amounts of carbon dioxide and is therefore an important factor in studies of global warming. Despite this, scientific knowledge of the processes taking place in nearshore waters and sediments (and even more so for deep waters and sediments) is still limited. This hinders solutions to questions about the capacity of the oceans to assimilate waste, which are central to all efforts to manage the oceans on a sustainable base for human use.

To manage the oceans as a sustainable resource requires the elimination of marine pollution. By implication, wastes which are potential pollutants should never be discharged into the seas, i.e. sources of pollution should be removed rather than relying on remedial measures to treat pollution that occurs. Closed systems of recycling waste are needed to eliminate operational discharges and accidental discharges require remedial treatment when they occur. Furthermore, large amounts of waste will continue to be produced and will require disposal. To dispose of them on land may create even more environmental problems than oceanic disposal. Marine pollution abatement therefore becomes a matter of balancing the costs of reducing pollution with the benefits to be derived from a cleaner marine environment. Pollution control incurs costs and the treatment of wastes to render all non-reusable residues environmentally harmless is prohibitively expensive in terms of resource consumption.

Even when there is awareness of the need for measures to control marine pollution, there is little agreement on the action necessary. What one coastal state allows in its coastal waters affects its neighbours and unilateral action by a single coastal state is bound to have only limited success. An 'international clean ocean act' may be highly desirable, but it is far from being politically feasible, even if scientific knowledge was sufficiently advanced to comprehensively identify the environmental impact of marine pollutants and to establish ambient standards. The 1982 UNCLOS provided clear guidelines for the future use of the oceans, both nationally and internationally. It delimited the political maritime zones and the powers vested in the coastal state with respect to the exploitation of marine resources rather than the control of marine pollution. Thirty-five per cent of the oceans now fall within maritime zones under the national jurisdiction of coastal states. These states can therefore adopt laws and regulations to combat pollution from ships to the outer boundary of the exclusive economic zone. The high seas and deep sea bed, the remaining two-thirds of the oceans, are recognized by UNCLOS as a common property resource. UNCLOS signatories accept an obligation to protect and preserve the marine environment but, as yet, the Convention does not command universal support. Even where states have supported the Convention there are problems ensuring implementation, which makes sustainable resource use less likely.

Whereas action at the global scale to combat marine pollution is limited, examples of more concerted efforts exist at the regional level. This is to be welcomed as the focus is on the most polluted parts of the ocean, i.e. coastal waters. In 1974 the United Nations Environment Programme announced a Regional Seas Programme which now covers ten areas—the Mediterranean, the Caribbean, West and Central Africa, East Africa, the Red Sea and Gulf of Aden, the Persian Gulf, East Asia, the South Pacific, South-east Pacific and South-west Atlantic regions. For each regional sea, action plans are agreed by the coastal states and a convention is signed which includes protocols dealing with waste dumping, the control of land-based pollution sources and co-operation in tackling pollution emergencies. Problems in other seas have been addressed independently, e.g. the North Sea is protected by 13 international conventions, 11 European Community (EC) directives, as well as numerous national laws. The North Sea became a 'special area' under the Marine Plastic Pollution Research and Control Act from February 1991, which means that no waste apart from food residues will be permitted to be dumped. Questions, however, remain about the effectiveness of such regional and national legislation. For example, in 1989 a quarter of the 440 UK bathing waters tested failed to meet the minimum safety standards established 15 years previously in the EC Bathing Water Directive.

Thus, although it is recognized that stewardship of marine resources and environments is required to ensure sustainability, in practice the approach adopted is one of the best practicable environmental options, in which the marine disposal of waste is managed to reduce pollution and mitigate its worst effects. It means that in some instances the disposal of waste into the sea, even though it damages the marine resources and environment, is accepted as preferable to any other means of disposal.

6.7 REFERENCES

Barnes, R. S. K. and Hughes, R. N. (1988) *An Introduction to Marine Ecology*. Oxford: Blackwell Scientific Publications.

Clark, R. B. (1989) *Marine Pollution*. Oxford: Clarendon Press.
Hinrichsen, D. (1990) *Our Common Seas: Coasts in Crisis*. London: Earthscan.
Ketchum, B. H. (1967) Man's resources in the marine environment. In: T. A. Olsen and F. J. Burgess (Eds) *Pollution and Marine Ecology*. New York: Interscience.
Meadows, P. S. and Campbell, J. I. (1988) *An Introduction to Marine Science*. London: Blackie.

6.8   FURTHER READING

Greenpeace (1987) *Coastline: Britain's Threatened Heritage*. London: Kingfisher Books.
Kinne, O. (Ed.) (1984a) *Marine Ecology V, Ocean Management Part 3, Pollution and Protection of the Seas: Radioactive Materials, Heavy Metals, and Oil*. Chichester: Wiley.
Kinne, O. (Ed.) (1984b) *Marine, Ecology V, Ocean Management Part 4, Pollution and Protection of the Seas: Pesticides, Domestic Wastes, and Thermal Deformations*. Chichester: Wiley.
Levinton, J. S. (1982) *Marine Ecology*. Englewood Cliffs, NJ: Prentice-Hall.
Marine Pollution Bulletin. Oxford: Pergamon. (A monthly journal with news and articles, many for a non-technical audience.)

SECTION II

GLOBAL ISSUES
(b) Change in Societies

CHAPTER 7

Population and Environment

M. S. Lowe and S. R. Bowlby

7.1 INTRODUCTION

In 1798 the English economist Thomas Malthus, in his '*Essay on the Principle of Population*', argued that eventually population growth would outstrip food production and hence lead to famine, war and human misery. Nearly two hundred years later, in 1990, Richard Ottaway, prospective Parliamentary candidate for Croydon South, writing for the Bow Group, once again drew attention to the relationship between population growth and the potential production of the land. He stated that 'at local, national, international and global levels . . . population growth goes hand in hand with increased pollution and environmental decay' (Ottaway, 1990). The longevity of this debate is testimony to the importance that policy makers have continued to place on population expansion as a significant environmental issue. Modern adherents of this view, the so-called neo-Malthusians, insist that there is a direct link between population growth and environmental degradation as, simply, more people will consume more resources and produce more waste. To others, however, the connections between environmental damage and population growth are more tenuous. This anti-Malthusian group see no necessary relationship between population growth and environmental dereliction, preferring instead to focus their arguments on issues such as the distribution and organization of people (as opposed to their absolute numbers), the use of inappropriate technology, overconsumption and inequality.

Despite a broad interest in the population–environment relationship students of population studies have been slow to recognize the specificity of the relationship between these phenomena in particular places and, more particularly, the central role that politics and ideology have to play in determining the population–environment equation in different localities. This chapter attempts to assess the complex relationship between population growth and environmental change and points out some of the many conflicts and contradictions involved in this important environmental issue.

The chapter begins with a general discussion of global population trends and a

Environmental Issues in the 1990s. Edited by A. M. Mannion and S. R. Bowlby
© 1992 John Wiley & Sons Ltd

review of the main schools of thought, neo- and anti-Malthusian, which have
contributed to the debate on the relationship between population and the environment. It then focuses on the linkages between population and the environment in two
specific countries. The first is Kenya, in the developing world, where the population
and environment relationship is most visible, and the second is Great Britain, in the
developed world, where the relationship between population issues and environmental change is not so immediately obvious, but is no less important.

7.2 GLOBAL POPULATION TRENDS AND ENVIRONMENTAL CHANGE

7.2.1 Global population trends: north and south, east and west

In 1830, the world population was about one billion (1000 million), by 1930 it was
two billion, by 1960, four billion and in 1990, approximately five billion. In the next
35 years it is expected that the world's population will grow by a further 3.2 billion to
reach 8.5 billion. However, the rates of population growth vary considerably between
continents (Fig. 7.1) and also within continents.

In the industrialized countries of North America, Europe and the Soviet Union,
death rates have been low and gradually declining throughout this century. Today
their death rates per thousand per annum are 9.0–10. Birth rates have also declined,
with typical figures of 13–16 per thousand per annum. As a result the populations of
these areas are roughly stable in overall size. Low birth rates and increased life
expectancy, resulting from improved living standards and medical care, mean that
the population of the developed world is an ageing population. In the year 2000
about 14% of the population of the developed world will be 65 or more years old and
this is considered to pose an economic problem of 'dependency' as the production of
those of working age now has to support an increasing number of people who are not
engaged in paid work or activities which are normally classed as 'productive'.

In the developing countries populations are still growing rapidly. In some instances,
such as in many of the countries of Latin America and East Asia, it is estimated by
the United Nations that population growth will stabilize by 2025–2050 (Fig. 7.1). In
South Asia and Africa, the population is not expected to stabilize until the 21st
century. Although there are important variations between countries in the Third
World, population growth is occurring because death rates have fallen to rates of
9–19 per thousand per annum while birth rates have stayed at around 35–47 per
thousand. Life expectancy in the developing world has improved during this century
(now between 54 and 66 years), but is still well below the figures for the industrialized
world. Thus, the populations of these countries are young, with the proportion of
people under 15 years old typically around 30–45%. Population growth is likely to
continue because a high proportion and high absolute number of people in these
countries soon will be of childbearing age. It means that these countries also have a
problem of dependency, but it is the problem of producing enough to support the
young, rather than the old.

Many commentators focus upon the immense size of the predicted population of
the world when discussing the relationship between population growth and the
environment. It is just as important to consider the income levels and production and
consumption patterns of the population. For example, although the developed world

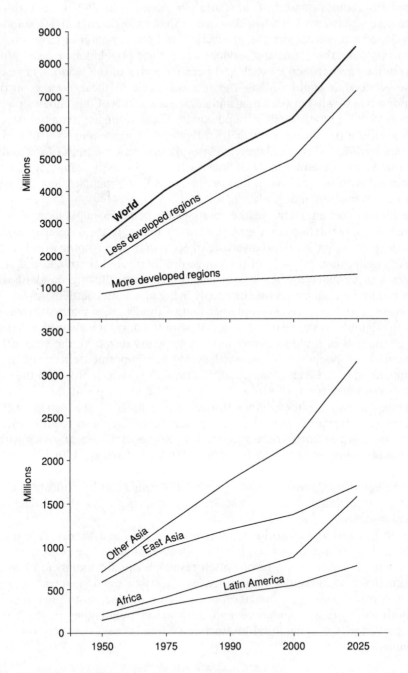

Figure 7.1 Population size and projections, 1950–2025. Upper panel, world total; lower panel, selected regions. Adapted from data in World Resources (1990), Table 4.1, p. 50. Data originally from UN (1989).

only contains about a quarter of the world's population, in 1985 it was estimated to produce three-quarters of the world's waste (Shaw, 1989). High living standards in the developed nations require the production and consumption of large amounts of energy to produce the goods and services which their populations expect. Although the gap in income between the rich and poor countries of the world is not expected to close—indeed it could widen—the income levels in these poor countries are expected to rise slowly between now and 2025. As a result of this, albeit slow, rise in incomes and the growth of their populations, Shaw estimates that the developing countries will be producing about half of the world's waste and 85% of new waste generation by 2025. This calculation assumes that no new methods of waste disposal are invented between now and 2025. Such figures can be used to suggest that it is not population growth but the rise in that population's living standards that threatens to accelerate environmental degradation.

A further important factor in the relationship between population growth and environmental degradation is the growth of urban living. In the developed world over 70% of the population live in urban areas, whereas in the developing world (excluding Latin America) about 60–70% of the population is rural. This situation is changing rapidly as people move to the cities in search of improved living standards and has already resulted a high proportion of people living in squatter settlements around the major cities. The United Nations predict that of the 24 cities that will have over 10 million inhabitants by the year 2000, 18 will be in the developing world and that 60% of the population in the developing world will be urbanized by the year 2025. The concentration of people in dense settlements has important effects on the local environment and on the environment of the rural areas which provision the city with material to produce food, shelter and clothing. The urban population also needs to earn money to pay for these items through some form of production, which will involve manufacturing as well as service production. They also require a transport infrastructure and motorized transport to service the exchange of goods within the city, between cities and between the city and rural areas. Urban populations, therefore, through their production and their consumption activities, generate large, localized, amounts of waste material ('residuals') which can be difficult to deal with without environmental damage, especially in countries with scarce financial and technical resources.

Also, in the developing world, the migration of men and women of working age from rural areas to the towns can have an adverse effect on the rural economy and environment. A lack of rural labour often results in the abandonment or neglect of environmentally sound agricultural practices, such as irrigation, terracing and crop rotation. Furthermore, the stimulation of cash crop production through the demand from both local urban populations and from urban populations in the developed world can distort the agricultural production system in ways that may contribute to environmental damage.

7.2.2 Review of the neo-Malthusian and anti-Malthusian debates

There are two main and opposing schools of thought on the relationship between population growth and environmental change. The first school, named after Thomas Malthus, is known as *Malthusian*. The second school opposes Malthus' viewpoint.

Anti-Malthusians believe that there is no direct relationship between population growth and environmental degradation.

Malthus' major arguments appeared in the seventh edition of his '*Essay on the Principle of Population*' in 1872. Malthus argued that population would always increase at a faster rate than food supply. Agricultural productivity could not continue to grow as the supply of land was finite and because there were declining returns to increased agricultural inputs. This imbalance between population growth and available resources was likely to result in 'powerful and obvious checks' on population growth such as famine, war or 'vice'—by which he meant contraception, abortion and infanticide. Unless people were willing to limit population through sexual abstinence, these 'checks' would result in an increase in mortality until population and resources reached a new equilibrium.

During the 20th century the ideas of Malthus have reappeared in a modified form and remain extremely influential. Under the head of neo-Malthusians, then, can be placed The Club of Rome whose *Limits to Growth* (Meadows *et al.*; 1972), based on a computer model, suggested that:

(a) If present growth trends in world population and resource depletion continue, the limits to growth on this planet will be reached within the next 100 years
(b) It is possible to alter these growth trends and to establish a state of global equilibrium; this could be designed so that the basic material needs of each person are satisfied and each person has an equal opportunity to realise their individual human potential.
(c) If the world's people decide to strive for this second outcome rather than the first, the sooner they begin working to attain it the greater their chances of success.

Also among the neo-Malthusians is Paul Ehrlich, whose 1968 book *The Population Bomb*, suggested that fast growing populations meant that the earth's capacity to feed people had almost been reached and that some minor fluctuation in agricultural productivity, for example, a failure of the monsoon in Asia for one year, would plunge the world into political chaos following famines and food shortages.

In the 1990s Malthusian ideas have been modified and expanded to include not only food shortages as a result of natural resources being depleted, but also other types of environmental damage which can result from there being too many people. Contemporary neo-Malthusian organizations include Population Concern, the United Nations Population Fund (UNFPA), the World Commission for Environment and Development, the Worldwatch Institute and individuals such as Prince Philip and Richard Ottaway. These share a common view that 'one fundamental reason for the damage being done to our planet, whether by pollution, deforestation, destruction of plant and animal life or by other factors is too rapid population growth' (Population Concern, 1989). They argue that the only way to curtail the damage being done to the environment by humans is to substantially reduce the numbers of people.

A variety of anti-Malthusian views are held by many commentators on population including Julian Simon, Barry Commoner and Frances Moore Lappe. Simon's text '*The Ultimate Resource*' (Simon, 1981) turns Malthusian theory on its head by suggesting that population growth presents no problem to the future of the earth. Rather, an increase in population leads directly to higher living standards and economic development through an increase in the invention of means to cope with the population increase because there are more potential inventors and because the pressures of population growth create the impetus for invention.

A different analysis is produced by Frances Moore Lappé and Rachel Schurman in their book *Taking Population Seriously* (Moore Lappé and Schurman, 1989). They take a power structures perspective to move beyond the neo-Malthusian arguments which they view as simplistic and deterministic. Instead, they focus on the causes of the population explosion by suggesting that a major factor determining the over-growth of developing world populations is the imbalance of power between men and women in society. This imbalance shapes people's reproductive choices (or lack of them), women's poor economic, social and cultural position often leaving them with few options other than reproduction.

Here an analysis of these perspectives is presented by examining the relationship between population and environment in a developing world country, Kenya, and a developed world country, Great Britain. Discussion will centre on three (of many) aspects of population issues: fertility and birth control, the ageing of the population, and migration.

7.3   KENYA: THE RELATIONSHIP BETWEEN POPULATION AND THE ENVIRONMENT

The population in Kenya is growing at one of the fastest rates in the world. Between 1965 and 1980 the total population has grown at 3.6% per annum and from 1980 to 1988 it grew at 3.8% per annum to reach a size of 22 million. Current estimates for Kenya's future population suggest a total population of 34 million in 2000 and 62 million in 2025. The rural population has been growing at an average annual rate of 4.0% in the 30 years from 1969 to 1990, whereas the urban population has been growing at 8.8%. Despite these rapid growth rates in the towns, three-quarters of Kenya's population is still rural and 81% of the labour force is employed in agriculture (World Resources, 1990).

During the 1980s Kenya's gross domestic product has grown at only 4.2% per annum (World Bank, 1990), which is only slightly greater than the growth rate of population. The growth of population and the increase of production necessary to achieve economic growth has resulted in an expansion of agricultural production of both food and cash crops and the development of modern, urban based industry. In such a situation the growth of population might be expected to create environmental damage through such processes as cultivation of marginal lands, overgrazing and overuse of fertilizers as well as by an increase in pollution resulting from the expansion of industry and the growth of large urban areas. Before examining this interrelationship, it is necessary to consider the social processes that contribute to population growth in Kenya.

7.3.1   Fertility and family size

The fertility rate in Kenya is one of the highest in the world. This rate was about 8.0 at the start of the 1960s and remained unchanged during the 1970s, although the Kenyan Government introduced a policy to reduce population in 1967. During the

1980s the fertility rate fell to 6.9; this is still very high. The government supports both public and private family planning clinics, but the take-up of contraception is low, although increasing slowly, at under 10% of married women of childbearing age in 1977–8 and 17% in 1986 (World Bank, 1984; 1990).

One important reason for the low take-up of contraception is that the desired family size is large. In 1987–8 only about 12% of women who were 'at risk' of becoming pregnant wished to have no more children. One factor influencing this desire for a large family is that in rural areas a large number of children is considered economically desirable. Children soon become workers on the family land and have therefore been regarded as a positive contribution to the family's economic wellbeing. They are also considered to be an investment for their parents' old age, when the children will be expected to care for them. Cultural factors are also important; for example, among some groups in Kenya children are seen as a way of carrying on the family or tribal name and parents may put pressure on their adult children to have a large family.

The social expectations and economic role of women are also very important to desired family size. Where women can make a cash contribution to the family's income through wage or market earnings, this contribution may be more valued than their indirect contribution to family income through bearing and rearing children. Women's power within the family is also important. Where women are in a subordinate role within the family they may have little say in the size of the family. If women have independent earnings this strengthens their position in the family decision making in general and in relation to deciding the size of their family. Thus the education and social status of women is closely linked to population growth. One reason for this is that women who have some education have better prospects of earning an income independently, either, in the rural areas, through marketing their own agricultural produce, or selling their own goods, or, in the urban areas, through having a paid job in manufacturing or services.

A Kenyan example quoted by Ottaway (1990) illustrates these points. In the Nyeri district, 150 miles north of Nairobi, a women's pig rearing co-operative was set up in the village of Mukurweini with initial funding from the Kenyan Family Planning Association. The co-operative was successful economically and for the first time the women had an independent source of income. The Association also provided education on the economic advantages of small families, information on contraception and mobile clinics providing contraceptives and some maternal and infant health care. After three years the birth rate began to fall. The acceptance rate of contraceptives among married women rose to 43% and infant mortality halved.

Infant mortality is another significant factor of great importance to desired family size. If parents expect several of their children to die they will tend to have more children to ensure that some survive into adulthood. Parents will err on the side of having more children than they 'want' rather than too few. In Kenya today infant mortality has fallen overall from 112 to 70 per thousand live births between 1965 and 1988, so this would be expected to lead to a decline in fertility.

Although the demographic figures presented give some idea of the current pattern of fertility in Kenya, they mask great regional variations. These are linked to the wide variation in income levels. The national average gross national product per capita was $370 in 1988 ($12 340 in the UK and $19 840 in the USA for the same year), but incomes in the major urban areas of Nairobi and Mombasa and the surrounding agricultural areas are far higher than those in some of the poorest rural areas where agricultural productivity is low. Similarly, infant mortality rates and contraceptive use vary strikingly between the richer and poorer areas. As in many developing countries health and educational resources have been channelled less readily to the poorer rural areas than to the urban areas, where the urbanized elite live, or to regions of commercial agriculture, which are also regions of political importance.

These disparities in income and in public resources are of great significance to the growth of populations as it is in those areas with low incomes and lack of public investment that fertility rates remain high. Without measures to increase incomes, the status, education and economic independence of women, and infant and maternal health care, this situation is likely to continue.

7.3.2   Age structure and dependence

Given the high population growth rates in Kenya it is not surprising that 51% of the population is under 15. The high proportion of young people in the population poses considerable strains on the education system and generates a high demand for rural and urban employment which must be met through continued economic growth. In addition, a large proportion of young people also means that there will continue to be a high proportion of people of childbearing age. If the fertility rate remains high, this fuels further population growth. In view of this situation the Kenyan Government finds it difficult to increase per capita expenditure on education and so to improve the skills and earning power of the population. Thus, it can be argued that the high percentage of young people is a factor that hinders the economic development of Kenya. Therefore it can also be argued that the youthful population structure makes it more difficult to reduce population growth rates as prosperity and education, especially for women, are such potent factors in reducing fertility rates.

7.3.3   Migration

Since independence Kenya has undergone major changes in the organization of its agriculture and industry which have been instrumental in the considerable redistribution of the population through migration. Before independence the best agricultural land was held either by independent European farmers in large farms or by commercial companies. During the 1960s, after independence, many of these farms were gained by the new Kenyan elite and remained in large profitable units. In the 1950s, during the Mau-Mau guerrilla war preceding independence, the Kikuyu (the dominant tribal group in Kenya) were moved for security reasons from the white farming areas and from their dispersed small farms on reserves into villages. Previously, the traditional form of Kikuyu landholding had been in several separate plots. This was considered to be an uneconomic form of land tenure, preventing the use of modern farming methods. As part of the process of integrating the Kikuyu into village life, the land was redistributed, consolidated and legal titles to land ownership were established. During the 1960s and 1970s there was continued population growth, but the new forms of land ownership meant that while those with title to both the large, previously European owned farms and the smaller (8–10 hectare) farms created by land consolidation were able to farm profitably, there was no means by which any of this land could be made available to the growing number of landless people. The disparity of resources available produced, and still produces, major inequalities in income.

The movement into the cities and the consequent rapid growth of the population has created a familiar set of environmental problems over the provision of housing, sanitation and services. Population growth is most rapid in the medium-sized towns rather than in the major cities of Nairobi and Mombasa, although absolute rates of growth in these cities remain high. However, fertility rates tend to be lower in the urban than in the rural areas. Families in the city depend on wage earning so that there is no longer the incentive to produce children as a contribution to the family's labour force. Furthermore, contraceptives, family planning advice and health care for children and mothers are more readily available in the cities than in the rural areas. Nevertheless, these facilities are far more easily available to the rich than to the poor.

7.3.4   Population and environment in Kenya

The anti-Malthusians and neo-Malthusians suggest three broad types of relationship between population and the environment. Neo-Malthusians suggest that population growth will, in the long term, damage the environment. Some anti-Malthusians suggest that population increase will stimulate further productivity through innovation and human creativity and that this will allow more people to use the environment without lasting damage. Another group of anti-Malthusians suggest that population growth is not the cause, but rather the symptom of economic and social relationships that are the real source of environmental damage. The Kenyan case provides support for all three positions as will be explained in the following.

Downing *et al.* (1990) argue that in the prime agricultural lands of central and eastern Kenya the considerable growth of population that has occurred through natural increases and, importantly, through migration has resulted in the intensification of agriculture and the greater integration of the rural economy with the urban economies of Nairobi and the medium-sized towns of Nyeri, Meru, Embu, Kangundo and Machakos. These are areas well suited to the growth of tea and coffee as cash crops and there is also a market for food and goods from the rural periphery in the towns. Thus, the rural areas help to support the urban population and some of the rural population have opportunities for annual or seasonal employment outside the agricultural sector. Agricultural production has been intensified, but this has not resulted in local environmental damage.

In contrast, in the more marginal semi-arid lands, the population has also increased, but there has been no increase in agricultural productivity. Farmers have expanded the cultivated area by cutting down trees, people have found supplementary sources of income such as charcoal production, or they have expanded their flocks to cope with years of poor rainfall. Unfortunately, these survival strategies damage the environment in ways that will undermine its future productivity. This situation appears to support Malthusian arguments.

These two examples can also be interpreted as showing that the underlying processes producing apparently successful agricultural intensification and environmental degradation are not those of population migration and increase. The processes responsible are those social and economic relationships involved in the transformation of a traditional peasant farming system to an agricultural and industrial economy that is fully integrated into the global capitalist economy. These processes involve the creation of a large group of landless people in addition to a small elite with large farms and many marginal farmers with too little land and capital and insufficient alternative employment to allow them to survive without farming. This implies that there is no straightforward correlation between population densities or rates of growth and environmental degradation, and that the relationship is much more complex.

7.4   GREAT BRITAIN: THE RELATIONSHIP BETWEEN POPULATION AND ENVIRONMENT

At a first glance the relationship between population and environment in Great Britain seems tenuous when compared with the obvious and direct linkages between these phenomena in Kenya. Indeed, many neo-Malthusians have suggested that it is unnecessary to go beyond the developing world to solve the population–environmental degradation problem. On further investigation, however, population change is as important a contributor to environmental change in the developed world as it is in the developing world. In this context it is important to examine the dynamics of population change. There have been profound shifts in the population size and composition of most places in Britain over the last two decades and these changes have had important environmental impacts.

The population of Great Britain currently stands at 56 million. In the century from 1875 to 1975 the population almost doubled, but since then it has been relatively stable. Population projections indicate that the population will peak at 58 million in 2025 and fall back to its current level between 2050 and 2075 (Joshi, 1989). Thus, the issues of most importance in the British case are those of population structure and location rather than absolute size. Although the large size and high standard of living of the population of Great Britain creates a large demand for resources in general, it is the way in which the population is organized which leads to specific environmental problems. This is because different segments of the population make demands for particular combinations of resources in different places. As in Kenya this can be shown by reference to the issues of fertility and family size, a changing age structure, and migration.

7.4.1   Fertility and family size

Notwithstanding the high fertility rate in the 19th century and the so-called 'baby boom' of the immediate post-Second World War years, it is generally accepted that high fertility is not a problem in Great Britain. Indeed, the reverse was true in the 1930s when there was a genuine concern about low fertility rates and suggestions that the population of England and Wales would fall below 10 million early in the 21st century. Following the 'baby boom', fertility rates have remained consistently low (about 1.8) since the 1970s. With a liberal policy on birth control, contraceptive services, freely available advice, and legalized abortion, the number of births declined by over a third from a peak of over one million in 1964 to less than 660 000 in 1977, although since then there has been a modest natural increase (Champion and Townsend, 1990)

This low and stable fertility rate, combined with the significant decline in mortality throughout the 20th century, are resulting in an overall population expansion, albeit on a small scale, which still has significant implications for the environment. Not only do more people consume more resources, but the rise of real incomes and, consequently, of consumption, that has occurred in Britain throughout the 20th century has had a multiplier effect on the per capita resource depletion (see Chapter 12). The consumption behaviour of the large British population is thus leading to all kinds of environmental difficulties, from the disposal of more and more industrial and domestic waste to the need for housing and recreation, which can place heavy demands on the countryside.

Moreover, the 19th century increases in the population of Great Britain were counteracted by an increase in emigration. Any population growth now has to be accommodated largely within Britain. In addition, during the last few decades there have been a number of significant alterations to household organization in Britain and the trends identified look set to continue into the next century. Such changes have a number of implications for the relationship between population and the environment.

The expansion of the elderly population (Section 7.4.2), combined with a falling birth rate, an increase in the number of young adults and the rising divorce rate have reinforced the trend to smaller household sizes (Champion and Townsend, 1990). From 1961 to 1971 the number of households in Great Britain increased by two million and by a further one million to 1981. These trends look set to continue. Marriage and childbirth are increasingly postponed and this, coupled with the growth in single parent families and divorce, has created an increased need for single person household units. Such trends demand different patterns of building and are exacerbated by the continued spatial concentration of the British population in south-east England as a result of migration (Chapter 18).

7.4.2 The ageing of the population

Stable fertility rates combined with improvements in life expectancy in the latter part of the 20th century have led to the ageing of the population of Great Britain. During the 1980s there was a higher proportion of the population over the age of 65 than had ever been recorded before. More specifically, there has been a significant increase in the number of very old people (in the 80 and 90 year old age group).

A clear implication of population ageing in Britain is a change in the demands for resources such as medical care, transport and leisure. For example, elderly people need different and, in general, more expensive types of medical care than younger people. They may need specialist accommodation such as sheltered housing or residential homes and they need higher heating levels than younger people. Many elderly people are unable to drive either because of illness or lack of money and are unable to use conventional public transport because of physical frailty. This increases the demand for accessible public transport, which requires public subsidy if it is to be provided effectively. Demand for recreation from elderly people is likely to place pressure on seaside resorts and on accessible areas in the countryside. Such leisure demands are likely to be coupled with those for housing in 'retirement' areas. The requirement for smaller housing units by an enlarged elderly population is likely to have an impact on open spaces.

One implication of the ageing of the population is that expenditure on care for the elderly, social security, sheltered housing and so on will compete with demands for increased expenditure on environmental initiatives and improvements. Furthermore, the specific demands of elderly people for housing and recreation have an impact on the use of land in scenically attractive areas of the country and on the type of urban infrastructure that a long-lived population requires.

7.4.3 Migration

During the 19th century industrial and urban growth in Britain lured increasing numbers of people to the towns and cities and resulted in a plethora of environmental problems. In the 20th century, and especially since the Second World War, there has been a steady but marked drift to the south and east of Great Britain, and from larger cities to smaller settlements and rural areas. As a result this area of Britian, and its more rural areas in particular, has experienced a number of environmental problems. Current disputes concerning housing land availability in the south-east, together with transport problems arising from the increasing use of cars by an increasing number of people (see Chapters 16 and 18) and recurrent debates over the loss of farmland and national heritage to housing estates and motorways, are some of the important environmental issues which arise as a result of this redistribution of population.

Moreover, this movement consists mainly of younger members of the population. Younger movers to the south and east and to the suburbs and rural areas demand smaller single person or suburban family units in these areas, while those who are unable to move, who tend to be old or poor or both, are concentrated in the inner cities and the north. This compounds the existing problems of poverty, lack of jobs and environmental dereliction resulting from industrial decline and poor maintenance of housing and residential environments in the latter areas.

7.4.4 Population and environment in Britain

During the post-Second World War 'baby boom', a number of people suggested that population growth in Britain should be controlled for the sake of worldwide resource availability as well as for the sake of this country. However, although Britain now has a population that is increasing very slowly, this has not meant that the environment in Britain is now relieved of pressure, nor that the British population are no longer damaging the global environment. If the population were to be reduced, it is not clear that this would result in less environmental pollution and damage. Indeed, the main environmental culprit in Britain is not population growth or distributional change, but the high standard of living that British people now enjoy.

This standard of living involves the use of large amounts of land for housing, transport, industry and commerce as well as requiring the availability of the countryside for recreation. It also means that people use products whose production and use involve the creation of polluting wastes. The spatial organization of the population and its age and social structure are clearly factors which influence the impact of these demands on the environment. However, it can be argued that the direct causes are not those of population dynamics, but the social and economic relationships that currently underpin an 'advanced' capitalist economy. From this perspective the crucial questions for policy towards the environment are not those of population control but of whether it is possible to maintain a sustainable capitalist economy (see Chapters 2 and 17). This is not to argue that policies to influence the impact of population growth or movement on the environment in Britain are not necessary, but to suggest that such policies are not the fundamental element in environmental protection.

7.5 POLITICS, ENVIRONMENT AND POPULATION IN THE 1990s AND BEYOND

The two case studies suggest that population change is only one of a number of factors which are currently leading to the degradation of the environment. They show that it is not sufficient to look to population growth to explain environmental degradation. Further, there are important differences between the developing and developed world in the links between population change and environmental damage.

In the developed world, countries with little or no population growth are big polluters because of their use of technologies that create pollution. The economy and way of life in these countries are predicated on the continued high consumption of goods and services. In contrast, in the developing world, the poor rural populations are forced to damage the environment in the long term to ensure their short term survival. Environmental degradation in these countries is often a symptom of underdevelopment and also a cause of further underdevelopment as it damages both the means of production (the environment) and the human beings who use that environment. In general, the contrasts between the developed and the developing world can be summed up by the suggestion that in the developed world the main force leading to environmental damage is the pressure to consume and to increase consumption, whereas in the developing world the main force is the pressure to produce to survive.

The World Commission on Environment and Development which reported in 1983 (the Bruntland Report) argued for a new approach to development in which the needs of the poor for jobs, food, energy, water and sanitation were given priority within an overall attempt to create 'sustainable development' (see Chapter 2). Although this has been widely accepted as a desirable goal, there has, as yet, been little evidence of the redistribution of wealth and power that would necessarily be involved in such an approach to development.

The relationships between men and women are central to any direct policies to influence population. Many people, including the authors of this chapter, would support the view that policies which allow women to make a freer choice about the number of children they have and which aim to provide appropriate medical care to both mothers and children should be supported as politically and socially desirable in their own right, whether or not they influence population growth. However, if a reduction in population growth is desired, such policies must be implemented. As such policies require substantial improvements in the social and economic status of women, they conflict with the interests of men, who are normally those in positions of political power, and therefore such policies can be very difficult to implement effectively.

7.6 REFERENCES

Champion, A. G. and Townsend, A. R. (1990) *Contemporary Britain: A Geographical Perspective*. Sevenoaks: Edward Arnold.

Downing, T. E., Lezberg, S., Williams, C. and Berry, L. (1990) Population change and environment in central and eastern Kenya. *Environmental Conservation*, 17 (2), 123–131.

Ehrlich, P. R. (1968) *The Population Bomb*. New York: Ballantine Books.

Joshi, H. (Ed.) (1989) *The Changing Population of Britain*. Oxford: Basil Blackwell.

Malthus, T. (1798) *An Essay on the Principle of Population and a Summary View of the Principle of Population*. A. Field (ed.) (1970). Harmondsworth: Pelican.

Meadows, D. L., Randers, J. and Behrens III, W. W. (1972) *The Limits to Growth: A Report to the Club of Rome's Project on the Predicament of Mankind*. New York: Potomac Associates.

Moore Lappé, F. and Schurman, R. (1989) *Taking Population Seriously*. London: Earthscan.

Ottaway, R. (1990) *Less People, Less Pollution: An Answer to Environmental Decline Caused by the World's Population Explosion*. London: Bow Group.

Population Concern (1989) *Annual Report*. London: Population Concern.

Shaw, R. P. (1989) Rapid Population Growth and Environmental Degradation: Ultimate versus Proximate Factors. *Environmental Conservation*, 16 (3), 199–208.

Simon, J. (1981) *The Ultimate Resource*. Oxford: Martin Robertson.

United Nations (UN) (1989) *World Population Prospects, Part 2*. New York: United Nations, Tables 1 and 2.

World Bank (1984) *World Development Report 1984*. Oxford: Oxford University Press.

World Bank (1990) *World Development Report 1990*. Oxford: Oxford University Press.

World Commission on Environment and Development (1987) *Our Common Future*. G. H. Brundtland (Ed.) Oxford: Oxford University Press.

World Resources (1990) *World Resources 1990–91: A Guide to the Global Environment*. World Resources Institute in collaboration with the UN Environment Programme. Oxford: Oxford University Press.

United Nations (UN) (1989) *World Population Prospects*, Part 2, Tables 1 and 2. New York: United Nations.

7.7 FURTHER READING

Moore Lappé, F. and Schurman, R. (1989) *Taking Population Seriously*. London: Earthscan.

Pacione, M. (Ed.) (1986) *Population Geography: Progress and Prospect*. Beckenham: Croom Helm.

Robinson, H. (1981) *Population and Resources*. London: Macmillan.

CHAPTER 8

World Energy Picture

J. G. Soussan

8.1 INTRODUCTION

The world's energy scene is changing rapidly. The oil crises of the 1970s put energy at the top of the international agenda and the equally dramatic events of 1986, when oil prices collapsed and Chernobyl blew up, have created an atmosphere of uncertainty in the world's largest industry. To these concerns can be added the increasing awareness of the environmental implications of energy production and use. The enhanced greenhouse effect, tropical deforestation, acid rain, the spectre of nuclear waste, the destruction of habitats by hydroelectric dams, air quality in cities and other environmental fears have led people to question the sustainability of a way of life based on high levels of energy use.

The last 20 years have been a period of unprecedented volatility in the world fuel markets. This followed a long period of stable real fuel prices and rapid growth in energy use from 1945 to 1973. During this period, the global consumption of commercial fuels rose by about 5% per annum, and demand for oil increased from around 380 million tonnes in 1950 to 2800 million tonnes in 1973. Cheap and plentiful energy seemed the natural order of things, and formed the basis of the reconstruction of the world's economic system after the Second World War.

The picture looks very different today. The oil crisis of 1973–74 resulted in a consensus of scarcity, which led to panic buying to build up oil reserves, more exploration efforts to develop new oil sources and the advocacy of a series of alternatives to oil. This view was eroded by the oil glut of the mid-1980s; a glut which led to the price collapse of 1986. The result is confusion. Are fossil fuels running out, or will oil, coal and gas continue to be available for future generations? What should be the role of nuclear power, and in particular will the dream of a world powered by the atom ever come to pass? What has been the impact of energy costs on the economies of different parts of the world? More fundamentally, what is the relationship between economic development and energy availability? What should be the price of oil? Can present levels of energy use be maintained without destroying the environment? These and many more questions have been thrown up since the

---

Environmental Issues in the 1990s. Edited by A. M. Mannion and S. R. Bowlby
© 1992 John Wiley & Sons Ltd

fateful Organization of Petroleum Exporting Countries (OPEC) meeting in October 1973, when the intransigence of the oil companies and Western support for Israel in the Yom Kippur war led to the price rises which so traumatized the world. Today, nearly 20 years on, there are still few satisfactory answers.

What has emerged is a better understanding of the importance of energy, and a clearer picture of the relationship between the political, economic and environmental forces which shape the world's energy picture. This chapter explores the differences in energy availability and use in different parts of the world and discusses their implications for the economic hopes and prospects of these regions. The main environmental challenges that energy presents are briefly considered, although a number of these issues (global warming, acid rain, deforestation and others) are considered in more detail in Chapters 3, 4, 5 and 11.

Table 8.1   Production and consumption of commercial fuels for selected countries in 1986. Adapted from World Resources 1988

	Production (PJ)	Consumption (PJ)	Per capita consumption (GJ)	Per capita gross national product (US$)
Africa	16 941	7109	12	—
Burkina Faso	—	6	1	150
Kenya	7	49	2	300
Libya	2279	395	106	—
Nigeria	3183	495	5	640
Zimbabwe	130	169	19	620
North and Central America	75 097	79 387	195	—
Nicaragua	2	30	9	790
USA	58 422	60 766	278	17 480
South America	11 431	8109	30	—
Argentina	1637	1580	51	2 350
Brazil	2152	3088	22	1810
Peru	433	310	15	1090
Asia	71 118	58 114	20	—
Bangladesh	122	188	2	160
China	24 349	21 771	21	300
Indonesia	4169	1392	8	490
Kuwait	3339	432	228	13 890
Sri Lanka	10	56	3	400
Vietnam	168	214	4	—
Europe	41 969	64 177	130	—
Czechoslovakia	1973	2860	183	—
West Germany	4545	10 097	166	12 080
Portugal	37	397	39	2250
UK	10 478	8858	157	8870
USSR	66 197	52 671	187	—
World total	288 367	273 201	56	—

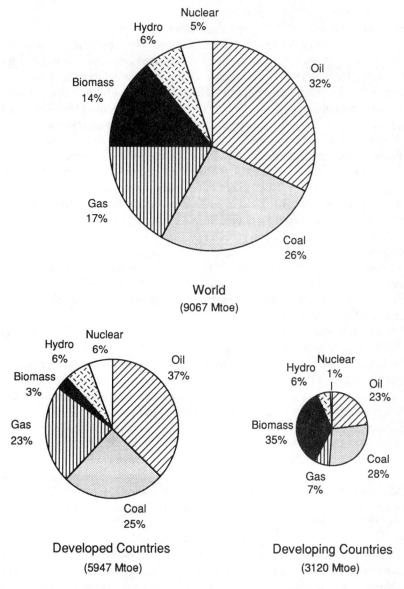

Figure 8.1 Global distribution of energy use in 1987. 1Mtoe (million tonnes oil equivalent) = 44 × 10⁶ GJ.

8.2 WORLD ENERGY DEMAND

Figure 8.1 summarizes the global distribution of energy use in 1987. The first point to notice is that the developed world uses far more energy than the developing world, nearly twice as much despite its far smaller population. Each person in the developed countries uses, on average, six times as much energy as a person from a developing country. There are, of course, huge variations in this (Table 8.1). The average

Canadian uses nearly 500 times as much commercial energy as someone from Burkina Faso (8945 as opposed to 18 kg oil equivalent in 1986). In contrast, the average energy use in Trinidad or Venezuela (both oil-producing countries) is actually higher than that in Italy, Spain or Ireland. Some developed countries (notably Japan, but including some European countries such as Sweden and Germany) have achieved rapid economic growth and now support a high standard of living using far less energy than others (especially the USA). In contrast, the Soviet Union and Eastern Europe have high levels of energy use relative to their economic performance. Despite this variety, what is clear is that there is a direct (though complex) relationship between economic development and energy use. For poor countries, increased energy supplies are a vital input into development, but this does not mean that they have to follow the profligate ways of the West. The developing world could, in theory, achieve Western living standards with only a 20% increase in energy consumption (World Resources, 1987). This would require huge amounts of capital investment, some significant technological breakthroughs and a major change in the way the world's energy economy works. This is a hopeful message; despite the claims of many commentators, the world community is not faced with a choice between poverty or eco-doom. None of the changes necessary to achieve the new energy path implied here are easy, but there is an alternative to repeating the mistakes of the past, however hard it is to achieve.

The second important point from Fig. 8.1 is that the fuels used by the developing world are different, and, in particular, biomass fuels are a major source of energy. These fuels are the energy of the world's poor; about half the world's people use little or no oil, electricity or any other commercial fuel and do not participate in global energy markets, even as consumers. They instead depend on biomass fuels, and as such are dependent on the local environment for their energy, as well as for the rest of their basic needs. In consequence, the energy concerns of developing countries, and especially the very poor, are very different to those of developed nations. They were badly affected by the oil price rises of the past, which dislocated their macro-economy and set back hopes of economic development. This they share with the developed world, only more so. Large parts of the developing world are also faced with severe problems which stem from fuelwood use and availability; problems which jeopardize the very survival of many of their people.

These differences in the amounts and types of fuels used in various parts of the world reflect the different purposes for which energy is used. Table 8.2 gives examples of patterns of energy demand from a number of countries. In very poor countries, such as Nepal and Somalia, almost all energy use is in the household sector. Most of this energy is used for cooking, and most of it comes from biomass fuels. This reflects the very limited development of these countries, and in particular the poverty of the people, which means that they cannot afford even the basic consumer goods which people in the developed world take for granted.

In developing countries commercial fuels, and in particular oil, are mostly imported, and the economic crisis that has developed through the 1980s and into the 1990s means that they are often unable to even maintain existing levels of use of commercial fuels. For example, in Somalia (one of the world's poorest countries) oil imports declined from 126 000 tonnes in 1986 to 55 000 tonnes in 1989. This leads to fuel shortages, which severely hamper their development prospects. In effect, such

Table 8.2 Energy demand balances for selected countries. Values given are 1000 metric tonnes oil equivalent (toe). Adapted from Soussan (1988); Department of Energy, Somalia (1990)

	Biomass	Petroleum	Others	Total (%)
Nepal (1980–81)				
Household	3089.3	30.4	6.6	3126.3 (95.1)
Industrial	2.1	4.8	39.4	65.3 (2.0)
Transport	—	69.1	1.2	70.3 (2.1)
Others	9.0	14.3	2.8	26.1 (0.8)
Total	3119.4 (94.9%)	118.6 (3.6%)	50.0 (1.5%)	3288.0 (100)
Somalia (1989)				
Household	1850.4	2.6	6.9	1859.9 (87.8)
Industrial	48	1.4	1.6	51 (2.4)
Transport	—	186.9	—	186.9 (8.8)
Others	—	19.8	0.2	20 (1.0)
Total	1898.4 (89.6%)	210.7 (10%)	8.7 (0.4%)	2117.8 (100)
Kenya (1980)				
Household	4058.6	173.2	45.8	4277.6 (60.0)
Industrial	1127.7	419.8	68.2	1615.7 (22.0)
Transport	—	997.3	1.1	988.4 (14.0)
Others	43.8	188.1	41.8	273.7 (4.0)
Total	5230.1 (73%)	1778.4 (25%)	156.9 (2%)	7165.4 (100)
Brazil (1980)				
Household	12 071.0	254.0	4960.0	17 285.0 (19.0)
Industrial	15 432.0	15 766.0	11 068.0	42 266.0 (46.0)
Transport	1 501.0	24 341.0	94.0	25 936.0 (28.0)
Others	508.0	3177.0	2339.0	6024.0 (7.0)
Total	29 511.0 (32%)	43 538.0 (48%)	18 461.0 (20%)	91 510.0 (100)

nations caught in a catch-22. Their economies cannot develop without the greater availability of energy, but they cannot afford to import the fuel until their economies are more developed. In economic terms, energy is a crucial and missing factor of production. Current patterns of energy use are consequently changing only slowly, if at all. Biomass fuels are the main fuel, and will continue to be so for the foreseeable future. As such, the most important energy issue in these countries is the ability of their land resources to maintain and increase the supply of biomass energy which is essential to the survival of their people.

Middle-income countries such as Brazil or the Philippines have a more diverse pattern of energy use (Table 8.2). Their populations are not so poor and are more urbanized (although, as in Brazil, they may also contain many very poor rural people who use little energy apart from biomass for cooking), urban based industry and services are more important, and transport and infrastructure facilities are more developed. These factors are reflected in the pattern of energy use found in these countries. Biomass fuels are still important, but other sources of energy (especially oil and electricity) are used widely and are of growing importance. In these countries the pattern of energy use is changing as the economy develops and the population urbanizes.

Table 8.3   Energy consumption in the UK in 1973 and 1983. Values given are million tonnes of coal equivalent (%). Adapted from Institute of Electrical Engineers (1985)

	1973	1983	Change (%)
Industry	103.2 (42.3)	67.8 (31.2)	−34.3
Transport	51.6 (21.1)	57.4 (26.4)	+11.2
Domestic	59.6 (24.5)	62.2 (28.6)	+4.3
Others	29.6 (12.1)	29.8 (13.7)	+0.1
Total	244.0 (100.0)	217.5 (100.0)	−11.1

The amounts of energy used *per capita* in the developed world are far higher than those of the developing world. This partly reflects the greater importance of industry in these economies, but energy use by individual consumers is also much higher. Table 8.3 shows the pattern of energy use in the UK for 1973 and 1983. In 1983, industry, transport and the domestic sector were all of similar importance, consuming around 30% of the energy used. The comparison with 1973 is revealing, as it shows that the total energy demand in the UK dropped by over 10% in the decade following the first oil shock, reversing decades of growth in energy demand. Table 8.3 also shows the changing structure of energy use, with industrial demand in particular falling and transport continuing to grow, despite the reliance of this sector on oil. This shows that the fall in energy demand in countries such as the UK is as much a function of the changing structure of their economies as it is a reflection of changing energy prices.

In the developed world each household contains many energy using domestic appliances. Private car ownership has also grown to a position where most households have at least one and many have two or more cars. This is reflected in the rapid growth of fuel use in transport. The more affluent people become, the more energy they use: two cars, two stereos or two televisions instead of one, larger fridge-freezers, bigger homes, central heating and so on. For the developed world, energy is an important issue, but is more about the maintenance of a wasteful way of life. For the world's poor, it is a matter of life and death.

8.3   WORLD ENERGY RESOURCES

The present picture of world energy resources looks far healthier than most commentators would have dared to suggest even in the mid-1980s, when predictions of oil costing $100 a barrel in the 1990s were common. In contrast, the optimistic forecasts from the 1950s of a nuclear-powered, electricity-based society before the end of the century now look like science fiction. The only certain thing about predictions of energy futures is that they will be wrong. What is clear is that fossil fuel resources are still relatively abundant, and will continue to be the main source of world energy well into the 21st century. They will, of course, become scarcer and more difficult to extract, and consequently more expensive, but this will be gradual. Their use also has disturbing environmental implications, and international efforts to limit the growth of fossil fuel burning are now a major focus of attempts to confront the threat of global warming. Despite this, it will be some considerable time before fossil fuels become so scarce that their cost produces major changes in the way societies use energy.

8.3.1 Oil

Odell (1986) has shown that oil has been discovered at a far faster rate than it is used, and much of the world has yet to be properly explored. Global oil consumption in the late 1980s was about 2900 million tonnes a year; a total not significantly greater than that of the early 1970s. Consumption in the developed world actually declined, from about 1900 million tonnes a year in the early 1970s to around 1600 million tonnes in the late 1980s.

The current best guess for the future of oil is that supplies, and consequently prices, will not be under serious pressure until the end of the century unless political factors intervene (the impact of the 1990–91 Gulf crisis again shows how real that possibility is). After that, production can certainly be maintained at a constant level until the middle of the next century and will then gradually tail off. The International Energy Agency gives a figure for the recoverable oil resource of 324×10^9 tonnes (Foley, 1987) or about 100 years' supply at current consumption levels. Some experts predict an even higher figure. The total amount of oil already consumed is about 60×10^9 tonnes. As such, oil will be a major source of energy for at least as many years into the future as it has been in the past. As Foley (1987, p. 82) says: 'The days of unimpeded growth in consumption are almost certainly gone for ever. But there is still a great deal of oil left'.

8.3.2 Natural gas

Consumption of natural gas has grown rapidly over the last 20 years, and, like oil, the resource future is bright. Proven reserves, at 98 trillion cubic metres, are equal to those of oil. Many geologists believe that the ultimately recoverable gas resource could be even greater than that of oil, enough to last for at least 100 years and probably far longer. A problem could come with its distribution, with 43% of current reserves in the USSR, but this unevenness is true for all energy sources. Consumption of natural gas has been growing more rapidly than that of any other fuel. Gas has many advantages as an energy source. It is a high quality fuel, is usable in many different sectors, and is particularly favoured by households and industry. It is often cheaper than competing sources of energy, and can be transported easily by pipeline. Above all, it is the cleanest of the fossil fuels, emitting lower levels of pollutants (including carbon dioxide) than either coal or oil. Many countries have plans to increase the use of natural gas. For example, there are plans to construct small gas-fired electricity stations in the UK and elsewhere, and developing countries such as Tanzania, Thailand and Bangladesh are actively developing their gas resources. Similarly, Japanese imports of natural gas are projected to rise sharply in the 1990s. The importance of natural gas is likely to grow significantly in the future; a development which is desirable for both economic and environmental reasons.

8.3.3   Coal

The picture for coal resources is even brighter. The exact extent of coal resources is controversial, but there is without doubt the equivalent of several centuries at current consumption levels. Coal use has declined in many developed countries such as the USA, Germany and the UK, but it is growing rapidly in some parts of the world. It is particularly important in India and China, both of which have sizeable coal reserves and an urgent need to increase energy use. The problems related to expanded coal use concern its suitability as a substitute for oil, transport costs, the dominance of a few large sources of coal reserves (with the Soviet Union, the USA and China in particular containing a sizeable proportion of the world's coal reserves) and the environmental implications of a major increase in global coal consumption.

Coal is dirty to handle, has a lower energy content proportional to weight and is far more polluting than either oil or gas. The air pollution of British cities in the 19th and early 20th centuries was largely a result of coal burning. The devastation of northern and eastern Europe's forests by acid rain is also clearly associated with coal-fired power stations (although the motor car is also a culprit). On the production side, the opencast mines of countries such as South Africa, Australia and Colombia devastate large areas of land. Above all, coal emits much higher proportions of greenhouse gases than oil or natural gas for each unit of energy produced. Many of these pollutants can be removed by technologies such as fluidized bed combustion, but as yet there is no technical solution to carbon dioxide emissions. These environmental effects are not reflected in the price of coal (or indeed other fuels). They are of major concern, but there is little doubt that coal use will continue into the future; it is far too cheap and plentiful, and is widely available to countries which have few alternative energy resources.

The abundance of coal, gas and oil means that the world will not be faced with an energy resource gap in the near future. This basic point sets the context for the discussion of any energy issue. There are powerful arguments for alternatives to the fossil fuels such as wind, solar, geothermal and even nuclear power. These arguments should be based on environmental and economic grounds (to show that the alternatives are cheaper and cleaner), and not some erroneous claim that the oil wells are about to run dry. Fossil fuels may not be renewable, but they are also not scarce.

8.3.4   Fuelwood

The most important energy source other than the fossil fuels remains biomass energy (wood, charcoal and crop residues). These fuels provide 14% of the world's energy consumption, and are the main energy source of the world's poor, providing an estimated 35% of all energy used in the developing world. Biomass energy is consequently the fuel of the poor, but despite (or perhaps because of) this, these fuels have received little attention from energy planners. Biomass fuels present a number of unfamiliar challenges to those concerned wth energy provision. The ways they are produced and used are very different from the more familiar international industries which characterize fossil fuels. These challenges need to be confronted if some of the most pressing energy problems in the modern world are to be addressed.

Despite the growth of energy use in other sectors, household consumption still dominates in many developing countries, and is particularly a feature of the poorer nations of Africa and South Asia. It constitutes over 75% of energy use in countries such as Nepal, Bangladesh, Ethiopia, Burkina Faso and even oil-rich Nigeria (Soussan, 1988). Most of this energy use is for cooking. In rural areas these fuels are gathered freely from the local environment (usually by women), and their production and use cannot be readily separated from other aspects of land resource management within rural economies.

Rural people rarely cut down trees for fuel use, and mostly depend on trees within close access to their homes. This means that trees outside the forest, within the agricultural landscape, are the main source of fuel for rural people. For the poor, in particular, there is no alternative to these fuels as a basic subsistence need, and problems associated with access to fuelwood can be considered as an integral part of the wider development crisis the rural poor face.

The land resources which provide fuel also provide for most other needs: rural people live in a biomass based economy and are largely dependent on the availability of land, and the plants which grow on the land, for survival. Meeting local energy (and other) needs presents few problems where land resources are plentiful. In circumstances where these resources are under pressure, competition between competing needs and different people for these resources emerges. The losers are invariably the least powerful (especially women and landless families), who do not control the land resources. Fuelwood use will be important for the foreseeable future, whatever happens to energy resources and prices at an international level. Problems are widespread, but solutions to these problems are complex, and no single, simple panacea exists. What is clear is that any interventions must work through the local community, providing solutions which meet their needs and reflect local resources and capabilites.

8.3.5 Nuclear power

Nuclear power is of minor importance in global energy terms. It provides no more than 5% of primary energy, and this is concentrated in a few developed countries. The USA, France, Japan and the Soviet Union accounted for over 65% of the total nuclear power generated in 1989 (British Petroleum, 1990), and less than 5% of all nuclear power was generated in developing countries. This is despite the huge sums which have been spent over the last 40 years on research and development into nuclear energy. In the period after the Second World War, nuclear power was presented as the energy source of the future, with claims that it would soon produce electricity that was 'too cheap to meter'. Many countries vigorously pursued nuclear energy programmes, supported by a powerful coalition of industrial, military and political interests. Despite this, a series of economic and environmental factors have undermined the nuclear dream to a position where it is more or less dead for the foreseeable future.

Orders for nuclear power stations have decreased dramatically since the 1970s, with countries such as Sweden deciding to phase out existing stations and others such as the USA, Brazil and the Philippines cancelling orders and abandoning nuclear plants which are under construction. The UK Government, a firm advocate of nuclear power, abandoned the fast breeder programme in 1988 and shelved plans for new pressurized water reactor stations in November 1989. Only France and, to a lesser extent, Japan among the leading industrial nations still cling to their nuclear programme (although a few other developing countries such as South Korea and Taiwan are still expanding nuclear power), and their main reason appears to be the strategic goal of not being dependent on imported fuels.

The main reason for abandoning nuclear power is economic. From the early days nuclear power has been shrouded in secrecy, but where the true costs of this energy source have been revealed (most notably in the USA and, more recently, in the UK as electricity production has been privatized and these costs have been exposed to commercial judgements) it has consistently been shown to be more expensive than alternative ways of generating power. In Britain, for example, nuclear electricity costs three times more to generate than coal-fired electricity, and the withdrawal of nuclear stations from the privatization of electricity in 1989 showed clearly that these stations cannot be sold, even at a huge discount.

There are also a range of environmental problems associated with nuclear power. The Chernobyl disaster in 1986 showed how catastrophic just one nuclear accident could be, with much of northern Europe covered with a radioactive cloud and huge areas of the Ukraine rendered uninhabitable. This was the worst nuclear accident to date, but the safety record of the nuclear industry is widely questioned, and other incidents such as that at Three Mile Island in the USA showed that similar disasters threaten all nuclear installations. The disposal of nuclear waste is also creating major environmental problems, necessitating the storage of thousands of tonnes of highly radioactive waste for hundreds of years. To date, no safe and publicly acceptable waste disposal method has been developed. The decommissioning of nuclear power stations is a further problem, involving substantial costs and creating further massive amounts of radioactive waste. These economic and environmental problems mean that nuclear power will continue to be a marginal fuel in the future, and is likely to decline from even today's low levels as a proportion of world energy supplies.

8.3.6 New and renewable sources of energy

New and renewable sources of energy come in many forms. Wind, solar, hydro, tides and waves, biomass technologies, geothermal and others present a theoretical potential which is almost infinite (Fig. 8.2). Many of these resources are technically difficult and expensive to harness, are limited in their range of applications (most are best suited for generating electricity) and are highly dispersed. These characteristics all limit the use of renewables in the modern world, but their potential for the future is tremendous, and it is likely that these energy forms will develop to replace fossil fuels during the 21st century and beyond.

The most widely used form of renewable energy is hydro power, which is important for electricity generation. The total amount of hydro power generated in 1989 was the equivalent of 526 million tonnes of oil, or over one sixth of total oil production. The present installed capacity is about 500 000 megawatts, and it provides about 23% of the world's electricity and around 6% of primary energy consumption (Foley, 1987). The theoretical potential of hydro power resources is an estimated 2 130 000 megawatts, so the current capacity is 23% of this theoretical total. The extent to which this will be developed is limited, however, as much of the remaining resource

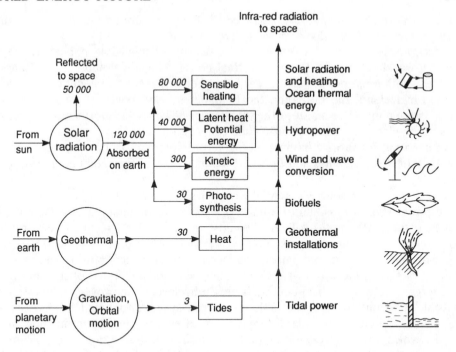

Figure 8.2   Global renewable energy flows. Units are terawatts (10^{12} W); global commercial energy consumption = 10.5 TW. Adapted from Twidell and Wear (1986).

is concentrated in developing countries, where demand is much lower and large dam schemes have run into a series of economic and environmental problems. Hydro power is fairly evenly distributed, with North America, Latin America, Europe and Asia all contributing between 15 and 25% to the total global production. It is the main source of electricity in some developed countries such as Canada and Norway, and is also extremely important in many developing countries such as Brazil, Ghana and Zambia.

Large-scale hydro power schemes are a cheap way to produce electricity, but this has often been achieved at a high environmental and social price. Adverse impacts are a major problem in developing countries, where the search for cheap power and irrigation water has led to big dam developments which are of questionable wisdom.

Massive schemes such as the Kariba dam between Zambia and Zimbabwe, Cabora Bassa in Mozambique, Volta in Ghana, Itaipu in Brazil (the world's largest dam), the Mahaweli scheme in Sri Lanka and Lam Pao in Thailand all produce cheap power, but insufficient investments have been made in distribution schemes and huge loans have added to debt in the host countries. They have also led to the flooding of forests or farmlands, the displacement of poor people, increased soil erosion, disrupted river regimes, an increased incidence of diseases such as bilharzia and other problems. Such problems can be minimized if enough care is taken in planning and constructing the schemes, but this increases costs and consequently makes the dams less attractive. Concern over these issues has led to a decline in investments in hydro power in developing countries. This trend is likely to continue, and the growth of hydro power will be far slower in the future.

Other forms of new and renewable energy are of negligible importance in the modern world. There are a couple of tidal power schemes, notably at La Rance in France; geothermal power (using heat from the earth's crust to generate electricity) is used in Iceland, the USA and the Philippines; wind power is used to generate electricity, pump water and for other uses; solar power is used for heating water and generating electricity using photovoltaic cells; and a series of more exotic schemes have been put forward to harness the perceived potential of these energy forms. In total, all of these schemes add up to a fraction of 1% of the total energy use of the world.

The theoretical resource potential of renewable energy sources such as solar or wind power (Fig. 8.2) is almost infinite (for example, the amount of solar radiation reaching the earth is 170 trillion kilowatts, the equivalent of 40 000 electric fires constantly burning for each person on earth). Most of the renewables also have a far less damaging environmental impact than fossil fuels or nuclear power. This increases their attraction as the world wakes up to the environmental effects of current sources of energy. These theoretical potentials are an irrelevance, however, in assessing the possible contribution that renewable resources can make to global energy use. The key point is to define the role different forms of energy can play in meeting energy needs, which is often very different to theoretical resource potentials. From this perspective the role of renewables looks far more humble, but can still be of great importance to the future of world energy development.

Many forms of renewable energy provide a different way of generating electricity, but are not suitable for large-scale, centralized power generation. The resources themselves, such as waves, wind and sunshine, are too dispersed and the costs are far higher than those of conventional power generation schemes. They are also often unreliable (there may be clouds in the sky or the wind may not blow), meaning that there is a need for back-up generation capacity from conventional sources if the lights are not to go out. This adds to the capital costs of the power system, and leads to the inefficient use of investment capital. Given these limitations, renewables can play a role as part of an integrated power system, but in practical terms are unlikely to be more than a minor component of the system where large-scale, concentrated electricity demand exists. The best bet in these areas is undoubtedly solar power, which can be installed at the point of use (in the home, office or factory) if integrated into the construction of the buildings. The cost of photovoltaic cells is falling rapidly, and the use of solar power is likely to grow.

What renewable sources of energy are ideally suited for is small-scale, dispersed power needs, and they have the added virtue of being more amenable to local control. This makes them ideal for isolated places such as islands, mountain regions and semi-arid areas, where providing conventional energy is a problem. They also have much to offer in many more densely settled parts of the developing world, where demand for electricity is not sufficient to make large-scale investments in conventional power generation viable. In these areas, harnessing renewable sources of energy can provide energy for local industries, water pumping, schools, health clinics, telecommunications and a host of other uses which can add greatly to the quality of life and which can act as a catalyst to a wider process of development. Seen in these terms, renewable sources of energy are unlikely to replace fossil fuels or, in developing countries, wood as the dominant sources of energy, but can provide for energy needs in circumstances where traditional sources of power are not available.

8.3.7   Conservation: the key to the future?

The discussion presented in Sections 8.3.1 to 8.3.6 makes it clear that there is no shortage of energy resources in physical terms, but that this should not lead to complacency, for two reasons. Firstly, there are many places in the world today where the price of fuels (or capital investments to provide these fuels) means that people are unable to meet their needs for energy. Secondly, the ways in which energy is used are leading to major and avoidable environmental impacts. These impacts occur at all scales. At a local level, air pollution in cities, leukaemia clusters around nuclear facilities, land degradation in the developing world and other problems are increasingly prevalent. At a regional scale there is acid rain, deforestation, the threat of nuclear accidents and so on. Globally, the spectre of global warming hangs over the future of the planet. These problems make a powerful argument for energy conservation, which means using less energy, or rather using energy more efficiently.

Energy efficiency has improved dramatically since the era of cheap energy ended in 1973. This is particularly so in the developed world, with most industrial countries using less energy to produce more goods than 10 or even 20 years ago. This is most dramatically true for oil, with consumption in the USA, Western Europe and Japan falling from 1848 million tonnes in 1979 to 1617 million tonnes in 1989. It holds true for total energy consumption in these countries as well, meaning that the amount of energy needed to produce a unit of economic output has fallen by, on average, about 2% per year since 1973 in the developed world. A leader in this field is Japan, which is the most energy-efficient economy in the world, but still sees steady improvements in its energy to output ratio. Between 1973 and 1987 Japan's gross national product (GNP) rose by an average of 4% a year, but primary energy consumption rose by only 1%, meaning that by the end of this period Japan was using 34% less energy to produce the same amount of goods and services. Other countries also improved: for example, over the period 1973–87 the UK used 25% less energy per unit of GNP and the USA 26% less. By 1987, Japan was using 0.26 tonnes of coal equivalent (tce), the USA 0.44 tce, the UK 0.43 tce and Germany 0.42 tce to produce $1000 of GNP.

There are three reasons why the energy efficiencies of developed economies have improved. The first is the effect of higher fuel prices since 1973, which have led to a greater energy consciousness and reduced energy use. The second is the changes which have taken place in the economies of the developed world, with traditional, energy-intensive industries such as iron and steel and engineering declining and new industries such as advanced electronics and service activities, which use far less energy, growing in importance. The third is investments to improve energy efficiency, in new technologies and better insulation. In the UK it is estimated that only 30% of the improvement in energy efficiency is attributable to such investments, and a great deal could still be done to encourage greater energy efficiency. For example, the use of more efficient light bulbs can save up to 75% of the electricity used for lighting, more efficient 'white goods' such as refrigerators and washing machines now available could save between 25 and 40% of electricity used in the household, and cars which are three times as efficient as those now on the road have been designed. Similarly, better designed buildings can reduce heat loss substantially, saving on household heating, which is one of the largest domestic uses of energy.

In part these savings will occur as old appliances are replaced with newer, more efficient ones, but there is a basic problem with this. The people who could benefit the most, those on low incomes, in old houses and so on, are usually the people least likely to buy new appliances. This is recognized, and some countries have active programmes to improve energy efficiency. The building standards enforced in Sweden, for example, mean that energy use for heating is far lower than in the UK, despite a colder climate. Similar standards enforced in Milton Keynes have reduced heating bills by 40% with only a 1% increase in building costs. Similarly, the USA has both 'energy labelling' policies, whereby the energy efficiency of appliances has to be clearly displayed on them, and rigorous appliance standards, which are legally enforceable energy consumption standards for domestic appliances. Japan has a system of compulsory energy auditing for all businesses, whereby they have to produce energy efficiency plans which are then approved by a government agency. The USA and Norway have similar schemes applied to electricity generation. Such active intervention by governments are resulting in a far more efficient use of energy than would be the case if energy savings were simply left to the market.

Improvements in energy efficiency have been concentrated in the developed world, but the effect of these savings is felt globally as fears about an ever growing energy demand are not realized. The improvements in the developed world are producing better technologies which will spread to the rest of the world, and reduced pressure on world energy resources will provide a breathing space in which the developing countries can gain access to the increased energy so vital to their development. The opening up of the economies of Eastern Europe is likely to result in equally dramatic improvements in their energy efficiency, and will provide a new incentive for further technical improvements.

Much has been achieved in improved efficiency, but there still remains a great deal to be done. It provides a key to the future whereby the benefits of greater energy use can be spread and resources can be more effectively used without the grave environmental effects which would be found if old, inefficient patterns of energy use were preserved. The levels of investment needed to realize the potential of energy conservation is particularly a problem for the developing world. It is doubtful whether these poor countries will be able to afford to pursue more energy efficient paths to develop unless they receive considerable help in gaining access to both the technologies and the finance to pay for them from the developed world, which controls both. They also need some respite from the development pressures (and above all the debt trap), which produce short-term approaches to economic change, if they are to build a more sustainable, energy efficient future. If this form of assistance is forthcoming, then there is hope; without it the developing world will follow the same disastrous path that the developed world has taken.

8.4 CONCLUSIONS

This chapter presents an overview of the different resources which provide energy now and into the future. In contrast with many people's assumptions, the modern world has abundant and varied energy resources. There is no 'limit to growth' from energy resource scarcity. The last two decades have also shown that humankind has, under the right conditions, an extraordinary capacity for technical change and innovation. Indeed, there are many technical potentials (such as solar photovoltaic cells) which are 'latent'; they are not used widely at present, but are available for rapid dissemination when conditions are right.

This optimistic assessment does not mean that there is no need to worry about the world's energy future, for mighty challenges face the global community. The era of cheap energy, which fuelled the post-war reconstruction, is over. These challenges are many and varied, but in the end come down to two basic issues. The first is whether individuals and nations can afford the economic cost of meeting their energy needs. The second centres on the environmental impacts of different forms of energy. These are the real limits to growth: access to finance (and technologies) and the capacity of local, regional and global environmental sinks to absorb the by-products of energy use. At the end of the day, like many of the issues discussed in this book, energy issues boil down to wider questions of development and the environment, and can only be understood in this context.

Above all, what is needed is a clear understanding of the actions needed to forge a sustainable energy future. Individual parts of the picture are present, such as the desirability of conservation or the problems associated with nuclear power, but these do not add up to a coherent package of desirable and viable steps. Given this, it is hardly surprising that the political will to challenge the powerful interests who benefit from the present energy economy and to forge a new pathway is absent.

What is clear is that the energy future will be very different to the energy past. The question is not whether the ways in which energy is produced and used will change, but what direction this change will take. The world's energy economy has been in a period of change and uncertainty since the first oil shock of 1973. The current era is one of flux, in which a simple and direct relationship between energy and economic development has given way to one which is complex and multi-dimensional. At the same time, environmental concerns have changed from concern with the localized pollution and hazards of energy exploitation to its regional and global impacts.

This volatility will last for some time, but over a period of time (perhaps as much as a hundred years) a transition from fossil fuels, which are cheap and convenient but ultimately scarce, to less convenient, less flexible but more abundant energy sources will occur. This will be accompanied by the development of an electrical society, a road which the developed world in particular is already travelling down. Speculation is dangerous, but solar power is probably the energy source which will fuel the future. The resources are vast and the technology (especially photovoltaic cells) is developing rapidly. This will produce a very different type of social organization of energy, with a move from a centralized energy economy, where the fuels are the main cost of power, to a dispersed energy economy where ownership of the technology gives free fuel from the sun. The implications of this are profound. These changes will occur gradually, and the abundance of fossil fuels means that they can take place without too much disruption if the global community can meet the economic and environmental challenges that fossil fuels throw up for the short and medium terms.

8.5 REFERENCES

British Petroleum (1990) *BP Statistical Review of World Energy 1990*. London: British Petroleum.
Department of Energy, Somalia (1990) *Somalia Energy Newsletter*. Somalia Mogadishu: DOE.
Foley, G. (1987) *The Energy Question*, 3rd edn. Harmondsworth: Penguin.
Institute of Electrical Engineers (1985) *Energy Supply and Demand in the UK: the Role of Electricity*. London: Institute of Electrical Engineers.
Odell, P. R. (1986) *Oil and World Power*, 8th edn. Harmondsworth: Penguin.
Soussan, J. (1988) *Primary Resources and Energy in the Third World*. London: Routledge.
Twidell, J. and Weir, A. (1986) *Renewable Energy Resources*. London: E. & F. N. Spon.
World Resources Institute and International Institute for Environment and Development (1987) *World Resources 1987*, Section 7, pp. 93–109. New York: Basic Books.
World Resources Institute and International Institute for Environment and Development (1988) *World Resources 1988–89*, Section 7, pp. 109–126. Published in collaboration with the UN Environment Program. New York: Basic Books.

8.6 FURTHER READING

Foley, G. (1987) *The Energy Question*, 3rd edn. Harmondsworth: Penguin.
Odell, P. R. (1986) *Oil and World Power*, 8th edn. Harmondsworth: Penguin.
Soussan, J. (1988) *Primary Resources and Energy in the Third World*. London: Routledge.
World Resources Institute and International Institute for Environment and Development (1987) *World Resources 1987*. New York: Basic Books.

CHAPTER 9

Biotechnology and Genetic Engineering: New Environmental Issues

A. M. Mannion

9.1 INTRODUCTION

Although the early 1990s will undoubtedly be remembered for media preoccupation with the prospect of global warming, they must also be acknowledged as a time when science is preparing for a new wave of environmental experiments. In the ensuing decades, the impact of biotechnology, and its sub-discipline of genetic engineering, will emerge as major environmental issues as their potential is translated into applications. Biotechnology already has many applications, but these have not prompted as much concern as the potential impact of genetically-engineered organisms. Surprisingly, the actual and potential impacts of these developments have received little attention in environmental texts. This is despite the fact that they may contribute to solving some of the world's environmental problems as well as providing a new range of potentially detrimental agents of environmental change.

Biotechnology is a general term that relates to the harnessing of living or dead cells, or cell components, to undertake specific processes with applications in medicine, industry, agriculture, conservation and the provision of food and fuel energy. Genetic engineering takes biotechnology a stage further by manipulating the genetic material that is contained within living cells. The objective is to isolate components of chromosomal DNA (deoxyribonucleic acid) which confer specific, preferred characteristics and transfer them into other species. These should then produce offspring that express the characteristics. Biotechnology and genetic engineering are thus more precise approaches to the plant and animal breeding programmes that owe much to the mid-nineteenth century work of Charles Darwin and Gregor Mendel who established the principles of heredity. Since then, the interbreeding of species has resulted in improved plant crops and animals with enhanced food or fibre value, or species that will flourish in what would otherwise be marginal

Environmental Issues in the 1990s. Edited by A. M. Mannion and S. R. Bowlby
© 1992 John Wiley & Sons Ltd

environments. Biotechnology is extending this process of plant and animal manipulation for human benefit by exploiting further ranges of organisms—the bacteria,
viruses and fungi. This, in turn, has led to a greater understanding of genetic
operation that is being developed to produce engineered organisms with advantageous traits. To date, most genetic engineering has been undertaken on bacteria,
viruses and plants, but inroads are being made into understanding the human genome
and the genetic basis of hereditary diseases. This has raised ethical and legal debates
about research on human embryos, which are outside the scope of this chapter.

From an environmental point of view, biotechnology and genetic engineering have
already had significant effects and both have enormous potential for further impacts.
Waste disposal technology, particularly that of industrial and sewage effluents, often
involves treatment with bacteria; biomining and resource recycling utilize the ability
of specific organisms to complex metals. In addition, biological materials can be
degraded by organisms to produce fuels such as ethanol, and the mass cultivation of
fungi and algae can produce protein-rich substitutes for animal and human food.
There are also a wide range of applications in agriculture, such as the engineering of
disease- and drought-resistant crops. The potential is exciting and is especially
apposite to the production of improved food supplies in the developing world. Such
innovations are, however, unlikely to occur without at least some adverse environmental effects. Concerns about the release of genetically engineered organisms are
not misplaced. This is uncharted territory with no precedents for guidance and society
can ill afford to risk further large-scale environmental damage.

9.2 BIOTECHNOLOGY

There are many applications of biotechnology, both actual and potential, which can
influence environmental quality. There is also a wide range of medicinal and industrial
applications, such as the production of enzymes and antibodies. These, like many
fermentation techniques that are used in the food industry, will not be discussed here
as they have no direct bearing on the environment. The agricultural applications
include crop improvement, the biological control of pests and the stimulation of
nitrogen fixation. Other ways in which biotechnology can influence environmental
quality include the treatment of sewage and industrial effluents. In a broader context,
biotechnology can contribute to resource recycling and biomining, and can be used
to produce food and fuel energy to supplement animal (and possibly human)
feedstocks and fossil fuels. All biotechnological enterprises require a high scientific
input and for each product specific industrial plant is necessary.

9.2.1 Applications in agriculture

Currently, agricultural and horticultural suppliers exploit the ability of many plants
to reproduce vegetatively (asexual reproduction) to produce fruit and vegetable
stocks and some garden plants from cuttings. This is particularly appropriate for
plants that already have advantageous characteristics derived from breeding programmes. Tissue culture techniques can also be used to produce useful plants and
plant products. Such techniques, which involve *in vitro* (in a growth medium)
production of plant cells and tissues, can also facilitate genetic engineering (Section

9.3). The ability of plants to regenerate from a single cell or tissue is due to a characteristic called totipotency. In theory, each individual cell has the potential to give rise to the entire range of cell types that comprise a given organism. This potential remains throughout the life span of individual plants, but in animals it is lost in the embryonic stages.

In the laboratory, tissue culture is used to produce individual plants via micropropagation, the techniques of which are discussed by Lindsey and Jones (1989). Auxiliary buds (which grow in the angle between leaf and stem), meristem (formative as opposed to permanent tissues) and callus cells, which consist of an undifferentiated mass of cells that are produced to repair damaged plant tissue, are the materials used. The resulting plants are clones of the parents and a wide variety of ornamental, woody, crop and vegetable species can be produced in this way, but with the notable exception of the major cereals. Apart from facilitating the mass reproduction of useful species, the technique also allows the production of disease-free species. Tudge (1988) discusses the example of cassava (*Manihot* spp.), a major crop in many African farming systems, which is often affected by a mosaic virus that reduces its food value. By selecting and then culturing uncontaminated cells, new disease-free seedlings can be produced rapidly. Tissues culture also offers potential for the improvement of slow-growing species such as oil palms, coconuts, rubber and banana trees. The most obvious application of such techniques in terms of environmental problems is their importance for increasing crop productivity in agricultural systems of the developing world. If initiated on a large scale and applied to a wide variety of crops, the resulting increase in food production and cash crops could help to halt the spread of cultivation into marginal areas.

Somaclonal variation can also be used for the development of improved crop plants. It occurs when clones derived from the callus of the parent plants exhibit genetic variation. Many somatic variants, or 'sports', are often not as useful as the parent because of the loss of desirable attributes but, conversely, such genetic variation can be used to produce new, improved varieties. Tudge (1988), for example, discusses their value for producing new varieties of the potato (Maris Piper), which is widely grown in the UK, that are less susceptible to virus attack and the fungal disease known as scab. Another crop which could benefit in this way is sugar cane. This is often badly affected by a fungal disease called red rot and somaclonal variants with resistance to this have already been identified. The potential of somaclonal variation for the improvement of crop plants (and ornamentals) is enormous and has already been recognized in wheat, maize, oats and rice as well as carrot, tomato, celery and tobacco.

New crop plants can also be produced by somatic hybridization, which requires the fusion of protoplasts (the cell without the cell wall) to give a novel cell with either two complete sets of chromosomes, two parallel sets of chromosomes or a mixture of chromosomes and foreign organelles. Organelles are plant cell components such as chloroplasts (responsible for photosynthesis) and mitochondria (responsible for respiration). This is not, strictly speaking, a form of genetic engineering because it does not alter chromosomal DNA. Nevertheless, the resulting configuration of chromosomes and/or organelles confers upon the plant carrier a new range of characteristics, some of which are advantageous for improving crop productivity. According to Lindsey and Jones (1989), there is a range of plant species of economic importance which can be produced in this way, e.g. tomato, tobacco and potato. The resulting hybrid may be produced by combining the protoplasts or protoplast components of the same or different species. It is possible, for example, to produce a hybrid by protoplast fusion between the wild potato (*Solanum brevidens*) and the commercially grown potato (*S. tuberosum*). The wild potato (and a high proportion of the resulting hybrids) is resistant to potato leaf roll virus and potato virus Y, while *S. tuberosum* is not. The technique thus has the potential for producing more efficacious crops than those already in use. Such hybrids are even more environmentally benign if they obviate the need to use crop protection chemicals.

Biotechnology also has applications in the biological control of pests. There is a range of bacteria, fungi and viruses which can produce fatal infections in many insect species and can supplement pesticide applications, or reduce them in integrated pest management strategies. For example, many strains of *Bacillus thuringiensis* produce insecticidal chemicals and are thus insect pathogens. Some strains are commercially available and are especially useful to combat insect pests that have developed resistance to other pesticides. Some of these strains are also used to control pests, such as mosquitoes, that present a health hazard to humans. Many bacteria also produce anti-fungal chemicals. One example is the common soil bacterium *Pseudomonas fluorescens* which, when experimentally inoculated into wheat seeds and potato tubers, led to increased yields of 27 and 70% respectively (Lindsey and Jones, 1989). Biotechnology is also used in the production of mycoherbicides (Ayres and Paul, 1990). This involves the large-scale culture of fungal spores (mycology is the study of fungi, hence the term mycoherbicides) from species that are natural pathogens of weeds. In the field crop, such spores will act as weedkillers. Among the requirements for the development of these agents are the identification of fungi that are host-specific but harmless to the crop, are rapidly effective and can be produced on an industrial scale. Two such mycoherbicides have been available in the USA for a decade. Collego® (a trade name) is effective against northern joint vetch which is a weed in rice and soya bean crops in the southern USA, whereas Devine® is targeted at milkweed vine, which affects citrus groves in Florida. These pests are not widespread, but the success of the mycoherbicides has prompted the development of further products to combat more common weeds. The mycoherbicide Casst®, for example, is targeted at sicklepod and coffee senna, which infest soya bean and peanut crops.

Harnessing the ability of nitrogen-fixing bacteria, notably strains of *Rhizobium* species which occur naturally in a symbiotic relationship with leguminous plants such as alfalfa and lucerne, is another aspect of biotechnology with much potential. The availability of nitrate in an agro-ecosystem is often limited and thus curtails crop productivity unless it is supplemented by nitrate fertilizers. As discussed in Chapter 11, nitrate fertilizers are a major contributory factor in cultural eutrophication so alternative means of prompting nitrate availability could be environmentally beneficial. The inoculation of *Rhizobium* spp. into soils and the seeds of leguminous crops has been shown to enhance crop productivity, especially in soils where there are no naturally occurring species and where nitrates are in short supply. There are, however, certain preconditions that are necessary before nitrogen fixation will occur. For example, nitrogen fixation by *Rhizobium* spp. will be inhibited if there is already an abundance of nitrate, and it is essential to select a strain of the bacterium that will readily form a symbiotic relationship with the desired crop, which must be a legume. The efficacy of leguminous species, which are components of many multi-crop systems in Africa, could thus be improved if rhizobial strains were also introduced.

9.2.2   Applications in biomining and resource recycling

Harnessing the ability of certain organisms to break down mineral-containing rock and mine wastes also has implications for environmental quality. This ability can be exploited to effect the recovery of useful materials, usually metals, from industrial waste water. Both applications are conservational in terms of resource use, and if initiated on a large scale they could reduce the need to exploit new mineral ores. Bacteria may also be used to desulphurize fossil fuels and thus provide an alternative to physical and chemical methods that are currently used in many North American and European power stations to combat the problem of acidification (Chapter 11).

In the context of biomining, the bacteria that effect metal extraction are known as chemolithotrophs. This is because they are capable of producing energy for growth by oxidizing inorganic compounds. Biomining consists of bacterial leaching, which renders metals soluble by bacterial activity on mineral ores or mined material. Sulphide ores, a major class of metal ores, are particularly amenable to bacterial leaching, which is used on a commercial basis to extract metals such as copper and uranium. The bacteria oxidize the sulphides to sulphates which are more soluble and from which the metals can be more easily extracted.

The bacteria which bring about this process are the thiobacilli, of which *Thiobacillus ferrooxidans* is the most important. It is active at low pH, which is advantageous because the process involves the formation of sulphuric acid, and it works at high temperatures (e.g. 80°C). Gray (1989) gives a range of examples which involve *Thiobacillus* spp. in metal recovery from low grade ores and mine tailings. These materials can be piled on to an impermeable base and sprayed with a leaching solution to dissolve the metals released by microbial oxidation. The metals can then be recovered by a variety of methods, including precipitation. Copper and uranium are the chief metals extracted in this way, but lead, arsenic, iron, silver, gold, chromium and cadmium can also be obtained. Gray (1989) states that about 10–15% of the annual copper production in the USA utilizes this method, notably from the Bingham Canyon mine in Utah. The leachate and the bacteria can be recycled and the overall operation of the plant requires only a small workforce.

Bioleaching has several advantages over the conventional recovery of minerals by mining. For example, it allows metal extraction from deep-seated ores without large-scale excavation, which is environmentally detrimental. It also facilitates metal extraction from low grade ores and mine wastes in addition to the removal of impurities such as arsenides. The resulting leachate must, however, be controlled to prevent it from polluting drainage systems and it is used to best effect if it can be recycled. Microbial cultures to enhance oil recovery from existing reservoirs and diffuse sources may also become important in the long term as oil reserves decline.

The ability of micro-organisms to scavenge metals also has a variety of applications in the treatment of waste water, although most are potential rather than actual applications. The bacterium *Sphaerotilus natans*, for example, will accumulate iron, magnesium, copper, cobalt and cadmium and could thus be used to recover these metals from waste water (Gray, 1989). Such techniques could also be used to treat radioactively-contaminated water produced by the processing of nuclear fuel as the yeast *Saccharomyces cerevisiae* and the bacterium *Pseudomonas aeruginosa* have been shown experimentally to concentrate uranium. Although not strictly a question of resource recycling, there is also evidence that microbial cultures could be used to break down chemicals, such as some pesticides and aromatic compounds, which are environmentally persistent. In fact, microbial cultures were used as one component of the clean-up operation after the oil spill from the *Mega Borg* that occurred off the coast of Texas, USA, in 1988 (Chapter 6).

9.2.3   Other applications

Of the many applications of biotechnology, the most environmentally relevant processes include waste water and sewage treatment and the production of food and fuel energy. There are also limited applications in wildlife conservation programmes.

Sewage and waste water treatment by organisms is one of the longer standing applications of biotechnology and has been in use since the early 1900s. As Table 9.1 shows, there are five stages in waste water treatment, of which the secondary

Table 9.1   Stages involved in waste water and sewage treatment

Stage 1	*Preliminary treatment* involves the removal of grit and coarse solids (or their maceration) in addition to oil and grease
Stage 2	*Primary treatment* involves the removal of solids, which collect as sludge after gravity separation and sedimentation
Stage 3	*Secondary treatment* involves the oxidation of suspended and colloidal organic material present in settled sewage that is the overflow from stage 2 after sedimentation. The process involves aerobic or anaerobic digestion by microbial populations so that the organic compounds are converted into microbial biomass. This is then precipitated as sludge.
Stage 4	*Tertiary treatment* is additional treatment of effluent from stage 3 to remove suspended solids, pathogens, nitrates, phosphates, bacteria, other toxic compounds. Effluent can then be released to inland or coastal waters
Stage 5	*Sludge treatment* involves water removal, stabilization and disposal

treatment process involves the mirobial fixing of organic materials that remain in the effluent. Many species of bacteria can oxidize the organic compounds to produce carbon dioxide and mineralized end-products that are discharged with the effluent. Alternatively, organic material can be biosynthesized by the bacteria, which produce new cells that are then precipitated in the sludge and can be recycled. Thus microbial populations help to purify the waste water and safeguard environmental quality. Unfortunately, the process is rarely adequate to remove all the potential pollutants.

If anaerobic conditions dominate the stage 3 process of secondary waste water treatment, one of the products of digestion is methane (as well as carbon dioxide). This is also produced in the anaerobic treatment of food processing wastes, animal manures, agricultural wastes such as straw, and organic-rich domestic wastes. The resulting methane, or biogas as it is often called, is an important source of energy that can be used to power the digesters in which the reactions take place. Thus it constitutes an alternative energy source to conventional fossil fuels. It is not extensively used in the developed world, but in China small-scale digesters are widely used to provide fuel for local domestic use. The microbial process is similar to that which occurs in landfill sites that are used for domestic rubbish disposal, the methane from which can be hazardous if it forms an explosive mixture with air. It is, however, possible to extract methane from landfill sites to produce energy on a commercial basis.

Biotechnology can also be used to produce other alternative sources of energy. The most well known is ethanol, which is widely used in Brazil, where it constitutes 30% of all fuel. The basis of the industry is sugar cane and cane molasses. The ethanol is released from the cane or molasses juice by a fermentation process involving yeast (Wayman and Parekh, 1990). In similar operations in North America, cereals such as maize, wheat and barley are also fermented to produce a variety of products, including potable and industrial alcohol as well as fuel alcohol. Wayman and Parekh (1990) point out that sugar beet, which is widely grown in north-west Europe, can also be used to produce fuel alcohol. The conversion of biomass to fuel alcohol could thus be part of the answer to reducing the world's consumption of fossil fuel. Fuel alcohol has a number of advantages. Firstly, it is not sulphur-rich and could, therefore, reduce the problem of acidification (Chapter 11), nor is lead required as an additive so another health hazard is eliminated (Chapter 2). The curtailment of methane production from waste materials is also to be welcomed as methane is a more potent greenhouse gas than the carbon dioxide produced by its combustion. Secondly, there are economic and political factors to consider. Biomass is ubiquitous and as long as it can be produced in amounts surplus to food requirements, it could provide nations that have few or no reserves of fossil fuels with an independent fuel source. This has implications for economic development, which is related to fuel consumption, and on a political front it reduces dependence on oil-producing nations and the ramifications of often volatile oil prices (Chapter 8). On a more parochial note, the production of biomass for fuel energy could be an alternative use for land that is currently producing food mountains in north-west Europe. It could thus be incorporated into extensification programmes that are being implemented as part of the European Commission's Common Agricultural Policy.

The use of microbial populations in waste water treatment has led to another development involving the production of food energy, notably the production of single-cell proteins (SCP). These consist of the accumulated, dried cells of single-celled organisms such as algae, yeasts, fungi and specific bacteria, which can be used for animal and/or human consumption (Lichfield, 1989). In Israel, for example, aerobic sewage treatment has been combined with the production of SCP for animal feed. The sewage provides the nutrients and the equable climate with year-round high light intensity is conducive to rapid algal growth, which can be harvested and thermally dried to remove pathogens. Algal biomass from lakes in many tropical countries is also harvested, dried and marketed as a health food. Most commercial production of SCP is via fermentation under anaerobic conditions. It involves the provision of a milieu with an appropriate mix of nutrients and an energy source to encourage the optimum reproduction of non-photosynthetic organisms. The best known examples of SCP produced in this way are Pruteen® and Mycoprotein®. Pruteen® was developed by ICI 20 years ago as an alternative to soya bean meal for animal feed. It consists of the bacterium *Methylophilous methylotrophus*, which is cultured in methanol, its energy source, with the addition of nutrients. Mycoprotein®, also known as Quorn®, derives from the fungus *Fusarium graminearum*, which is fermented in a mixture of glucose, the energy source, and ammonia as the nitrogen source. The end-product consists of fungal mycelia (thread-like strands) that are fibrous, resembling the texture of meat. The product Quorn® is marketed by Rank Hovis McDougall, with a variety of flavourings, as a substitute for meat.

A further aspect of biotechnology, which has limited applications to date but is nevertheless likely to become increasingly important, is its role in the conservation of wildlife species. As Conway (1989) states, the use of biotechnology in this context is a 'crisis response to the threat of extinction'. He uses the term biotechnology in a broader sense than that adopted here (Section 9.1) and includes biological manipulation at the whole organism rather than at the cellular level. Examples of this are the propagation of species in zoos and, in some instances, their reintroduction to the wild. Propagation techniques can involve natural breeding, as well as fostering and artificial incubation, which have been successfully used to increase specific bird and reptile populations. There are also instances where artificial insemination can be effective, such as in the breeding of the giant panda. It has also proved possible to transfer embryos between species; the domestic cow, for example, can successfully carry embryos of the Indian wild ox.

All of these methods can be used to ensure the preservation of small populations and to increase their gene pool. They are ways of curtailing extinction and maintaining biotic diversity, but it is unlikely that they will do more than preserve just a few species that are threatened with extinction by human activity. The collections of animals, especially of threatened species, in zoos are also gene banks, as are plant collections in botanical gardens, seed collections and germplasm stores (where genetic material rather than whole organisms are stored). These are of vital importance, not only as a resource for genetic engineering (Section 9.3), but also for ensuring that field extinction does not eradicate species. Unfortunately, existing plant and animal gene banks contain only a minute proportion of the genetic reserves that are contained within the earth's flora and fauna.

9.3   GENETIC ENGINEERING

Genetic engineering involves the manipulation of DNA and the transfer of gene components between species to encourage the replication of desired traits. It is variously known as recombinant DNA technology, gene cloning, and *in vivo* (in cell) genetic manipulation. It consists of the insertion of foreign DNA, containing the genetic code for the desired characteristic, into a vector which is transferred to the host. The host will then interpret the inserted genetic material so that its offspring will exhibit the required trait.

Such technology has already been used to produce substances with medical applications including interferon, an anticancer agent, and human insulin. Most work to date has focused on micro-organisms and plants, but the potential of genetic engineering is enormous. In relation to the environment, there are potential advantages and disadvantages. The engineering of new crop plants, for example, provides opportunities to improve food production, but there are concerns that some engineered organisms could constitute an environmental threat or a threat to public health. Could such techniques contribute to the development of weapons for biological warfare? How might engineered microbes affect global biogeochemical cycles? Is it possible that organisms engineered for their pathogenicity towards particular crop diseases could prove equally fatal for other beneficial organisms? Questions such as this are legion and, as few altered species have, so far, been released into the environment, there are few definitive answers. However, such doubts emphasize the need for strict international controls on laboratory and field experiments prior to marketing.

9.3.1 Applications in agriculture

Some of the most important applications of genetic engineering relate to agriculture and seek to improve crop productivity. This can be directly enhanced by improving the ability of specific crops to fix carbon in photosynthesis. The production of frost- and pest-resistant species, in addition to species capable of fixing atmospheric nitrogen, would also result in increased productivity. Such developments, as well as helping to solve the world's food supply problems, could have other environmental advantages. For example, it may not be necessary to cultivate marginal areas, thus avoiding further soil erosion (Chapter 14) and desertification (Chapter 15). The use of artificial fertilizers and crop protection chemicals, such as pesticides, could be reduced, preventing the associated environmental problems such as cultural eutrophication (Chapter 11) and the contamination of food by pesticide residues (Chapter 12). All such developments depend on the identification of the gene components in cultivated or wild plants that confer these particular advantages, and devising methods for transferring the genetic code into host organisms. As the science involved is chiefly a product of the 1980s, the overall possibilities remain to be investigated. This makes it even more imperative to prevent the extinction of plant and animal species and thus preserve as large a gene pool as possible.

In the context of improving the productivity of existing crops, most research has focused on ribose biphosphate carboxylase-oxygenase, a protein known as Rubisco, which is involved in carbon fixation. As Lindsay and Jones (1989) discuss, the biochemical pathways involved in photosynthesis are so complex that genetic manipulation to produce more efficient carbon fixation is not yet feasible. Nevertheless, it could become a reality in the future and the ability to manipulate photosynthesis would indeed revolutionize food production. The possibility of genetically engineering disease- and pest-resistant species is closer to fruition. It is now possible, for example, to engineer tobacco and tomato plants with resistance to the tobacco and tomato mosaic viruses. Buck (1989) states that, in field tests, normal infected tomato plants showed a decrease in yields of between 25 and 36%. Transgenic (genetically altered) plants were not affected. Tomato plants have also been engineered so that they produce fruit which softens more slowly after harvest and is thus easier to market. This is achieved by introducing a gene to suppress the production of the enzyme responsible for softening.

Advances have also been made in the genetic engineering of crops for insect resistance, as discussed by Hilder et al. (1990), who also detail the advantages (Table 9.2) of such changes. At present, genetic engineering can lead to resistant plant species only if a single gene product (e.g. a protein), rather than a multi-gene

Table 9.2 Advantages of genetically-engineered insect resistance. Adapted from Hilder *et al.* (1990)

Continuous protection is provided, eliminating the need for accurate predictions of insect pest outbreaks

Protection is provided *in situ*, exactly where it is required, and covers all plant tissues, including those underground which are not easily treated with conventional insecticides. This would be particularly advantageous in the treatment of below-ground pests such as nematodes.

Protection is provided for known pests, thus beneficial and harmless insects would not be adversely affected

Protection is confined within the plant, so minimizing the possibility that crop protection chemicals could adversely affect the wider environment

Protection of this kind is much cheaper to develop than conventional crop protection chemicals

---

product, confers the resistance. The potential is exemplified by recent work (Hilder *et al.*, 1990) on the cowpea, a grain legume which is an important crop in West Africa and South America. It is liable to substantial losses during storage due to the bruchid beetle *Callosobruchus maculatus* F. As a result, a programme was established at the International Institute of Tropical Agriculture in Nigeria to develop a resistant species. It involved the screening of thousands of specimens of cowpea, of which only one proved to have significant natural resistance to the larvae of the beetle. This was found to be due to the presence of higher than average amounts of an enzyme inhibitor that interferes with part of the digestion processes in the insect. The resistance is, therefore, a result of the antimetabolic property of the chemical. Laboratory trials using a variety of insect pests, including the tobacco budworm, corn earworm and armyworm, showed the chemical to be an effective insecticide. Hilder *et al.* (1990) have also engineered tobacco plants containing the cowpea enzyme inhibitor which show an improved resistance to the tobacco budworm. The next stage is to engineer the major crop plants, although the cereals have not proved as easy to manipulate as other crops.

The production of insect-resistant plants by genetic engineering is also harnessing the abilities of specific bacteria to make insect toxins. For example, the bacterium *Bacillus thuringiensis* can produce crystalline spores which are natural insecticides. These are known as Bt toxins. The gene that controls Bt production can now be cloned into tobacco plants, conferring resistance to tobacco budworm and the large white butterfly. Advances have also been made in engineering crop plants that are resistant to herbicides (Buck, 1989). This would facilitate the treatment of field crops with a broad-spectrum herbicide that would kill the weeds but not the crop. There is also the possibility that crops could be engineered to combat environmental hazards such as frost, drought and high salinity. Tudge (1988), for example, reports on efforts to produce a salt-tolerant wheat by transferring the relevant genes from a salt-tolerant grass. In relation to frost damage, Lindow (1990) has developed an 'ice-minus' bacterium to protect crops. In many instances frost damage occurs because bacteria, such as *Pseudomonas syringae*, which live on the foliage produce a protein that encourages ice nucleation in the temperature range 0 to $-2°C$. Lindow developed genetic engineering techniques to remove the gene that codes for this protein and also methods to produce the new bacterium in large amounts. Field trials showed that frost damage to sprayed tomato and potato crops was considerably reduced.

Much effort has also been expended to improve the fixation of nitrogen from the atmosphere by bacteria. Most fixation is undertaken by free-living bacteria in the soil or by bacteria that occupy root nodules and form a symbiotic relationship with higher plants, specifically legumes. As Ligon (1990) discusses, nitrogen fixation could be improved in a variety of ways. Firstly, genetic engineering could be used to increase the number of genes that code for nitrogen fixation and thus increase efficiency. Experimental work on the bacterium *Rhizobium meliloti* indicates that this is now possible. Secondly, bacteria that are not able to fix nitrogen or to promote root nodulation in leguminous species could be engineered to do so. Thirdly, it should eventually be possible to engineer crop plants which do not naturally develop a symbiotic relationship with nitrogen-fixing bacteria to develop such a relationship. Engineering cereal plants to do this would be especially advantageous, but the goal remains a long way off. Nevertheless, the environmental benefits of all these possibilities could be considerable, provided, of course, that the engineered organisms are not themselves environmentally harmful. One major advantage would be the reduced need for artificial fertilizers and thus curtailment of cultural eutrophication (Chapter 11).

9.3.2 Other applications

Many of the applications of biotechnology examined in Sections 9.2.2 and 9.2.3 could be improved by genetic engineering. Making them more efficient, to reduce costs, could encourage their more widespread use to improve environmental quality. This is particularly relevant to waste water treatment, resource recycling and biomining. It is unlikely that the microbial recovery of metals from ores will replace conventional mining techniques in the foreseeable future, but if more efficient organisms could be engineered and produced on an industrial scale then existing metal ores and oil deposits could be exploited more effectively. Such scavenging organisms could also improve recycling programmes so that they become more economically attractive than primary sources as a means of obtaining useful substances. Treating waste water, for example, in this way would also curtail pollution. In addition, the engineering of organisms to break down non-biodegradable substances, such as polychlorinated biphenyls and dioxin, could help to solve disposal and pollution problems.

The reclamation of derelict land could benefit from biotechnology and genetic engineering. Most waste tips, often derived from mineral exploitation and coal combustion, are deficient in nitrogen, which impedes colonization by vegetation. Seeding with nitrogen-fixing bacterial cultures would help mitigate this problem and encourage revegetation, thus aesthetically improving despoiled landscapes and making them more attractive for development. The genetic engineering of bacteria to more efficiently fix nitrogen in a variety of relatively hostile substrates could also enhance the reclamation process. There are already large-scale operations to produce biomass fuels (Section 9.2.3). Engineered organisms could make these more efficient and possibly extend the range of substrates from which biomass fuels can be produced. This would reduce the pressure on fossil fuel reserves and provide a more environmentally benign substitute with fewer pollution implications.

9.3.3   Genetic engineering: the environmental risks

In addition to the potential benefits of genetically-engineered organisms (Sections 9.3.1 and 9.3.2), there are equally important potential drawbacks and, as stated in the introduction, the 1990s will be the temporal setting for a new wave of environmental experiments. Since the advent of genetic engineering, scientists and politicians have aired their concerns about the possible adverse effects of transgenic organisms on public health and the environment. In fact, from 1974 to 1975, scientists in the USA involved with genetic engineering imposed a voluntary moratorium on aspects of their work to allow the US government to establish 'watchdog' committees that could formulate safety guidelines and monitor progress. As a result, safety procedures were laid down which detailed the level of containment necessary for specific experiments, and laboratories were designated according to the perceived risks of their experiments. These guidelines have been widely adopted by nations engaged in such work. Essentially, there are two types of containment: biological containment which requires that the engineered organism is unable to survive outside the laboratory, and physical containment which prevents escape from the laboratory. The latter is obviously more difficult to enforce and monitor and opponents of genetic engineering believe that, sooner or later, an escapee will cause 'biological havoc'. This is despite the fact that engineered organisms are generally less efficacious than their natural counterparts. However, such a statement is more rhetorical than factual and is less easy to substantiate for bacteria than it is for engineered higher plants such as crop plants. There remains the possibility that microbial engineering could be used for developing biological weapons. Here there are parallels with the development of nuclear power for peaceful purposes and its use as a military force.

Similar parallels can be drawn between the release into the environment of transgenic plants and agents of biological control, such as bacteria and viruses, with other plants and animal species that have been introduced into non-endemic environments. Some of these, such as the introduction of cane toads, rabbits and water buffalo into Australia, have caused severe ecological problems. The release of transgenic organisms could cause equally severe problems. There is thus a need for rigorously controlled and monitored field experiments prior to general release. This is a difficult task and proposals to undertake such field experiments have, understandably, met with considerable opposition. Proposed field trials of the 'ice-minus' bacterium (Section 9.3.1), for example, were the subject of litigation in the USA in 1986. This failed, and the bacterium was subsequently tested in the field in 1987 with no apparent adverse environmental effects.

This example raises a plethora of questions relating to adequate controls and the roles of government, via legislation, and regulatory bodies. To date, there are no internationally agreed guidelines and in some countries, e.g. Germany, there is a moratorium on field testing. The formulation of regulations is indeed difficult and field experiments will always involve an element of risk. The risk assessment procedures must, therefore, be as comprehensive as possible and are easier to implement for some transgenic species than for others. Lindsey and Jones (1989), for example, consider that there are few risks associated with transgenic higher plants and that these risks can be assessed in the laboratory. The propensity of such plants to produce novel toxic compounds or to develop weedy tendencies can be determined prior to release. However, the possibility that engineered genes could be transferred by cross-pollination to produce a vigorous weed species is less easy to ascertain in the laboratory, although there is no precedent for this from conventionally bred species.

The release of transgenic bacteria and viruses poses a similar set of questions. It is, however, much more difficult to determine whether such species will proliferate uncontrollably in the environment and provide competition for naturally occurring species that undertake ecologically significant processes. Bacteria, for example, are important components of biogeochemical cycles. Nor is it easy to establish the potential for gene transfer between transgenic and naturally occurring species which could give rise to toxin-producing species with repercussions for plant, animal and public health.

Although there are regulations for field experimentation in some countries, and moratoria in others, there is a clear need for stringent controls, reinforced if necessary by legislation. The risk element must be minimized and this will require close co-operation between industry, governments and regulatory organizations. Some form of international agreement would also be appropriate, at least to prevent the uncontrolled release of trangenic organisms by unscrupulous companies in countries, notably developing countries, where few or no formal regulations exist.

9.4 CONCLUSIONS

The applications of biotechnology discussed in this chapter have been developed for economic reasons rather than out of a concern for environmental quality. Nevertheless, they illustrate some of the ways that microbial cultures can be used to mitigate pollution and to more efficiently utilize resources by recycling useful substances such as metals. In these contexts biotechnology is an instrument for influencing the flux rates of biogeochemical cycles (Section 9.2.2). The agricultural applications of biotechnology also include aspects of biogeochemical flux manipulation, notably the enhancement of nitrogen fixation by soil and root-nodule bacteria. In addition, the use of biotechnology to produce new and improved crop plants and SCP represents a means of manipulating energy flows (Section 9.2.3).

Genetic engineering opens up a new and vast range of possibilities for influencing biogeochemical cycles and energy flows. It provides a more sophisticated approach to plant and animal breeding programmes that have been in operation for millennia (Section 3.3.3). Most significantly, it affords the basis for a novel approach to agriculture because it provides the potential for engineering crop plants (and animals) for specific environments. This contrasts with past and present agricultural practices that concentrate on tailoring the environment to the crop, by the application of fertilizers, crop protection chemicals and irrigation. There is now the possibility of producing nitrogen-fixing cereals, disease- and insect-resistant crops, and drought- and salt-tolerant crops with high carbohydrate, protein and fibre values. No such transgenic species are so far available, but the technology to translate this potential into reality is rapidly being developed. Such crops could make a valuable contribution to the world's food and fuel supply problems, in addition to contributing to the mitigation of some aspects of environmental degradation. In these contexts biotechnology and genetic engineering have significant contributions to make to achieving sustainable development (Chapter 2), although inevitably the technology is being developed where it is least required. Technology transfer, so that the poorer nations benefit, thus becomes an economic and political matter.

Such developments are, however, unlikely to occur without some adverse environmental repercussions. Assessing the risks of transgenic species in terms of plant, animal and public health is a difficult undertaking. It is incumbent on governments, regulatory bodies and scientists to ensure that adequate regulations are established to minimize the risks. This, plus the experimental release of transgenic organisms and wider applications of biotechnology in waste treatment, will be very important environmental issues in the 1990s.

9.5 REFERENCES

Ayres, P. and Paul, N. (1990) Weeding with fungi. *New Scientist*, **127** (No. 1732), 36–39.

Buck, K. (1989) Brave new botany. *New Scientist*, **122** (No. 1667), 50–55.

Conway, W. G. (1989) The prospects for sustaining species and their evolution. In: D. Western and M. C. Pearl (Eds). *Conservation for the Twenty-first Century*. New York: Oxford University Press, pp. 199–209.

Gray, N. F. (1989) *Biology of Wastewater*. Oxford: Oxford University Press.

Hilder, V. A., Gatehouse, A. M. R. and Boulter, D. (1990) Genetic engineering of crops for insect resistance using genes of plant origin. In: G. Lycett and D. Grierson (Eds). *Genetic Engineering of Crop Plants*. London: Butterworths, pp. 51–66.

Ligon, J. M. (1990) Molecular genetics of nitrogen fixation in plant-bacteria symbioses. In: J. P. Nakas and C. Hagerdorn (Eds). *Biotechnology of Plant–Microbe Interactions*. New York: McGraw-Hill, pp. 145–187.

Lindow, S. E. (1990) Use of genetically altered bacteria to achieve plant frost control. In: J. P. Nakas and C. Hagerdorn (Eds). *Biotechnology of Plant–Microbe Interactions*. New York: McGraw-Hill, pp. 85–110.

Lindsey, K. and Jones, M. G. K. (1989) *Plant Biotechnology in Agriculture*. Milton Keynes: Open University Press.

Litchfield, J. H. (1989) Single-cell proteins. In: J. L. Marx (Ed.). *A Revolution in Biotechnology*. Cambridge: Cambridge University Press, pp. 71–81.

Tudge, C. (1988) *Food Crops for the Future*. Oxford: Blackwell.

Wayman, M. and Parekh, S. R. (1990) *Biotechnology of Biomass Conversion*. Milton Keynes: Open University Press.

9.6 FURTHER READING

HMSO (1990) *Developments in Biotechnology*. London: HMSO.

Marx, J. L. (Ed.) (1989) *A Revolution in Biotechnology*. Cambridge: Cambridge University Press.

Nossal, G. J. V. and Coppel, R. L. (1989) *Reshaping Life: Key Issues in Genetic Engineering*, 2nd edn. Cambridge: Cambridge University Press.

Wheale, P. and McNally, R. (Eds). (1990) *The Bio-Revolution: Cornucopia or Pandora's Box?* London: Pluto Press.

CHAPTER 10

Environmental and Green Movements

S. R. Bowlby and M. S. Lowe

10.1   INTRODUCTION

The late 1980s witnessed what is often thought to have been an unprecedented upsurge in environmental and 'green' political activity. This movement has influenced not only government action but also industry and culture. In Britain, in 1990, the government published the first major 'Environment' White Paper for more than a decade, while supermarkets embraced 'environmentally friendly' products and packaging which were until recently tucked away in specialist health food shops or wholefood co-operatives. The emergence of the recycled toilet roll and of magazines such as *Lifestyle* are testimony to the contemporary impact of green thinking. Such movements are not exclusive to the developed world. Indeed, they cross nations, class and culture. For example, pop stars have united with Amazonian tribal chiefs to fight for the rain forests and native land rights.

   Contrary to popular opinion, the ideas which inform the contemporary environmental and green movements in the West have a long history. This chapter begins by examining the historical development of ideas about the environment and nature in the West. The next section discusses the contemporary environmental and green movements and gives examples of green political activity in the developed and developing worlds. The final section briefly explores the motivations behind the current revival of these movements.

10.2   A BRIEF HISTORY OF IDEAS ABOUT THE ENVIRONMENT

This historical survey of changing ideas about the environment focuses on Europe (in particular, Britain) and the USA. These are the areas where economic and social change initiated the world-wide industrial developments that are now thought to be threatening the natural environment.

   In the 16th and 17th centuries the dominant view among Europeans and American colonists was derived from the then prevalent Christian teaching that nature was made by God to serve humanity. Animals and plants were provided either for

Environmental Issues in the 1990s. Edited by A. M. Mannion and S. R. Bowlby

practical human use (as labour or food) or to serve as moral lessons or stimuli. For example, in 1728 a Virginian suggested that horse flies were created 'that men should exercise their wits and industry to guard against them' (quoted in Thomas, 1983). One important aspect of this conception was the belief that as animals and plants had no souls, humans need not be concerned about any pain they might cause to them. Indeed, many contended that animals and plants had no feelings. However, it was also recognized that the boundary between animals and humans was indistinct and that humans had 'animal' aspects and appetites. It was believed that these had to be subdued by religion and morality for people to become fully human. Animal-like behaviour—such as dirtiness, lust or greed—was, by definition, immoral. This view was used to justify the appalling treatment of people who were thought, conveniently to those in power, not to be fully human. For example, the poor, negroes, American Indians, the Irish and women were all often referred to in terms otherwise reserved for animals.

In the case of women, it was suggested commonly that their biology rendered them too close to nature to be fully human. Great stress was laid on the animal aspects of childbearing and breast-feeding, which were held to be indications of women's dangerous links with nature, witchcraft and sin. As women were identified with nature, which was presented in contemporary religious thought as something given to man for his use, Christian views about nature could be used to justify women's social domination by men.

Even these orthodox ideas about humans and nature were not held universally. It is possible to find evidence of concern for animals and plants dating back as early as the medieval period (Thomas, 1983) and, later, of more enlightened views about the humanity of women, slaves and the poor. How then did these alternative ideas begin to become more important? One influence was the growing interest in the application of human reason to the scientific study of nature. Initially this study was undertaken to further the understanding of God's creation and to gain greater control over nature to enhance human livelihood. Scientists such as Linnaeus (1707–78), who were motivated by deep reverence, began to investigate systematically the number and characteristics of different species. Explorers of the New World and the Orient brought back exotic plants and animals not only for their potential economic value, but increasingly as objects of scientific and popular interest. By the end of the 17th century botany and the study of animals were becoming popular pastimes among the middle classes of England, France and Germany. By the 18th century the study of natural history had become a highly fashionable activity and in England several clubs and societies were founded to promote it. This activity led in turn to the growth of interest in nature conservation and the preservation of rare or important species.

Another influence was the increasing acceptability of sentiment among the educated classes. From the 17th to 18th centuries there was an increasing willingness to acknowledge the importance of feelings and emotions as justifications for action. The sensitive and tender-hearted man became a person to respect rather than to despise. The sensibilities of women could be cited as a reason for their moral superiority to men rather than as evidence of their inferiority. The increasing acceptance of sensibility helped to create a new belief that 'unnecessary' cruelty towards animals and any delight in inflicting pain and suffering on them was morally wrong. Moreover, instead of seeing wild and uncultivated areas as aesthetically unpleasing because they had not yet been tamed by human endeavour, they were increasingly seen as evidence of the magnificence of the Divine power. Visits to mountains became fashionable and the new taste for wilderness was reflected in the fashion for informal 'wild' gardens. The belief that communion with nature was morally and spiritually uplifting became widespread.

During the same period, major shifts in economic life were changing the relationship between people and nature. Although the massive redistribution of population from the country to the city did not take place until the 19th century, the movement that did occur during the 18th century helped to bring about a new appreciation of the countryside as a place for recreation and the enjoyment of natural beauty. In Britain, the proportion of the population living in towns with more than 5000 inhabitants increased from 13 to 25% between 1700 and 1800. The new enjoyment of the countryside developed not only because of the physical problems of urban living—the towns were heavily polluted by coal burning and by industrial processes— but also because the new merchant class sought to emulate the nobility by purchasing country estates or houses. The countryside had become fashionable. However, the countryside itself was being transformed as the cultivated areas were extended and also enclosed between neat and orderly hedges. Thomas (1983) argues that the 18th century taste for romantic, uncultivated landscape was, in part, a reaction to this shift in the nature of the cultivated landscape.

The move to the towns and the transformation of agriculture were themselves the result of the shift in the economy, first to the increased importance of trade and, subsequently, to the gradual development of capitalist industries. By the end of the 18th century writers frequently complained about the impact of new roads, canals and industrial buildings on the urban and rural environment. Here, again, the beginnings of contemporary concern with the conservation of rural environments can be discerned.

The 18th century was thus a period when new ideas about nature and the relationship between human beings and nature were being formed and debated. Worster (1977) suggests that there were two main traditions in attitudes towards the environment: the arcadian and the imperial. The *arcadian* is typified for him by Gilbert White (1720–93), a quiet country parson in the village of Selborne who spent his days studying the natural history of the area. His motives were to understand the wonders of creation and to help society to use that creation wisely to its own advantage. His book on the *Natural History of Selborne*, published in 1789, became a major influence. In particular, by the mid-19th century his life and writings had come to symbolize a lost pastoral haven in which all creatures, from human beings to the smallest beetle or ant, lived in harmony with one another. His scientific work was held by his admirers to show the necessity of studying the whole interrelatedness of nature rather than of investigating parts in isolation.

The *imperial* tradition of thought dates back at least to the writings of Francis Bacon (1561–1626). Bacon believed that human beings should develop and apply science to conquer and organize nature. Through the application of rigorous science humans would be able to create and rule over an ideal world. This imperial tradition is one that has been dominant in the West and which inspired much of the scientific advance of the 19th, and indeed the 20th, century. However, both the arcadian and the imperial traditions encouraged the objective study of plants, animals and their environment and so helped to change popular ideas about the natural world.

During the 19th century the speed of social and economic change in Europe and America intensified. Urban areas mushroomed in size, workers moved from field to factory, and for most of them everyday contact with a rural environment was lost. Many people felt that these developments were moving humanity in the wrong direction. They looked back to the past and idealized the world of Gilbert White. Others tried to develop the arcadian tradition to find a relationship with nature that allowed for the new science about the earth and its life forms, but which also acknowledged the spiritual importance of contact with nature to the human soul.

One of the most influential of these *Romantic, transcendental* thinkers was the New England writer and naturalist Henry Thoreau (1817–62). Thoreau believed that 'nature' is a single organic entity in which all living organisms are related to one another. He wrote in 1851, 'The earth I tread on is not a dead, inert mass, it is a body, has a spirit, is organic, and fluid to the influences of its spirit, and to whatever particle of that spirit is in me' (quoted in Worster, 1977). He therefore believed in the spiritual need for people to keep in contact with nature.

Such Romantic ideas informed the movements that were then developing in America and Britain to campaign for the preservation of wilderness and the conservation of natural habitats. However, it is important to remember that there is not a simple polarization between the arcadian tradition and environmental conservation on the one hand and the imperial tradition and environmental exploitation on the other. It was, in fact, people working within the latter tradition who were most important in establishing many of the agencies of resource conservation in America. They did so to ensure adequate resources for the future. Moreover, Thoreau himself, and many others with similar ideas, were also concerned to use the earth's resources for the benefit and economic well being of humanity.

Romantic thought was often allied with radical political views. Thoreau expressed his ideas about an ideal way of life in practice, living a simple and ascetic existence, trying to minimize his material wants to develop both his inner self and to gain a greater understanding of social morality through studying the world of nature. Thoreau, like other Romantics, believed that true democracy could be attained by imitating what 'they understood as the lesson of nature—the pursuit of self-actualization and creative diversity within mutually sustaining communities' (O'Riordan, 1981).

This was but one response to the tremendous social and economic upheavals of the Industrial Revolution. Others tried to find different ways of combating what they saw as the increasingly dehumanizing conditions of life in cities and the growing, impersonal power of large bureaucracies. One strong tradition of thought argued for the establishment of small self-sustaining communities. For example, the anarchist Peter Kropotkin argued for the establishment of decentralized communities in which agriculture, industry and the arts were all practised so that the inhabitants would not suffer the alienation from their work brought about by industrialization and the division of labour. Kropotkin's ideas and the arguments for small independent communities have been important influences on the tradition of town planning in Britain, through the work of such people as Ebenezer Howard, Patrick Geddes and Patrick Abercrombie. Similar ideas were put forward in the 1960s by Schumacher in his book *Small is Beautiful* and have been important influences on the modern green movement.

Another tradition of thought that grew out of Romantic ideas is that of *bioethics*.

This sees a link with nature as vital for the moral and spiritual health of human beings. Moreover, society must not merely conserve or preserve but has an ethical duty to ensure that 'nature' is free to follow its own course. Thus, human beings must act within the ecological boundaries set by nature. This strand of thinking has been important in the development of 'Gaianism' and 'deep ecology' (see Section 10.3).

During the early 20th century a diverse range of pressure groups and environmental activists in the West helped to establish ways to preserve or conserve habitats and species, to preserve the built environment and to ensure access to 'wilderness' for urban dwellers. However 'the environment' did not become a high profile political issue until the late 1960s, when the modern environmental movement began to take shape. This grew out of two developments: the wave of quasi-radical, student-centred, 'flower-power' political activism and the increasing doubt over the future of humanity, raised by the existence of nuclear weapons and the Cold War of the 1950s.

The first brought back into popular consciousness many of the concerns of Romantic transcendentalists and some of the social ideas of the 19th century anarchists. It is no surprise that many of the radicals of the 1960s sought to establish communal living or to retreat to the simple life in a rural or 'wild' environment. The second gave rise to the peace movement and also represented, but in a new and more urgent form, the mistrust of society's ability to use technology wisely. This had been a theme of some of the debates over the impact of the Industrial Revolution on human communities in the 19th century. The peace movement, therefore, has always had close links with the environmental and green movements. Feminism was another important movement that grew out of the various radical movements of the sixties. Like the peace movement this has had a significant influence on the modern green movement (see Section 10.3.3).

The 'environmental movement' of the late 1960s came to political prominence with the publication in 1972 of two major documents: *Blueprint for Survival*, which was published by the British *Ecologist* magazine (Goldsmith *et al.*, 1972) and the so-called 'Club of Rome's' *Limits to Growth* (Meadows *et al.*, 1972). Both suggested that without radical social and economic change, overpopulation, resource scarcity and pollution would bring about world-wide economic collapse. The ensuing popular concern led to the establishment not only of several important environmental pressure groups, but also the various European 'green' political parties. During the late 1970s and the early 1980s, the recession and the consequent failure of the Organization of Petroleum Exporting Countries (OPEC) oil cartel to induce enduring widespread scarcity of oil in the West removed environmental issues from newspaper headlines. Nevertheless, popular interest in these issues continued, fuelled by continuing small-scale local environmental change, reports of massive and life-threatening environmental changes in some developing countries, and by a number of major ecological near-disasters caused by industry or disasters such as Three Mile Island, Bhopal and Chernobyl. In the late 1980s this popular concern reemerged and now, in the 1990s, green issues are once more at the top of the political agenda.

10.3 ENVIRONMENTAL AND GREEN MOVEMENTS TODAY

This section starts with a discussion of current environmental ideologies and then gives three examples of the ways in which such ideas are linked to environmental politics.

10.3.1 Environmental ideologies

One simple classification of Western environmental ideologies is proposed by O'Riordan (1989). He divides them into the *technocentric* and the *ecocentric* (see Table 10.1). Technocentric ideologies only value the natural world insofar as it provides for human needs and is related to the imperialist tradition (Section 10.2). In contrast, the ecocentric approach espouses the arcadian view that the natural world has value in its own right.

Table 10.1   A typology of attitudes to the environment. Adapted from O'Riordan (1989)

Ecocentrism		Technocentrism	
Gaianism	*Communalism*	*Accommodation*	*Intervention*
Faith in the rights of nature and of the essential need for co-evolution of human and natural ethics	Faith in the co-operative capabilities of societies to establish self-reliant communities based on renewable resource use and appropriate technologies	Faith in the adaptability of existing institutions and approaches to assessment and evaluation to accommodate environmental demands	Faith in the application of science, market forces and managerial ingenuity to intervene in nature to create economic growth and overcome environmental problems
Demand for redistribution of power towards a decentralized, federated economy with more emphasis on informal economic and social transactions and the pursuit of participatory justice		Belief in the retention of the status quo in the existing structure of political power, but a demand for more responsiveness and accountability in political, regulatory, planning and educational institutions	

Technocrats consider that technical innovation and science will be able to solve any environmental problems arising from resource scarcity and pollution. Most people in the West today subscribe to some version of the technocentric approach. In contrast, ecocentrics believe that humanity will not always be able to solve environmental problems through technical innovation and scientific advance, and must radically restructure their economic and social organization. They think that the current high standard of material consumption in the West must be reduced, and decentralized patterns of living and devolved political power must be created.

Ecocentrists can be divided into two groups. Communalists believe that harmony with nature comes from the establishment of small self-reliant communities. Their ideas are derived from those of Kropotkin and later thinkers such as Schumacher (Section 10.2). Gaianists derive their ideas from the recent theory of James Lovelock, who argues that the biosphere operates as if it were a single living entity which he calls Gaia, after the Greek Goddess of the earth (Chapter 1). The Gaia hypothesis suggests that the organisms of the earth are not merely dependent on their environment but actively change it into conditions optimal for the survival of life. If humans interfere with this mechanism too much, Gaia will survive, but in a form in which human beings will have no part. The Gaia hypothesis has been invoked by those adhering to bioethics and by the so-called *deep ecologists* to argue that the rights of the non-human world must be given parity with, or even priority over, human welfare.

Some extreme adherents of deep ecology in the USA, who are members of an organization called Earth First! have become involved in illegal 'guerrilla warfare' to protect various natural environments. Their tactics include spiking trees to ruin the blades of chain saws and putting sand in the engines of the machines used by developers. Deep ecologists have been criticized for their emphasis on population control without a corresponding concern with the problems of development and inequality that beset so many of the world's people (see Chapter 7).

Mention of the guerrilla tactics of Earth First! is a reminder that people also differ in their views of what sort of political action is needed to deal with environmental issues. At one end of the spectrum are those who believe that the necessary change can and should be achieved through established political parties and processes. Others adopt the view that new political groupings and institutions must be established to bring about radical changes in social and economic behaviour. Lastly, there are those who consider that change can be brought about through example, that by adopting ecologically sound lifestyles themselves they will persuade others to do the same.

Only a minority of people actually take political action over the environment, for example, by belonging to an environmental pressure group, by joining an environmental political party, or by adopting an ecologically sound lifestyle with other like-minded people. A larger group of people may make modifications to their lifestyles by buying 'green' products or may vote for a 'green' party. The minority are political activists (the word political is used here in its broad sense and not in the narrow 'party political' sense).

People who are politically active over environmental issues belong to a political 'movement'. The broad *environmental political movement* includes both ecocentrics and technocentrics. The *green movement* comprises those environmental activists who believe that to save the environment major changes to political and economic institutions must be made. They are therefore drawn from the ranks of the ecocentrists.

10.3.2 Green political parties

Some people in the green movement are active in green political parties (Porritt and Winner, 1988). The first green party was the Values Party of New Zealand, which was established in the late 1960s. Although the Values Party collapsed in the early 1980s after internal disagreements, its principles were an inspiration to the founders of green parties in other countries. Its proposals included: 'the need for a steady state population and economy, . . . ecological thinking, soft-path energy systems, decentralization of government, equality for women, and rights of native peoples, as well as . . . valuing the traits traditionally considered feminine: cooperation, nurturing, healing, cherishing, and peace-making' (Capra and Spretnak, 1984).

A number of other green parties and green political alliances have since been established in several Western countries. There is no green party in the USA, despite the vast range of 'green' pressure groups and the existence of several 'networks' of communication linking these groups. One reason for this is that the voting system in the USA and the powerful organization and massive funds of the two main political parties, the Democrats and the Republicans, leaves little room for minority parties to make an impact. Green parties have been most successful in gaining political representation in mainland Europe, where the voting systems allow some green politicians to be elected despite their small share of the vote. These electoral successes have allowed the various green parties to put a variety of environmental issues on to the political agenda and have been one factor in persuading larger political parties to take environmental issues seriously. However, their success in doing so may now threaten the survival of the green parties themselves.

The United Kingdom Green Party has not enjoyed the political success of the other European 'green' parties, perhaps in part because of the UK's 'first past the post' system of election, which means that Green Party candidates have never been elected to parliament. It was founded as the People Party in February 1973 and initially based its manifesto on the four principles of the Club of Rome's *Blueprint for Survival* (Goldsmith *et al.*, 1972). These principles were: (a) minimum disruption of ecological processes; (b) maximum conservation of materials and energy; (c) a population in which recruitment equals loss; and (d) a social system in which the individual can enjoy rather than feel restricted by the first three conditions.

Following a change of name to the Ecology Party in 1975, the focus of the party shifted to issues such as nuclear disarmament and wildlife protection. It was not until 1985 that the party that we now know as the Green Party came into existence. Its major principles are listed in Table 10.2 and show that the concerns of the party have shifted again. There is a greater awareness of the issues of basic needs and development in the Third World (Principles 5 and 10) and a clearer commitment to equal rights, individual freedom and human fulfilment (Principles 6, 8, 9 and 12).

In the UK during the 1970s, the Green Party (then the People Party) only gained a small share of the vote: 1.8% in February 1974, 0.7% in October 1974 and 1.5% (as the Ecology Party) in 1979. Its share of the vote fell back to 1.0% in 1983. The early 1980s saw links being formed with green parties in Europe and the green's success in national elections in Europe gave the UK Green Party some small increase in political credibility. Its share of the vote remained low throughout the 1980s, reaching 1.4% in the 1987 General Election.

The UK Green Party's most significant success to date was in the 1989 European Elections, when the Green Party achieved an unprecedented 15% of the vote. Although this rise in support has been dismissed as a protest vote against the Conservatives, it had the effect of giving more prominence to Green Party policies and politicians in the national media. It also brought to the fore disagreements within the UK Green Party concerning the need for individual leaders and a hierarchy of control, a problem not confined to the UK Greens and one which seems to reflect deeper divisions in the green movement.

During the 1980s the mainstream British political parties began to pay attention to green issues. For example, the Labour Party brought out a Green Manifesto in October 1990, shortly after the Conservative Government's Environment White Paper. The Conservative Government has claimed to be adopting green values, although its critics suggest that it has pursued these with rather limited, if not negligible, success (British Association of Nature Conservationists, 1990). How far this shift in thinking is attributable directly to the Green Party's activities is difficult to determine, but clearly it has been a significant element in keeping environmental

Table 10.2 The principles of the UK Green Party

1 Humankind is especially responsible for the care of the planet, holding it in trust for all other living things and for other generations	9 Increased personal freedoms and rights must be balanced by increased personal responsibility
2 No activities which may cause irreversible damage to our environment or planet should be undertaken	10 Economic activity and work should be personally fulfilling and geared to the needs of all the World's peoples, not just the wants of a few
3 In a world of finite resources, uncontrolled economic growth cannot continue indefinitely	11 All land belongs to the community occupying it, never to individuals
4 A society which is dependent upon finite natural resources is unsustainable	12 Progress should be measured in terms of quality of life for all the World's peoples: personal freedom, human fulfilment and spiritual growth rather than centralized power, uniformity and material wealth
5 Basic human needs should be met first: food, shelter, clothing, health and education for all the World's peoples	
6 The equal rights of people should be protected, irrespective of nation, colour, creed, sex, sexuality, age, physical or mental ability	13 The values of caring, co-operation, nurturing and sharing must be encouraged to redress the balance with the values of competitiveness, domination and aggression which have characterised our society in the past
7 As valuable resources become scarce, competition for those resources grows: competition which often leads to aggression, conflict and even war; we should work for world peace rather than prepare for war	
8 Laws must be agreed which protect the planet, which increase rather than diminish individual freedoms, rights and choices, and which, when necessary, mediate between people where those freedoms and rights conflict	14 Our policies must reflect the interdependence of all living things and the interconnectedness of all political and social activity
	15 Urgently needed reforms must be set within long-term strategies which attack the root causes of our society's ills rather than just the end results: prevention rather than cure

© The Green Party (1990). Reproduced by permission of the Green Party, 10 Balham High Street, London SW12 9YY.

issues in the public eye. Moreover, this shift could well mean the end of the Green Party as an effective political party (as opposed to a lobby group), as environmentally concerned voters may think that a vote for one of the major parties will be more effective than a vote for a minority party.

10.3.3 Ecofeminism

Ecofeminism started as a Western green movement. However, it seeks to make connections with women in the Third World who are involved in environmental movements and to emphasize the connections between the need to improve the social position of women in Third World countries and action to create a sustainable economy.

Ecofeminists believe that women need to organize together politically if a sustainable economy is to be created. They argue that women have greater concern for the environment than men and suggest that the control and exploitation of nature and of women by men have many features in common. For example, they point to the long-standing tendency in Western thought to treat mind and body, culture and nature as separate and opposing entities, with body and nature as inferior to mind and culture.

They also emphasize the links between views of nature and of women as objects for male exploitation and use (Section 10.2). To understand the ideas that inform ecofeminism it is necessary to look back briefly at the history of current feminist thought.

As discussed in Section 10.2, one popular explanation for women's social inferiority in Western thought was that women were closer to nature than men. Consequently, women were close to the Devil and were morally dangerous, tempting men to sin. However, like the rest of the natural world, women were also thought to have been created for the benefit of men.

Just as the dominant attitudes to animals and plants began to change from the 1600s onward, so, gradually, did attitudes towards women. In the 19th century strong feminist movements developed throughout the Western world. These movements attacked the prevailing ideas about women. In doing so they deployed two opposing types of argument. One refuted the view that women's biology rendered them intellectually and morally different to men; the other used men's own arguments, based on biology, to suggest that women were not only different but superior. These two opposing strands of feminist thought have continued to characterize the feminist movement. However, it is the view that women are different that has been the strongest influence on ecofeminism.

Many ecofeminists believe that women are closer to nature than men because women bear children and are, by nature, less aggressive and more caring than men and therefore more opposed to war and its threat to the survival of the planet. They also believe that women's scientific knowledge (e.g. their knowlege of herbal medicine and of midwifery), which was suppressed by men during the period in which modern science was developing, was and is more in tune with the creation of a sustainable and caring economy than the exploitative 'imperial' male science on which the Industrial Revolution was founded. They argue for a move away from 'male' emphasis on individual leaders, hierarchical power structures and lack of tolerance of difference. This aspect of ecofeminism clearly has strong links with the ideas of those in the green movement who argue for small communities and the minimization of centralized hierarchies of power. It is important to make clear that not all ecofeminists agree with the 'biological difference' view. Some argue that, if women are more caring and less aggressive and power-hungry than men, it is because of social conditioning rather than a result of innate differences.

A further important element in ecofeminism derives from research carried out since the 1970s which emphasizes that, globally, most of the work involved in the production of the means of survival and of future generations is carried out by women either as paid or unpaid work. Women also play a central role in organizing domestic consumption. Their actions as both producers and as consumers are thus very important to policies for creating a sustainable economy (see Chapter 2). Thus ecofeminists argue that it is women who must organize politically to change society's behaviour towards the environment and to create a sustainable way of life. One example of such ecofeminist political organization is the UK Women's Environmental Network, which lobbies for, and provides information about, ecologically sound methods of consumption and production. Ecofeminists have had a major influence on the green movement, as is indicated by the UK Green Party's principles (Table 10.2).

There is, however, a great deal of controversy within the women's movement concerning some ecofeminists' claims that women's nature is biologically determined. Ecofeminism, then, is a part of the green movement which, like most other political movements, incorporates a variety of viewpoints and has internal as well as external opponents. We, the authors, while having great sympathy with many of the proposals of ecofeminists, cannot agree with arguments which suggest that there are innate moral differences between men and women.

10.3.4 A green movement in the developing world

Having discussed the history and development of the environmental and green movement from a Eurocentric perspective, this subsection counters this bias through a brief description of one example of green politics in the developing world.

There are, as yet, no green parties in developing world countries, but this should not be taken to mean either that people in these countries do not care about environmental issues or that such issues are of less importance. Indeed, damage to the environment in these countries often poses a more immediate and evident threat to the survival of the population than it does in the West. For example, soil erosion (see Chapter 14) may cause the starvation and death of many of the local people, whereas in the West often the worst outcome is a loss of jobs. Because, in the developing world, so many people depend directly on the land, pollution and environmental degradation can threaten the livelihood of many poor people, However, because they are poor and lack economic and political power, it is often difficult for them to take effective action.

Despite these problems there are many examples of political action to protect the environment in developing world countries. In some instances, such as in the Brazilian rain forests, local people who are taking action to protect their environment have managed to enlist the aid of people outside their own country to bring international pressure to bear on their national government and the private companies involved to adopt improved policies and practices. Whether this strategy will work in the Brazilian example remains to be seen. There are also examples of 'green movements' that have been successful without such outside assistance. A particularly famous and impressive example is the Chipko movement in Uttar Pradesh State in northern India (Weber, 1989). The word 'chipko' means literally 'to hug' and one of the movement's principal weapons was for groups of protesters to hug trees to prevent them being felled.

The movement grew out of local opposition to the discriminatory practices of government and private companies towards local hill people. Although the hill people depended on the forest's products for fuel, fodder and tools, the government favoured large commercial companies when awarding contracts for felling and for the use of timber and resin. The first occasion on which trees were hugged to prevent felling was in 1973. The motivation for this action was to prevent a large private company from felling trees which had been denied to a local co-operative which made wooden agricultural implements for local use. The original aims of the movement were, thus, not primarily ones of environmental protection. However, these aims rapidly shifted from the protection of forests from commercial felling so that the trees could be felled by local people, to the protection of the forests to preserve the forest cover, prevent soil erosion and landslides and create forests that could be used in sustainable ways. Women have played an important part in the direct action of the Chipko movement. One reason for this is that the women suffered more immediately than the men from the destruction of the forest. Women are responsible for collecting firewood and fodder from the forest and as the forest was felled, they had to walk further and further in search of them.

The aims of the movement now include reforestation and the development of new sustainable agricultural methods among the local people. The growth and spread of the movement have been achieved by the leaders of the movement making long marches on foot (paydayatras) from village to village to spread information. The movement has now become famous internationally and has led to the establishment of similar movements in other parts of India, most notably in Himachal Pradesh, Rajasthan and Karnataka. From being a movement that was despised by the government it is now recognized as an important political force.

Those people involved in various environmental and green movements in the West have taken much of their inspiration and ideas from the Western tradition of thought and action in relation to the environment. In a similar fashion the Chipko movement has drawn its inspiration from the Indian tradition and in particular from the philosophy of Gandhi. Thus, the Chipko movement has emphasized non-violent protest, equality, simple lifestyles and local self-sufficiency.

10.4   CONTEMPORARY REVIVAL OF ENVIRONMENTAL AND GREEN MOVEMENTS

Why has there been a revival in environmental and green movements at the beginning of the 1990s? Cynical commentators argue that values associated with being 'green' are also those bound up with an emphasis on individual choice and individual freedom developed by the 'new right' in the 1980s. The still more cynical suggest that the explosion in green consumerism is part of a master plan by capitalists to capture more markets by expanding into new profitable territory. Others suggest that the universal expansion of green and environmental movements is the result of the recognition that something must be done to halt people's destruction of their environment. These perspectives are examined in the following.

10.4.1   Green times

There is no doubt that the 1980s were marked by a change in political values. From a post-war consensus in favour of the expansion of universal rights and of the welfare state, the shift to the so-called 'new times' of the 'new right' presented a different value system of individuality, diversity and choice. Such values can fit in with a green political scenario (Steward, 1989). In particular, personal responsibility for the consequences of one's actions, self-sufficiency and new patterns of work and consumption for individual satisfaction nestle neatly within a framework of green thinking. The recent growth of green consumption suggests the creation of a green future through individuals choosing the products they wish to buy and hence influencing what is available in the future.

10.4.2   Green consumption

Green consumption has another facet. There can be few people who have not considered the idea that the growth of green consumption, including in particular the publication of volumes such as the *Green Consumer Guide* 'designed to appeal to a "sandals to Saabs" spectrum of consumers' (Elkington and Haines, 1988), is attempting to expand the profitability of a variety of producers through exploiting and influencing green values. This is not likely to be a development in tune with the ideas of the green movement. First, it may well encourage people to buy more rather than to adopt a low consumption lifestyle. Second, there is no assurance that all the products that are marketed as 'green' are, in fact, environmentally benign. Methods of regulating the claims of 'green' producers are being developed in several countries, but at present it does appear that consumers are often misled about the environmental impact of the products they buy. Furthermore, although a variety of commercial and business organizations are proud to advertise their environmental credentials, consumers may be right in feeling sceptical about this new awareness. Supermarkets may peddle new environmental products while at the same time building new, more profitable stores on locations of environmental value. Firms who advertise the help they give to environmental groups or the environmental value of their products may be damaging the environment through their production processes.

10.4.3   Green politics

The green movement would argue that the revival of interest in green and environmental issues is neither part of new rightist ideals nor a master plan by capitalists to reach fresh heights of profitability. These commentators would suggest that the current perceptions of the threat of environmental destruction have led directly to the green considerations of the present. The obvious effects of air, river and sea pollution in the first few decades of this century tended to be limited in their geographical extent. However, in the post-war period the increased globalization of economic relations combined with the speed of technological advance has resulted in observable damage to the global environment. The effects of disasters such as Chernobyl, and the evidence which suggests that real global damage is being brought about by the enhanced greenhouse effect (see Chapters 3 and 4), stand testimony to this fact. Equally, such commentators would view the new green ideals as part and parcel of the emergence of new values of collectivism, universalism and social purpose rather than the advent of new conservatism (Steward, 1989).

Clearly, green and environmental issues have emerged from a variety of sources and it is difficult to select between these different interpretations of their origins. Whatever their basis, at present green and environmental issues are high on the political agenda, but whether they will remain so is not clear. If they do not spring from widespread and durable environmental concern and are simply a temporary consumption fad or a minor facet of new right thinking, they may well be short-lived.

10.5   CONCLUSIONS

This chapter has shown that environmental concerns have a long history and that there has been a gradual shift of ideas in the West towards the view that humanity must learn to live with rather than to exploit nature. The new recognition that there may be major threats to the survival of the human species unless there are significant changes in economic and social organization appears to have produced an environmental movement that has succeeded in bringing about shifts in the political agenda in a number of countries. The recognition of potential environmental problems has also led to new initiatives in international politics, with attempts to reduce global levels of the greenhouse gases. As is made clear in Chapter 2, if sustainable development is to be achieved there will need to be more fundamental changes in both national and international politics. If the gradual change in people's attitudes towards the environment is deep seated, it should support the development of such political changes. If not, the future of the green movement, and perhaps of the planet, is doubtful.

10.6 REFERENCES

British Association of Nature Conservationists (1990) *A Report on the Prime Minister's First Green Year*. Newbury: British Association of Nature Conservationists.

Capra, F. and Spretnak, C. (1984) *Green Politics*. London: Hutchinson.

Elkington, J. and Haines, J. (1988) *The Green Consumer Guide: From Shampoo to Champagne*. London: Victor Gollancz.

Goldsmith, E. (1972) *A Blueprint for Survival*. London: Penguin.

Green Party (1990) *Our Political Principles*. London: The UK Green Party.

Meadows, D. H. *et al.* (1972). *The Limits to Growth*. London: Potomac Associates and Pan Books (1974).

O'Riordan, T. (1989) The challenge for environmentalism. In: Peet, R. and Thrift, N. (Eds). *New Models in Geography*. London: Unwin.

Porritt, J. and Winner, D. (1988) *The Coming of the Greens*. London: Fontana.

Steward, F. (1989) Green times. In: Hall, S. and Jacques, M. (Eds). *New Times: The Changing Face of Politics in the 1990s*. London: Lawrence and Wishart.

Thomas, K. (1983) *Man and the Natural World*. London: Allen Lane.

Weber, T. (1989) *Hugging The Trees: The Story of the Chipko Movement*. Calcutta: Penguin.

Worster, D. (1977) *Nature's Economy: A History of Ecological Ideas*. Cambridge: Cambridge University Press.

10.7 FURTHER READING

Davis, K. (1990) What is ecofeminism? *Women and the Built Environment*, **13**, 3.

The Ecologist (1992) Special Issue on Feminism, Nature, Development, **22**, 1.

Lowe, P. and Goyder, J. (1983) *Environmental Groups in Politics*. London: Allen and Unwin.

O'Riordan, T. (1981) *Environmentalism*. London: Pion.

Paehlke, R. C. (1989) *Environmentalism and the Future of Progressive Politics*. Newhaven: Yale University Press.

Prentice, S. (1990) What's wrong with ecofeminism? *Women and the Built Environment*, **13**, 3.

SECTION III

LOCAL IMPACTS AND REACTIONS
(a) Change in the Physical Environment

CHAPTER 11

Acidification and Eutrophication

A. M. Mannion

11.1 INTRODUCTION

Although environmental issues at the beginning of the 1990s are dominated by the prospect of global warming (Chapters 3 and 4), the 1980s witnessed the emergence of acidification and eutrophication as environmental concerns and political bones of contention. Acidification and eutrophication are processes that occur naturally in the environment as a response to changing nutrient status in soils, the run-off from which influences the pH and nutrient loading of drainage systems. However, during the last two decades there has been increasing concern about accelerated rates of these two processes. It is now established that this acceleration is mainly anthropogenic.

These processes have become environmental issues because of the adverse impact they have had on environmental quality. Accelerated acidification, for example, has caused lakes, and possibly forests, to decay. Cultural eutrophication, a term used to distinguish it from natural eutrophication, has adversely affected aquatic ecosystems and aquifers. The causes of the two processes are well established and reflect society's ability to disturb the biogeochemical cycles of elements such as sulphur, phosphorus and nitrogen. Like so many pollution problems, accelerated acidification and cultural eutrophication are inadvertent repercussions of scientific advancement and both are the products of fuel-powered urban–industrial systems (Section 1.2.2).

Since it was first recognized by Scandinavian ecologists in the 1960s, accelerated acidification has been the focus of much scientific endeavour. Pollution histories derived from lake sediments in Europe and North America have linked it with fossil fuel consumption and as a consequence it has become a contentious international political issue. Cultural eutrophication is also a contentious issue. It is the nutrient enrichment of drainage systems, coastal regions and aquifers that creates environmental degradation. As Fig. 1.3 shows, cultural eutrophication results from high concentrations of people that produce a range of waste products, notably nitrates and phosphates, in sewage and waste water. In addition, large non-food producing populations require the manipulation of agricultural systems to produce food surpluses. This involves the intensive use of nitrate and phosphate fertilizers which, in

Environmental Issues in the 1990s. Edited by A. M. Mannion and S. R. Bowlby
© 1992 John Wiley & Sons Ltd

conjunction with inappropriate cropping systems, also contribute to nutrient enrichment in aquatic environments.

Accelerated acidification and cultural eutrophication impair environmental quality, although in both instances mitigating technologies and strategies are available. These should thus be integral components of sustainable development policies (Chapter 2) if future generations are to inherit as broad a resource base as is enjoyed by the present generation. Understanding the processes involved in these issues and determining the underpinning causes are precursors to the implementation of preventive policies which lie in the remit of politicians and policy makers who are involved in power relations (see Section 1.2.4).

11.2   ACIDIFICATION

The term 'acid rain' appears widely in the scientific and popular press. It is, however, something of a misnomer because even uncontaminated rain has a pH below 7 and is, therefore, acid. The addition of sulphurous and nitrous gases to the atmosphere causes precipitation to become even more acid as these gases combine with water. The natural pH of rain-water is about 5.6, but the rain occurring in the eastern USA, north-west and central Europe has an average pH of 4.0 to 5.0, and as the pH scale is logarithmic a decrease of one pH unit represents a ten-fold increase in acidity. Rain with such increased acidity can have environmental repercussions, even before it reaches the ground (Fig. 11.1). The resulting aerosols, for example, can influence

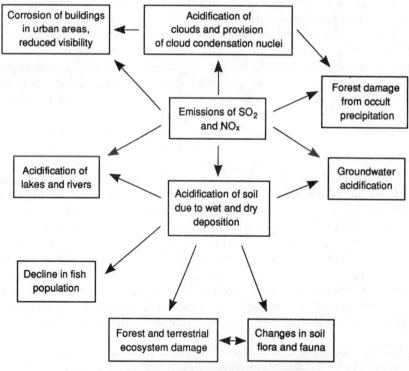

Figure 11.1   Environmental impact of acidification.

cloud formation by providing cloud condensation nuclei. This, in turn, can influence local and regional climates as the amount of cloud cover affects the amount of incoming and outgoing radiation. However, the major impact of acidification has been on freshwater ecosystems in the acidic bedrock regions of the northern hemisphere. The impact of acidification on terrestrial ecosystems is more controversial; forest health may be one casualty, but that could also be affected by increasing tropospheric ozone. Where, however, acidic emissions have been reduced and where mitigating measures have been employed, there is evidence that ecosystems can recover.

11.2.1 Chemistry and distribution of acidification

As Fig. 11.2 shows, the production of acids in the atmosphere involves the combination of sulphur dioxide (SO_2) and nitrogen oxides (NO and NO_2, which are collectively known as NO_x) with hydroxyl radicals (OH^-) or monatomic oxygen. These reactions require the presence of sunlight, and the resulting acids are entrained within clouds. Depending on the prevailing wind directions, these acids can be deposited hundreds of kilometres from the source area in the process of wet deposition. Occult deposition can also occur when low cloud and mist cause deposition on vegetation and ground surfaces. In addition, the dry deposition of sulphur and nitrogen oxides as particulates, gases or aerosols can occur near the source area. Wet and dry deposition can both have an environmental impact, which is usually gradual as specific areas receive acidic precipitation over long periods of time. Surges of acidified water, due to cloud bursts or rapidly melting winter snowfall, can also cause immediate ecological damage. Although such events are comparatively rare they can have devastating effects on fish stocks.

Once it reaches the ground the impact of acid precipitation also involves chemical reactions. For example, acidic soils and peatlands which receive such precipitation will gradually decline in pH as hydrogen ions (H^+) accumulate in the system. The sulphate and nitrate anions (SO_4^{2-} and NO_3^-) combine readily with nutrients such as sodium and potassium, and the resulting compounds are easily washed out of the substrate. The milieu will thus become even more base impoverished and will constrain vegetation growth, leaving soils and peats vulnerable to erosion. Such nutrient impoverishment may also be a cause of widespread forest damage in polluted areas. Aluminium is also more soluble at low pH and it too will be removed from soils or peats into lakes and rivers where high concentrations can have an adverse impact on fish populations. A combination of a pH of 5.0–5.5 and high concentrations of aluminium in calcium-deficient waters causes physiological changes in fish, especially in immature fry, that result in death. Aluminium can also combine with phosphorus to form chemical complexes. This removal of phosphorus from active circulation can contribute to a decline in primary productivity (this is the amount of organic material that green plants produce by photosynthesis). The nitrogen cycle is also sensitive to acidification. Where hydrogen ion accumulation occurs to depress the pH to 5.4–5.7, the activity of nitrifying bacteria is restricted so that ammonia, which is usually oxidized to nitrate by these bacteria, accumulates in the system. This can then be used by certain groups of algae in a complex process that results in even higher H^+ accumulation to compound acidification.

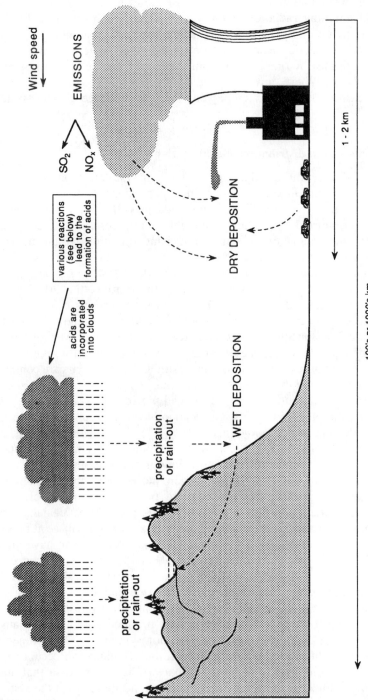

A. Sulphurous and sulphuric acids

SO_2 is emitted from natural and anthropogenic sources and dissolves in cloud water to produce sulphurous acid:

$$SO_2 + H_2O \longrightarrow H_2SO_3 \rightleftharpoons H^+ + HSO_3^-$$

Sulphurous acid can be oxidized in the gas or aqueous phase by various oxidants

$$SO_2 \xrightarrow{\text{oxidant}} SO_3$$

Aqueous sulphur trioxide forms sulphuric acid:

$$SO_3 + H_2O \longrightarrow H_2SO_4 \rightleftharpoons H^+ + HSO_4^- \rightleftharpoons 2H^+ + SO_4^{2-}$$

B. Nitrous and nitric acids

N_2O is emitted by the process of denitrification and although relatively inert it is a greenhouse gas. NO and NO_2 (collectively designated as NO_x) are produced by combustion processes and lightning. They are involved in many chemical processes, some of which damage the ozone layer in the stratosphere:

$$O_3 + NO \longrightarrow NO_2 + O_2$$

Other chemical processes may generate ozone in the troposphere causing photochemical smogs:

$$NO_2 \xrightarrow{\text{light}} NO + O$$
$$O + O_2 \longrightarrow O_3$$

In addition, nitric and nitrous acids may be produced:

$$2NO_2 + H_2O \longrightarrow HNO_3 + HNO_2$$

These acids are components of acid rain along with sulphurous and sulphuric acids

Figure 11.2   Processes involved in the formation and deposition of acid pollution.

Acidic deposition, then, is not just a straightforward case of hydrogen ion accumulation. Rather it creates a series of positive feedbacks by influencing the biogeochemical cycles of the major nutrients that are essential to plant growth. Thus, while some nutrients are limited in supply, others accumulate to concentrations that are detrimental. The overall result is a reduction in species diversity at the primary producer level which influences the composition of heterotrophic communities (animal communities that cannot produce their own food and are thus dependent on green plants).

Acid rain varies spatially and temporally on a global basis in relation to emission sources and prevailing wind directions. Most acidification has occurred in the industrialized northern hemisphere where emissions of sulphur and nitrogen oxides are highest. It is manifest where high precipitation rates occur in areas of acid bedrock that offer little buffering capacity. Alkaline bedrock, in contrast, has a neutralizing effect. The areas so far identified as having been most severely affected by acid precipitation are Scandinavia, northern Europe including the USSR west of the Urals, parts of China, the north-east USA and eastern Canada. In the next few decades additional areas will probaby become affected by acidification as developing countries industrialize and increase their consumption of fossil fuels.

The problem of acidification has also had political repercussions, mainly due to the fact that acid-producing nations are not always those that suffer the greatest damage. As Table 11.1 shows, sulphur dioxide emissions are highest from the USSR and the USA, but the amount that these nations receive is comparatively small. Conversely, countries such as Sweden and Norway which produce relatively low amounts of sulphur dioxide actually receive most of their acid deposition from other European sources. Similarly, Canada receives much of its acid deposition from the USA. The same trends occur in relation to nitrogen oxides. Thus remedial measures to curb

Table 11.1   Annual sulphur dioxide emissions and deposition in selected countries. Adapted from McCormick (1989). Dates given are approximate

Country	Sulhur dioxide emissions (10^3 tonnes)		Sulphur deposition (%)		
	1980	1983–86	Foreign	Domestic	Unknown
USSR	25 000	24 000	32	53	15
USA	23 200	20 800	?	?	?
China	?	13 000–18 000*	—	90	10?
Poland	4100	4300	52	42	6
East Germany	4000	4000	32	65	3
Canada	4650	3727	50	50	—
UK	4670	3540	12	79	9
Czechoslovakia	3100	3050	56	37	7
West Germany	3200	2400	45	48	7
France	3558	1845	32	54	14
Finland	584	370	55	26	19
Sweden	483	272	58	18	24
Norway	141	100	63	8	29

* Estimates vary.

Table 11.2   Annual nitrogen oxide emissions from selected countries. Adapted from McCormick (1989). Dates given are approximate

Country	Nitrogen oxide emission (10^3 tonnes)	
	1980	1983–86
USA	20 300	19 400
European USSR	2790	2930
West Germany	3100	2900
UK	1916	1690
France	1867	1693
Canada	1725	1785
Czechoslovakia	1204	1100
Poland	—	840
Sweden	328	305
Finland	280	250
Norway	—	215
China*	4400	4130

* From World Resources 1990–91.

acid deposition are required on an international basis, but because energy consumption is related to economic prosperity it is difficult (Section 11.2.5) to find a mutually acceptable formula.

11.2.2   Historical perspective: pollution histories

Investigations of lake sediments in Europe and North America have revealed temporal trends in lake water pH during the last 200 years. The main methods used are diatom and pollen analyses and the determination of heavy metal and soot particle concentrations in conjunction with dating techniques. In North America such work is being undertaken under the auspices of the Palaeoecological Investigation of Recent Lake Acidification (PIRLA) project. A similar project is underway in the UK, Norway and Sweden, which is called Surface Waters Acidification Programme (SWAP). These and other projects (reviewed in Mannion, 1989a, 1989b) have shown that acidification is widespread. They have also been instrumental in identifying the major cause of acidification as fossil fuel combustion, although afforestation and natural interglacial soil impoverishment (see Chapter 3) may also contribute to acidification.

These studies also indicate that the degree of acidification and its timing have varied. The research utilizes the fact that particular species of diatoms, unicellular green algae that are widespread in aquatic habitats, are very sensitive to water pH. Any changes that have occurred in the composition of diatom communities will be reflected in the fossil assemblages preserved in lake sediments. These will also contain a geochemical record of heavy metal deposition and soot particles, both of which are the products of fossil fuel combustion.

Much of the research on UK lakes has focused on the Galloway area of south-west Scotland and is exemplified by a study on the Round Loch of Glenhead and Loch Enoch (Battarbee et al., 1989). The diatom assemblages indicate that the two sites have become acidified since about 1850 AD by about one pH unit, from 5.5–5.7 to 4.5–4.7. The two lakes occur on granite bedrock in catchments that are dominated by peats or peaty podzols but which have not been afforested. The sediment records also indicate an increase in the concentrations of lead, zinc, copper and soot particles in the post-1930 sediments. These data point to increasing fossil fuel combustion as the primary cause of acidification, especially as other factors such as land-use changes can be discounted on the basis of historical records and pollen assemblages (which would indicate any vegetation change) within the sediment. Moreover, the Galloway region receives rainfall that is particularly acid, with a pH of 4.4–4.6. There is also an annual sulphate loading of 59 kg, which is ten times the average background value in the northern hemisphere and which originates from England and Europe.

The results of similar work in North America have been summarized by Charles (1990), who states that diatom-inferred pH histories are now available for at least 150 lakes. Many of these, in the Adirondack Mountains of New York State, northern New England, Ontario, Quebec and the Canadian Atlantic provinces, show pH decreases of between 0.5 and 1 pH unit. This contrasts with control sites in the Rocky Mountains and Sierra Nevada, western USA, which show little or no change. As with the Galloway lakes, acidification has occurred in the post-1850 period in response to industrialization and increasing use of fossil fuels.

11.2.3 Acidification of freshwaters

In lakes that have experienced accelerated acidification the ecological impact has been considerable. Enhanced concentrations of hydrogen ions can adversely affect the ionic regulation of freshwater organisms, causing an often fatal release of sodium cations and chloride anions. The impact of increased concentrations of metals, notably aluminium, and disruptions to the phosphorous and nitrogen biogeochemical cycles (Section 11.2.1) have ramifications for plant and animal growth and their reproductive success. In combination these factors impair primary productivity and reduce algal diversity so that only the most acid-tolerant species survive. Observations on acid lakes show that acid-tolerant algae frequently form a dense mat on the surface of the lake sediment and on debris. This, although it enhances water clarity, restricts microbial decomposition and further inhibits nutrient cycling. Such algal species are not especially palatable to many herbivores (plant-eating animals), which subsequently become stressed due to an inadequate food source as well as polluted water.

From an economic point of view, most interest has focused on fish stocks which are particularly vulnerable when the pH is less than 5.0. Although acidified conditions may not always cause high mortality, unless the acidification occurs in rapid pulses, fish stock depletion is mainly due to recruitment failure. This is due to the adverse effects of increased aluminium concentrations (Section 11.2.1). The situation is exemplified by a review of the status of fish stocks in Norway (Hesthagen *et al.*, 1989) which is summarized in Table 11.3. The most severely affected stocks are those of brown trout, especially in the counties of Vest-Agder, Rogaland and Aust-Agder in southern Norway, where fish populations are almost extinct. Hesthagen *et al.* (1989) also state that catches of Atlantic salmon in ten rivers of Aust-Agder and Vest-Agder have declined from about 53 tons in the 1880s to less than one ton in the 1980s. These

Table 11.3 Lost and affected populations of five fish species due to acidification in different counties in Norway. N = number of stocks; L = number of stocks lost; A = number of stocks affected. Reproduced from Hesthagen et al., 1989, by permission of the British Library

County	Number of lakes examined	Brown trout			Perch			Arctic charr			Roach			Other species		
		N	L	A	N	L	A	N	L	A	N	L	A	N	L	A
Østfold (1975)	425	175	34	97	305	11	149	5	2	2	80	11	37	232	12	92
Oslo/Akershus	171	66	31	20	151	6	18	11	6	2	38	2	8	54	2	9
Hedmark	129	100	27	37	102	1	18	18	2	13	25	0	5	94	2	25
Oppland	212	182	20	57	130	3	7	78	12	23	0	0	0	87	4	11
Buskerud	973	855	93	207	342	20	62	82	5	22	3	0	0	186	5	26
Vestfold	107	95	4	65	78	2	26	2	0	1	2	0	0	62	0	2
Telemark	1020	931	340	428	186	9	34	87	3	38	0	0	0	118	8	64
Aust-Agder	1034	917	591	105	257	80	37	1	0	0	0	0	0	39	1	0
Vest-Agder	1021	625	419	124	99	50	13	22	7	3	—	0	0	56	0	5
Rogaland	941	849	277	173	0	0	0	116	15	18	0	0	0	97	1	2
Total	6033	4795	1836	1313	1650	182	364	422	52	122	148	13	50	1025	35	236

data reinforce the widespread view that accelerated acidification remains one of the major environmental problems of the 1990s.

11.2.4   Acidification of terrestrial environments

How accelerated acidification has affected terrestrial environments is more equivocal than that discussed in the previous section for aquatic environments. This is because the enhanced production of ozone in the troposphere is also thought to contribute to ecosystem degradation. However, there is no doubt that acid precipitation affects soils. Soil pH also influences plant growth and only a limited range of species will tolerate acid conditions, particularly at pH <5. Nutrient cycling in the soil is influenced by the pH because of its effect on microbial activity. As the pH declines, fungi take over from bacteria as the dominant decomposers. Less mixing and less aeration occur due to reduced soil faunal activity. Such processes can, however, occur naturally, especially where there is a high rainfall causing water movement down-profile for most of the year. Some of the cations, such as calcium and sodium, removed in this process will be replenished as bedrock weathering occurs, but over a

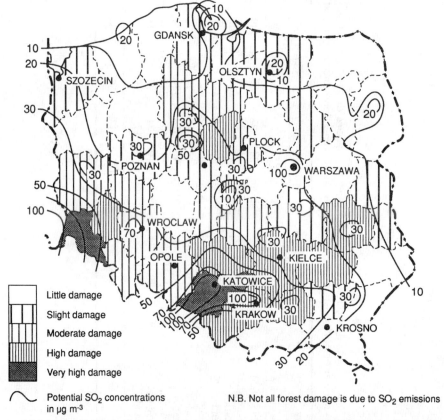

Figure 11.3   Distribution of potential forest damage in Poland in relation to sulphur dioxide emissions. Adapted from Mazurski (1990).

period such as an interglacial (see Fig. 3.5) there will be gradual impoverishment. However, soil acidification can occur rapidly, especially under coniferous plantations that produce an acid litter from which organic acids leach into the substrate. In many parts of Europe soil acidification caused by conifers masks the amount of acidification due to acid precipitation. As, in Britain for example, most afforestation takes place in upland areas of acid bedrock in receipt of acidic precipitation, it is probable that both the conifers and acid precipitation are contributing to acidification. Nevertheless, acidification is also accelerating in areas where there is no afforestation. Skiba *et al.* (1989) have shown that modelled wet and dry acidic deposition in Scotland correlates with highly acidic peats (about pH 3) with a low base status and which are unaffected by conifers.

Declining forest health throughout Europe has also been associated with acidic precipitation, although the relationship is not straightforward. The deposition of heavy metals, also the result of fossil fuel combustion, can contribute to forest decline and there is undoubtedly damage when occult deposition occurs. Each particular case must be considered on its own merits. As an example, Mazurski (1990) has examined the extent of forest damage in Poland in relation to sulphur dioxide deposition. Not only is Poland a high producer of sulphur and nitrogen oxides (Tables 11.1 and 11.2), but it also receives a further 50% from its industrialized neighbours. As Fig. 11.3 shows, the most damaged forests are located where the sulphur dioxide concentrations are the highest, notably in the south-west region. Mazurski (1990) states that overall some 7000 km^2 are affected by air pollution, including acidification, with an annual timber loss of 3×10^6 m^3 in addition to deterioration in quality and losses in game management. Together these losses total 200 million US dollars, and as industrialization continutes the areal extent of damage and its economic cost will increase. This illustrates how ecological damage can be translated into economic damage.

11.2.5 Mitigating measures

There are two aspects of recent research which point to the fact that acidified ecosystems can recover. Firstly, the liming of lakes on an experimental basis has resulted in marked increases in pH. The addition of powdered limestone also increases calcium concentrations and depresses aluminium concentrations. However, unless liming is repeated regularly the ameliorative effect is short-lived. The same is true where catchment liming is undertaken. The only really effective long-term treatment is to curb the emissions of acid-producing gases. This can be achieved by reducing the amount of fossil fuel combustion, by turning to alternative energy sources, by using less sulphur-rich fossil fuels and/or by removing the pollutants before they reach the atmosphere. The latter, for example, is now being applied in several developed nations where flue gas desulphurization units have been fitted to coal-fired power stations. There is also the possibility that biotechnology (and genetic engineering) could be used to desulphurize fossil fuels (Chapter 9).

Desulphurization has largely been prompted by the recognition that acidification is not confined by national boundaries and that high acid-producing nations are not only polluting themselves but also their neighbours. Acidification has thus become a major international political issue and has been the source of acrimony between acid-producing and acid-receiving nations. This has, however, prompted international action. The convention on Long-range Transboundary Air Pollution was signed in November 1979, in Geneva, by 35 countries. Although only a token treaty initially, it at least represented a general consensus of industrialized nations that acidification required collective ameliorative action. By 1983 the so-called '30% Club' had been established, resulting in a new protocol by 1985. The signatory countries, which did not include the UK or the USA, agreed to reduce sulphurous emissions by at least 30% (of 1980 levels) by 1993. Several other nations agreed to reduce emissions by 40 or 50%, e.g. Canada, West Germany, France and Norway. In 1988 a second protocol was signed, including the UK and USA this time, which incorporated an agreement to curb emissions of nitrogen gases. The result of these agreements has been a substantial reduction in the ouputs of sulphur dioxide and nitrogen oxides from most industrialized nations. Such policies are important adjuncts of sustainable development (Chapter 2) because they reduce the environmental costs of economic growth.

Whether the magnitude of these reductions is sufficient to secure the recovery of already acidified ecosystems remains to be seen. It is, however, encouraging that lakes, for example in the Sudbury region of Canada, in Scotland and in Sweden are showing evidence of recovery in response to reduced acidic emissions. Although emission reductions will help to curb acidification, there is the probabilitiy that present sinks for sulphurous emissions, such as peatlands, will release their accumulated pool into the atmosphere. This may already be occurring and release would be accelerated under enhanced greenhouse warming. There is also the likelihood that acidification will shift in the 1990s and beyond to newly industrializing countries, particularly China.

11.3   CULTURAL EUTROPHICATION

Unlike acidification, which involves nutrient depletion, eutrophication occurs in response to nutrient enrichment. Where that enrichment is due to human activity the result is cultural eutrophication. The nutrients involved are phosphorus and nitrogen in the form of nitrates and phosphates which emanate from fertilizer use, sewage output and urban run-off. The recreational use of lakes, notably boating, can also contribute to cultural eutrophication by bank erosion and sewage addition. Cultural eutrophication has received media attention in recent years because of the degradation of freshwater and coastal habitats, and the contamination of aquifers by high nitrate concentrations which may prove detrimental to human health.

11.3.1   Process and distribution of cultural eutrophication

The accelerated nutrient enrichment of aquatic environments results from anthropogenic disruption of the biogeochemical cycles of phosphorus and nitrogen. Phosphates have long been mined to provide fertilizers, and since the development of the Haber process in the early 1900s nitrate fertilizers have been artificially produced by fixing nitrogen from the atmosphere. Human intervention is thus short-circuiting these biogeochemical cycles by accelerating natural flux rates and creating new ones (Chapter 1). The use of nitrate fertilizers, for example, replaces or supplements the nitrifying effect that specific types of bacteria produce in soils. The mining of phosphate far exceeds the flux of phosphorus from sedimentary rocks by weathering which would occur in undisturbed conditions. Specific agricultural practices also promote loss of nitrogen from the soil, while urban run-off and sewage effluent (including the use of sewage sludge in agriculture) contain concentrations of nitrate and phosphate that would not occur naturally. Consequently high concentrations of these nutrients cause pollution.

Cultural eutrophication manifests itself in lakes by stimulating the ageing process. This involves sediment accumulation within which biological remains are incarcerated. It usually takes thousands of years for a once open water body to develop into a terrestrial ecosystem. However, the accelerated addition of nutrients due to human disturbance in lake catchments speeds up this process, because algal growth rapidly increases to take advantage of the nutrient supply. This, as Table 11.4 shows, has implications for water quality, as well as for plant and animal communities. Sediment input also increases in response to catchment disturbance and silting thus accelerates.

Table 11.4   Characteristics of lakes experiencing cultural eutrophication

Biological factors:

(a)   Primary productivity: usually much higher than in unpolluted water and is manifest as extensive algal blooms
(b)   Diversity of primary producers: initially green algae increase, but blue-green algae rapidly become dominant and produce toxins. Similarly, macrophytes (e.g. reed maces) respond well initially but due to increased turbidity and anoxia (see below) they decline in diversity as eutrophication proceeds
(c)   Higher trophic level productivity: overall decrease in response to factors given in this table
(d)   Higher trophic level diversity: decreases due to factors given in this table. The species of macro- and micro-inveterbrates which tolerate more extreme conditions increase in numbers. Fish are also adversely affected and populations are dominated by surface dwelling coarse fish such as pike and perch

Chemical factors:

(a)   Oxygen content of bottom waters (hypolimnion): this is usually low due to algal blooms restricting oxygen exchange between the water and atmosphere. Oxygen deficient (anoxia) conditions develop, especially at night when algae are not photosynthesizing. Thus seasonal and diurnal patterns of oxygen availability occur. The decay of algal blooms also produces anoxia
(b)   Salt content of water: this can be very high and a further restriction on floral and faunal diversity

Physical factors:

(a)   Mean depth of water body: as infill occurs the depth decreases
(b)   Volume of hypolimnion: varies
(c)   Turbidity: this increases, as sediment input increases, and restricts the depth of light penetration which can become a limiting factor for photosynthesis. It is also increased if boating is a significant activity

Water uses:

(a)   Water quality for domestic and industrial uses: this is usually poor
(b)   Amenity use: this can be severely impaired due to the production of noxious odours and loss of floral and faunal attractions

In lake ecosystems and coastal regions the outcome is visually obvious in the form of algal blooms. In aquifers, the problem is one of high nitrate concentrations which have no visual impact.

The problem of cultural eutrophication is globally widespread. It is, however, prevalent in the developed world where agriculture has intensified, and where urban spread has increased sewage output.

11.3.2   Historical perspective: pollution histories

As in the example of acidification, lake sediments contain a record of cultural eutrophication (reviewed in Mannion, 1989a, 1989b). In North America, for example, the cultural eutrophication of lakes in New England and the Mid-West followed European settlement and associated land-use changes during the late 1700s and 1800s. The changes registered in the diatom assemblages and the geochemical record reflect enhanced nutrient inputs, initially due to forest clearance and the inception of logging practices followed by arable cultivation. Ploughing also accelerated soil removal from the catchments into lake basins. In some instances, sedimentation rates increased ten-fold, which reflects the magnitude of colonial landscape disturbance. Similar events are also recorded in the sediments of the Great Lakes, indicating the considerable human impact that occurred in North America as migration from Europe increased in the post-independence period.

For Europe, too, there are numerous examples of the cultural eutrophication of lakes during the last 200 years. In many instances, nutrient enrichment is due to sewage effluent which, even if treated, remains high in nutrients, especially phosphates. The discharge of untreated or partially treated sawmill and woollen mill effluent and sewage into lakes, as well as dairy waste, and industrial and domestic detergents which are phosphate-rich, all contribute to cultural eutrophication.

11.3.3   Cultural eutrophication of freshwaters and aquifers

The ten-fold increase in the use of nitrate fertilizers which has occurred, mainly in the developed world, since the inter-war period is one of the major reasons why the rate of cultural eutrophication of aquatic environments has increased. As Table 11.3 illustrates, enhanced nutrient inputs into freshwater ecosystems, including reservoirs, can have marked chemical, physical and biological effects.

In many of the world's major waterways, for example the Nile and the lower Mississippi, the water hyacinth (*Eichhornia crassipes*) has become a major problem. Introduced from South America, it has responded to increasing nutrient inputs (and lack of natural predators) to such an extent that it causes oxygen deficiency and thus high fish mortality. Moreover, its rate of reproduction is so high that it rapidly recolonizes after clearance. The kariba weed (*Salvinia molesta*) constitutes a similar problem and measures to curb the growth of these two species have so far been unsuccessful. The proliferation of these species can impair the economic use of aquatic environments. Lake and river ecosystems can lose their amenity value, with a consequent loss of revenue for associated services. Reservoirs may have their working life shortened by excessive siltation that requires expensive remedial measures.

In regions that are underlain by porous sedimentary rocks intensive fertilizer use and cropping methods have influenced the chemical quality of groundwater which is used for domestic water supply. Throughout Europe and the USA high nitrate concentrations are a major pollution problem of the 1990s. Directives from the World Health Organization (WHO) and the European Community (EC) recommend that nitrate concentrations in domestic water should not exceed 50 mg l^{-1}. This limit is imposed because of increasing concern that high nitrate levels can adversely affect human health.

Nitrates themselves are harmless but once inside the body they can form nitrites or nitrosamines. Infant methaemoglobinaemia, or 'blue baby' syndrome, can occur in young children when nitrites produced in their digestive systems pass into the blood stream and combine with haemoglobin. This prevents the haemoglobin from performing its usual function of carrying oxygen and in acute cases a blue skin coloration develops. It is a reversible condition and rarely fatal; of fourteen cases reported in the UK since 1945 only one has been fatal. There is also the possibility that high nitrate concentrations may be linked to stomach cancer. Nitrate ingestion may be linked to the production in the human body of N-nitrosamines, some of which are known carcinogens. Although epidemiological research has failed to establish unequivocally that such a link exists, the imposition of limits on nitrate concentrations is a justifiable precaution. This is an example of a proactive response to a potential problem (Chapter 1).

However, meeting the targets for nitrate concentrations is another matter. As Dudley (1990) discusses, the problem in the UK is particularly acute in the south and east of England where agriculture has intensified in response to the EC's Common Agricultural Policy. The water authorities most severely affected are Anglian and Severn-Trent, where 14–15% of water supplies will have nitrate concentrations in excess of 50 mg l^{-1} by the mid-1990s. Similar problems are being experienced in other intensively cultivated parts of Europe. The problem may also become more acute before it improves as there is much controversy surrounding the rapidity with which nitrates enter aquifers. The nitrates emerging in groundwater today may be the product of the last few years of fertilizer use and agricultural practice, or they could be the result of practices of 30 or 40 years ago. If the latter is true, then nitrate concentrations are set to rise for some time to come.

11.3.4 Cultural eutrophication of coastal areas

Drainage from agricultural land, sewage effluent and urban run-off can also cause the cultural eutrophication of coastal waters where enrichment is most severe in shallow shelf areas and enclosed seas and estuaries. Pollution problems arise directly from waste input and because there is restricted exchange of water with the open oceans.

As well as increased concentrations of nitrates and phosphates, there is an additional problem created by organic waste, mostly from sewage, that is subject to decomposition by bacteria. This involves the oxidation of organic molecules using oxygen dissolved in the water and creates a biochemical oxygen demand. Where organic inputs are high, oxygen in the water is not replaced sufficiently rapidly from the atmosphere and anoxic (oxygen poor) conditions ensue. Under these circumstances anaerobic decomposition occurs, producing noxious gases such as hydrogen sulphide, ammonia and methane. The stimulation of algal reproduction also causes blooms to form. These can produce surface waters rich in oxygen (as a product of photosynthesis), but deeper waters are anoxic due to the inhibition of oxygen diffusion from the atmosphere. As a result, and depending on the regularity of the discharge and its content, diurnal and seasonal variations in the amount of dissolved oxygen present can occur.

Marine cultural eutrophication is a global problem, but in recent years the plight of the Mediterranean has been highlighted in the media. This sea is particularly susceptible because it occupies an enclosed basin with restricted exchange of water with the Atlantic Ocean. It also receives drainage from some of the most densely populated and intensively cultivated nations of Southern Europe, the Middle East and North Africa. High concentrations of nitrates and phosphates promote the growth of massive algal blooms that consist of diatoms and dinoflagellates. In extreme circumstances the dinoflagellate growth is so prolific that so-called 'red tides' are created which can extend more than 100 kilometres and can adversely effect tourism and fishing industries. Fish mortality can be locally high where anoxic conditions occur and the production of noxious odours can reduce amenity values. There may also be a significant hazard to humans as some shellfish, for example clams and oysters, can accumulate human pathogens from sewage wastes. Where red tides occur shellfish can become contaminated with toxins, which, if eaten, can cause nausea, convulsions and even death.

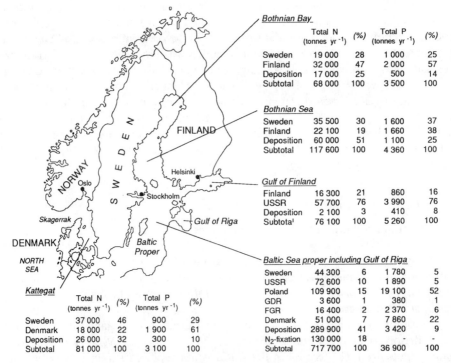

Figure 11.4   Inputs of nitrogen and phosphorus to the Baltic Sea and Kattegat. Adapted from Rosenberg *et al.* (1990).

The problems created by marine cultural eutrophication are also illustrated by the examination by Rosenberg *et al.* (1990) of the Swedish coastal areas of the Baltic Sea and Kattegat. As Fig. 11.4 shows, external supplies of nitrogen and phosphorus to the Baltic Sea are considerable. About 50% of the nitrogen is due to atmospheric input, notably nitrogen fixation, whereas about 90% of the phosphorus derives from land sources. The values represent substantial increases during the last century, with a four- to six-fold increase in nitrogen and more than an eight-fold increase in phosphorus. In the Kattegat there has been an eight-fold increase in nitrogen, but a slight decline in phosphorus due to the introduction of phosphorus removal from sewage in the 1970s.

Rosenberg *et al.* (1990) also examined the interplay that occurs between nitrogen and phosphorus as regulators of primary productivity. For example, species of nitrogen-fixing algae are limited by low phosphorus availability and have locally declined as a result of improved sewage treatment. A mosaic of variations in primary productivity and primary producers results, depending on whether nitrogen or phosphorus is the chief limiting factor. To generalize, blooms of blue-green algae can occur in the Baltic Sea during the summer and an increase in filamentous bottom-dwelling algae has had a detrimental impact on herring. Fish catches have, however, increased overall in the Baltic Sea, indicating that cultural eutrophication can have some beneficial effects. There is also evidence to show that primary production has increased in the Kattegat, but increases in toxin-producing algae have caused high mortality in littoral communities of bivalves and fish. Rosenberg *et al.* (1990) also describe the development of hypoxic conditions in the Baltic and Kattegat, in waters deeper than 70 metres. This means that oxygen-impoverished waters prevail, reducing bottom faunas and causing periods of high fish mortality.

11.3.5 Mitigating measures

There are many measures that can be taken to treat the symptoms of cultural eutrophication, as have been discussed by Croll and Hayes (1988). For example, suppliers of domestic water can mix sources so that water with a low nitrate content will dilute water with a high nitrate content. This is already necessary in some parts of the UK to comply with WHO and EC regulations. Other methods, such as denitrification (see Table 9.1), can also be used but are costly. Similarly, sewage can be more effectively treated to remove phosphorus and nitrogen. To reduce the amount of phosphate in waste water due to detergents requires the use of non-phosphate detergents coupled with regulations to limit industrial effluent.

The control of such point sources of sewage and detergent-rich industrial effluent is easier than dealing with diffuse sources, especially those related to artificial fertilizer use. In this instance the most effective way of curbing cultural eutrophication involves treating the cause rather than the symptoms. This requires a massive reduction in the use of artificial fertilizers and the adoption of arable farming practices that are conducive to nitrate retention within the soil. The recent changes to the Common Agricultural Policy of the EC, which require agricultural extensification to reduce food mountains, could help to combat the problem of cultural eutrophication in Europe if specific areas were selected for extensification, notably areas of intense cereal growing. As extensification is presently occurring on a voluntary basis, it is likely that more marginal, usually more upland, regions will be taken out of agriculture first, thus preserving the status quo elsewhere. Even if fertilizer applications were reduced forthwith, it is unlikely that the rate of cultural eutrophication in aquatic environments would decline immediately because of the recycling of nutrients from biomass and sediments. In the short term remedial measures will be necessary, but the only viable long-term solution is to tackle the primary causes of cultural eutrophication by limiting fertilizer use and employing adequate sewage treatment.

11.4   CONCLUSIONS

Acidification and cultural eutrophication are chiefly products of the post-Industrial Revolution period. They are the environmental price that society has paid for increasing energy consumption, high technology agriculture, rapid population growth and urbanization. Both are important political issues, as well as environmental issues, because they affect that most vital of resources—water. The two problems also create economic and amenity repercussions and both can be mitigated, to a certain extent, by short-term remedial measures. Where such measures have been adopted, there is evidence that acidification and cultural eutrophication can be halted and, in some instances, reversed.

Mitigation of these two problems requires longer-term measures that address the underpinning causes. Such measures are in keeping with sustainable rather than unsustainable, development (Chapter 2). Acidification, for example, will only be reduced if fossil fuel consumption is curtailed (energy policies are also required for sustainable development, Chapter 2) and if desulphurization (and denitrification) techniques are widely utilized in power-generating plants. Desulphurization is being implemented in parts of the developed world where economic growth can be continued without further increases in energy consumption. In fact, energy consumption has declined since the oil crisis of the 1970s. However, the developing world is beginning to industrialize and its use of fossil fuels is increasing so the efforts at energy saving, and by association a reduction in acidic emissions and greenhouse gases, in developed nations may be counteracted. This is inevitable as little effort has been invested in developing alternative energy sources. Thus the globally significant enhanced greenhouse effect (Chapters 3 and 4) is unlikely to be mitigated and the problem of acidification will shift into vulnerable parts of the developing world.

For cultural eutrophication there are two significant measures that can be instigated to curb and reverse the problem. Firstly, the use of nitrate fertilizers needs to be reduced. This has economic and hence political ramifications. However, in view of the excessive food production which occurs in the developed world and the policies of extensification that are in operation in the USA and the EC, it should not, if the political will is there, be too difficult to combine the two objectives. Whether this will occur remains to be seen and, as for so many of the environmental issues addressed in this book, the 1990s are likely to witness confrontation between environmental pressure groups on the one hand and politicians and policy makers on the other. Reconciliation between the two would be a significant improvement on the acrimony of the 1970s and 1980s. It is essential for the improvement of environmental quality, which is a major objective of sustainable development in the 21st century.

11.5 REFERENCES

Battarbee, R. W., Stevenson, A. C., Rippey, B., Fletcher, C., Natanski, J., Wik, M. and Flower, R. J. (1989) Causes of lake acidification in Galloway, south-west Scotland: a palaeoecological evaluation of the relative roles of atmospheric containment and catchment change for two acidified sites with non-afforested catchments. *Journal of Ecology*, **77**, 651–672.

Charles, D. F. (1990) Effects of acidic deposition on North American lakes: palaeolimnological evidence from diatoms and chrysophytes. *Philosophical Transactions of the Royal Society of London*, **B327**, 403–412.

Croll, B. T. and Hayes, C. R. (1988) Nitrate and water supplies in the United Kingdom. *Environmental Pollution*, **50**, 163–187.

Dudley, N. (1990) *Nitrates: The Threat to Food and Water*. London: Green Print.

Hesthagen, T. *et al.* (1989) The effects of acid precipitation on freshwater fish in Norway. In: Longhurst, J. W. S. (Ed.). *Acid Deposition: Sources, Effects and Controls*. London: British Museum Technical Communications, pp. 117–142.

Mannion, A. M. (1989a) Palaeoecological evidence for environmental change during the last 200 years. I. Biological data. *Progress in Physical Geography*, **13**, 23–46.

Mannion, A. M. (1989b) Palaeoecological evidence for environmental change during the last 200 years. II. Chemical data. *Progress in Physical Geography*, **13**, 192–215.

Mazurski, K. R. (1990) Industrial pollution: the threat to Polish forests. *Ambio*, **19**, 70–74.

McCormick, J. (1989) *Acid Earth: The Global Threat of Acid Pollution*. London: Earthscan.

Rosenberg, R., Elmgren, R., Fleischer, S., Jonsson, P., Persson, G. and Dahlin, H. (1990) Marine eutrophication: case studies in Sweden. *Ambio*, **19**, 102–108.

Skiba, U., Cresser, M. S., Derwent, R. G. and Fulty, D. W. (1989) Peat acidification in Scotland. *Nature (London)*, **337**, 68–69.

World Resources (1991) *World Resources 1990–91*. Washington DC: World Resources Institute.

11.6 FURTHER READING

Acid Magazine is issued twice yearly, free of charge. Contact: The Swedish Environmental Protection Agency, S-171 85 Solna, Sweden.

Follett, R. F. (Ed.) (1989) *Nitrogen Management and Ground Water Protection*. Amsterdam: Elsevier.

Saull, M. (1990) Nitrates in soil and water, Inside Science, 37. *New Scientist*, **127** (1734).

CHAPTER 12

Pollution and Development

M. S. Lowe and R. D. Thompson

12.1 INTRODUCTION

World-wide concern over pollution is currently being expressed through international treaties on acid rain and the enhanced greenhouse effect and the increasing availability of 'CFC-free' and 'environmentally friendly' household products. However, pollution problems are by no means exclusive to the late 20th century. For example, Shakespeare was aware of the 'foul and pestilent congregation of vapours' (*Hamlet*, Act Two, Scene Two) that characterized the atmosphere of Tudor times.

In this chapter, the sources and types of environmental pollutants are examined and three (of many) major contemporary pollution issues are emphasized. The chapter concludes with a summary of the necessary strategies for action which are essential to minimize the impact of pollution. These strategies are largely issues of green politics with international implications.

Pollution is defined as the deliberate or accidental contamination of the environment with waste from human activities. It includes the release of substances which harm the quality of air, water and soil, which destroy or perturb biogeochemical cycles (linking people to animals and plants) and which damage the health of humans (taking decades or generations to produce terminal diseases). Pollution is usually aggravated by growing populations and increased economic growth. These are the so-called 'bases of development' in society and are controlled largely by political decision-making and investment. Since the Industrial Revolution, environmental pollution has been regarded either as the 'price of progress' or as the result of developments to improve society. Consequently industrialization and urbanization have accelerated with a complete disregard for the well-being of the physical environment. In the 1990s, the costs of this disregard are evident, for example, in the form of 'holes' in the ozone layer and in the occurrence of 'dead' seas and rivers. It is apparent that the time is right for society to reassess its abuse of the environment although, before this can be done, it is important to understand contemporary pollution issues. This chapter examines this understanding, although limited space forces a concentration on the developed world and especially the UK. However, it

Environmental Issues in the 1990s. Edited by A. M. Mannion and S. R. Bowlby

must be remembered that the Third World is not immune from such environmental abuse (for example, Bhopal, India in 1984). Furthermore, the associated problems will be accentuated in future decades as these regions become more industrialized and urbanized.

Throughout history human beings have regarded the air as a depository for gaseous and particulate waste products. In the last century or so, the land surface and water bodies have been regarded as additional dumping grounds. Of course, the atmosphere has always been polluted to some degree with natural emissions, ranging from the injection of volcanic aerosols on a global scale to the localized foul-smelling air over swamps and marshes. Pollution due to human activities was initially a local problem following the invention of fire for cooking and heating in caves and buildings. By the 13th century, air pollution had become a serious regional problem with the smoke from burning sea coal over urban areas (such as London). In 1306 a Royal Proclamation prohibited artificers from using sea coal in their London furnaces because of its 'deleterious effects on health'.

Despite this early concern for the quality of air over European cities (and especially London), the situation deteriorated on a regional scale over the next three centuries. In 1661 John Evelyn (one of the first apostles of clean air) submitted his text *Fumifugium or the Smoke of London Dissipated* to King Charles II. It contained some graphic descriptions of pollution levels observed in London, comparing the city with Mount Etna, the Court of Vulcan and the suburbs of Hell. However, once again and despite the documented awareness of a chronic pollution problem, conditions continued to get worse. Furthermore, the progressive deterioration of air quality was intensified by the vast emissions of smoke and fly ash which accompanied the Industrial Revolution in north-west Europe.

Nineteenth century Britain attempted to control the emissions of offensive gases, smoke, grit and dust from specified industries (e.g. the Alkali Act of 1863, which was the first attempt at a comprehensive clean air act, and the 1875 Public Health Act). However, the severity of air pollution grew exponentially in the current century due to a number of major developments. The first of these was population expansion, as this generally increases the demand for fuel power and manufactured products. The second was industrial and technological growth as aerosol emissions increase with the expansion of existing plant capacity and with the establishment of new manufacturing activity to meet increased local demand and feed the vital export drive. The evolution of new technology such as nuclear reactors has also created new pollution hazards. Finally, an increased standard of living in recent decades (at least in the developed world) has resulted in greater manufacturing and energy consumption, which has increased the emission of particulate and gaseous pollutants. Also, the expansion of urbanization has concentrated people and industrial activity in close proximity so that the producers of pollutants sometimes exist alongside the receptors.

Today, pollution is evident on a global scale, particularly in the atmosphere where the enhanced greenhouse effect presents global warming hazards (Chapter 4). Acid rain also remains an immense problem, and was discussed in Chapter 11. Until recently, human beings have developed societies which completely abuse their physical environment. With experience and with continued population and industrial growth, the realization has dawned that society must regulate and protect the land, food and water supplies. However, despite the fact that the atmosphere is the real essence of life, the control of the quality of this resource has been neglected. Indeed, people have been misled by the fallacious concept that the atmosphere is vast and limitless and has an infinite capacity for the dilution and dispersion of endless waste emissions. Since the 1950s, increasing concern about environmental pollution has been mainly associated with the public outcry following disastrous events such as the infamous London smog of December 1952, which killed 4000 people, mainly from chronic bronchitis. The British Government immediately set up the Beaver Committee on Air Pollution, whose reports formed the basis of the 1956 Clean Air Act, which subsequently transformed the face and air of Britain.

Environmental pollution covers a wide range of solids, liquids and gases which can be described as disagreeable, noxious and toxic. Generally it includes any emission (natural or anthropogenic) of a kind and quality which is considered as objectionable and harmful. However, atmospheric contamination has received the greatest attention because aerosols and gases traverse national boundaries and place global environments in a precarious state (Chapter 4).

12.2   SOURCES AND TYPES OF ENVIRONMENTAL POLLUTANTS

12.2.1   Air pollution

It is apparent that atmospheric contamination has been an important local issue for about 2000 years. However, its serious implications on regional and global scales have only been recognized in recent decades, particularly since the 1970s when exponential projections of fossil fuel consumption predicted unacceptable concentrations of air pollution over the next century. Emission controls are vital to protect global environments. Unfortunately, successful controls have been difficult to introduce due to the complexity of the air pollution problem which results from the wide variety of sources and emissions involved in the production of pollutants (Maunder, 1969). There is a great deal of controversy related to the major sources of pollutants, particularly between domestic and industrial contributions.

Industry is a conspicuous and toxic offender which is most readily subjected to control in the design and siting of industrial plants and associated discharging points. For example, in the acid rain debate, coal-fired power stations (and associated sulphur dioxide emissions) are the main targets for significant reductions, whereas the emission of nitrogen and sulphur oxides from petrol engines continues unabated. It is now apparent that most of the suspended particulate matter (smoke) is emitted from inefficient domestic combustion, whereas over 90% of sulphur dioxide is from industrial and power station contributions.

The 1956 and 1968 Clean Air Acts in the UK were aimed at all coal-burning sources and consequently smoke production declined from 2.36×10^6 tonnes to 0.6×10^6 tonnes between 1951 and 1973 (Fig. 12.1). It should be noted that this significant decline took place despite a 10% increase in population and a 17% rise in annual gross energy consumption over the period. This indicates the success of the Clean Air Acts, which replaced domestic coal burning by cleaner gas and electricity. The sharp decline in industrial smoke (Fig. 12.1) was due to controlled emission *per se*, through new furnace designs, and reduced local concentrations through approved chimney heights. The Clean Air Acts were also responsible for a significant reduction

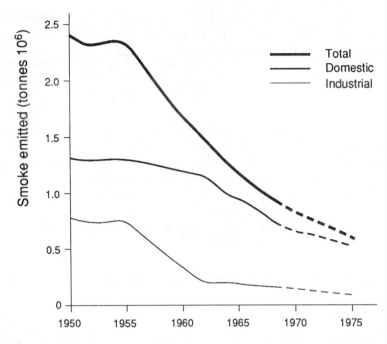

Figure 12.1   Emission of smoke in the United Kingdom, 1950–70. Adapted from Smith (1975) by permission of dti.

in sulphur dioxide emissions (about 20% between 1963 and 1975). However, emissions have decreased even more dramatically in the last two decades or so, from 3.0×10^6 tonnes in 1970 to 1.77×10^6 tonnes in 1984, although emission increased again to 1.93×10^6 tonnes in 1987 (Mason, 1990).

Petrol-powered motor vehicles are also important sources of air pollution and contribute about 35% of the total UK emissions (compared with 60% in the USA) due to the inefficiency with which the engine burns fuel. The main pollutants are carbon monoxide (92% of the total emitted), followed by hydrocarbons (4.5%), nitrous oxides (3%) and sulphur oxides (0.5%). Air pollution is composed of a complex range of emissions made up of mainly gases (90% of the total pollution) and particulate or liquid materials. The major types of pollutants are as described in the following sections.

Sulphur compounds are the most common form of pollution as sulphur is an impurity of coal and oil which is released into the air through combustion. However, two-thirds of the global total are produced by natural sources such as sea spray and especially bacteria and organic processes. Also, more than two-thirds of sulphur dioxide emissions occur in the northern hemisphere as more than 90% of combustion releases are located north of the equator. Sulphur dioxide has an atmospheric life span of only a few days as it is rapidly removed by acid rain wash-out (Chapter 11 and Mason, 1990) and absorption by vegetation.

Carbon monoxide is a colourless, odourless and lethal gas produced by the incomplete combustion of carbonaceous material, especially petrol-powered motor vehicles which emit about 80% of the total gas production. Carbon monoxide is a very stable, inert gas which is removed from the atmosphere by oxidation to carbon dioxide (i.e. it has an indirect greenhouse effect), bacterial metabolism and absorption by the oceans.

Carbon dioxide is a heavy colourless, odourless gas and is not truly a pollutant as it is essential for all life processes. Most (80%) of the gas is produced naturally by animal respiration, whereas the remainder is produced by the combustion of fossil fuels. It has been estimated that over 300×10^9 tonnes of the gas have been added to the atmosphere since the Industrial Revolution, leading to an enhanced greenhouse effect (Chapter 4). About half of the gas produced remains in the atmosphere, whereas about 14% is absorbed by sea water (the oceans are assumed to contain about 60 times as much carbon dioxide as the atmosphere) and 36% enters the biosphere, mainly into biomass and soil humus.

Hydrocarbons are mainly produced naturally from decomposing swamps and marshes (1.6×10^9 tonnes annually), compared with a mere 90×10^6 tonnes per year from motor vehicles. However, the latter reactive hydrocarbons (especially alkenes and benzene compounds) undergo important photochemical changes when exposed to solar radiation and create photochemical smog. This is a relatively new air pollution problem (first observed in Los Angeles in 1942) and forms peroxyacetyl nitrate (PAN). Figure 12.2 shows the distribution and effects of photochemical pollution in California between 1961 and 1963.

Oxides of nitrogen form a pungent, harmful gas when oxidized to nitrogen oxide. The oxides are produced naturally by soil bacteria, lightning and volcanoes, and artificially by fertilizer and explosives factories and especially from power stations and the exhausts of motor vehicles. Nitrogen oxide is an important ingredient for the development of photochemical smog and PAN (along with reactive hydrocarbons) and is a secondary contributor to the acid rain problem (Mason, 1990). Nitrous oxides also contribute to the enhanced greenhouse effect (Chapter 4).

Particulate matter comprises about 10% of atmospheric pollutants in the form of solid and liquid particles. Coarse particles (larger than 10 μm) consist mainly of fly ash and dust, which act as a shield to incoming solar radiation (Chapter 4). Smaller particles (less than 5 μm) remain suspended in the atmosphere as smoke and haze, although they can combine with radiation fog to produce the infamous 'pea-soup' smogs.

12.2.2 Pollution of land

Since the Industrial Revolution, pollution of the landscape has been associated with the accumulation of unwanted waste materials and major environmental disturbances resulting from accelerated urbanization and industrialization. There are also aesthetic changes such as the spoiling of the natural beauty of the landscape with the gross ugliness of structures such as warehouses, storage tanks, cooling towers and motorways. This section will concentrate on the environmental contamination associated with rubbish dumps and spoil heaps, although a wide range of waste (of varying toxicity) is involved from industrial, agricultural and domestic sources.

The annual production of domestic rubbish in the UK exceeds 300 kg per person and totals over 18×10^6 tonnes, plus another 20×10^6 tonnes from commercial and industrial sources. Since the 1950s, domestic waste has changed in content, with the

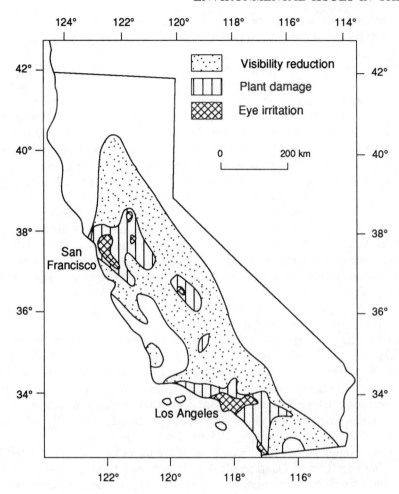

Figure 12.2   Distribution and effects of photochemical pollution in California, USA, 1961–63. Adapted from Leighton (1966) by permission of the American Geographical Society. Copyright © 1966 AGU.

disposal of less cinders and ash but more packaging materials (especially plastics). The weight and density of the rubbish has decreased but the total volume has increased by over 45% since 1955. This means more dustbins, bigger containers and more frequent collections, resulting in spiralling costs at a time of fiscal expediency in local authorities' budgets. A major disposal problem is also associated with the dumping of over 700 000 cars annually in the UK, particularly the 70 000 or so vehicles which are simply abandoned in gardens, streets and lay-bys. These become local eyesores and contravene the 1967 Civic Amenities Act.

Currently, 90% of all household rubbish is tipped as landfill in controlled and uncontrolled sites. The former are in a minority, but comply with a strict Department of the Environment code of practice on the depth permitted (2.4 m or 1.2 m if pulverized refuse), the gradient of tip faces (less than one in three) and the sealing required at the end of each day (faces and flanks to be covered with 23 cm of suitable sealant). The tipping of effluent is also controlled so that underground water supplies are not contaminated. Subsoil filtering assists natural purification, although toxic leachate can travel considerable distances and cause severe pollution of groundwater in chalk and fissured rock (see next section). At the end of a finite tipping period, the area is restored to amenity use or agricultural production. For example, between the early 1950s and early 1970s, 11 000 hectares of rubbish dumps in the UK were restored to these uses.

12.2.3   Pollution of water

Water pollution is a serious issue discussed in detail in Chapters 6 and 13. It will clearly remain a problem in the 1990s and beyond, and can only be tackled by massive investments in water cleansing programmes, which will involve considerable costs to water consumers. Water pollution involves both thermal and chemical changes, although the latter pollutants have traditionally received the most attention. Thermal pollution is unwanted heat accumulation in a lake or river when water (used to cool heat exchangers at power stations) is returned to the natural water body. For every one kilowatt hour of electricity generated in a modern coal-fired power station, about two kilowatt hours must be dissipated by water used to cool the heat exchangers. This increases the water temperature by 5–8°C when the cooling water is returned to a river or lake. Nuclear power plants are even greater offenders, using about three kilowatt hours for every kilowatt hour generated, which increases the water temperature by over 10°C.

The effects of adding heat to water are very complex, but increases of about 2°C are known to seriously affect fish and other aquatic life. Increasing water temperature also raises the metabolic rate in organisms (which doubles with every 10°C increase in body temperature) and also increases their oxygen requirements, although there is less oxygen available at higher temperatures. For example, dissolved oxygen decreases by over 17% with an increase in water temperature from 20 to 30°C. However, the consequences of thermal pollution appear invisible and innocuous when compared with the more obvious disturbances caused by chemical pollutants in lakes and rivers. The sources of chemical pollution are very varied and include domestic sewage, industrial effluent, acid drainage and toxic salts from mines, pesticides and fertilizers, and farm effluent (especially from slurries and silage).

As with air pollution, water contamination by human activities is not a new problem and the first UK legislation to deal with it appeared in 1388. However, for many centuries river pollution was the accepted consequence of urbanization and industrialization as the first effective legislation giving powers to control the dumping of sewage and industrial effluent was the River Pollution Act of 1876. This Act remained in force for the next 75 years and further acts were then passed in 1951 and 1961 which prohibited the discharge of poisonous, noxious or polluting matter into non-tidal rivers and specified tidal estuaries. However, pollution control was mainly ineffective, due to the decentralized administration of the water industry into 1700 public and private undertakings. Consequently, pollution legislation was strengthened by the 1974 reorganization of the water industry into ten regional authorities which unified control, reduced the conflict of interest and injected greater efficiency into water management. Today's regional privatization of the water industry will (hopefully) continue the anti-pollution pledge of all concerned and raise water standards to European Community (EC) requirements. For example, an EC directive on dangerous substances in abstracted river water limits lindane (used in industrial wood processing) to 0.1 μg l^{-1}. Failure to comply with all EC directives will result in heavy fines. However, the high cost of improvements in quality will be passed on to the consumer, with sharp rises in water rates expected in the 1990s.

12.3 CONTEMPORARY POLLUTION ISSUES

12.3.1 Lead poisoning

The main supply of lead in the atmosphere is associated with car exhaust emissions originating from the anti-knock agents added to petrol. Secondary anthropogenic sources include coal burning and oil smelters producing ferrous and non-ferrous metals; natural emissions from volcanoes produce negligible amounts of total lead. Each gallon of petrol contains about 2 g of lead and between 25 and 50% of this becomes airborne in the form of fairly stable lead halides and oxides. In 1970, car exhaust emissions of airborne lead in Britain approached 10 000 tonnes, compared with about 100 tonnes from coal burning. In addition, more than 200 tonnes of this petrol-derived lead were particularly dangerous organo-lead compounds.

The lead content of city dust tends to be about 1%, although this can approach 5% in busy urban areas. This is a serious situation as the absorption of lead in the bloodstream harms the central nervous system and is particularly likely to cause brain damage in children. Adults in cities are also at risk from exhaust fumes, which can cause toxic lead encephalitis (inflammation of the brain) which causes depression, headache and fatigue. Fortunately, treatment for lead poisoning with calcium EDTA rapidly leads to a cure or improvement in 85% of patients tested. In the 1970s, research showed evidence of airborne lead poisoning in the UK and caused considerable anxiety, leading to a number of important lead-free petrol directives in the 1980s. Consequently, atmospheric lead has been reduced significantly since peak levels were recorded in 1971.

Lead in drinking water is another serious environmental issue, particularly associated with long lengths of lead piping in older (pre-1944) dwellings. Between 1975 and 1976 the Department of the Environment in Britain initiated a study of daytime water samples from lead pipes in rigorously selected households. It revealed that the 1975 EC limit of 0.05 mg l^{-1} was exceeded in 10% of the sampled households, equivalent to 1.9 million households across the country.

12.3.2 Aquifer contamination

This is a serious hazard in the densely populated south-east of England, which relies heavily on groundwater for its public water supply. The sources of contamination are varied and include sewage effluent from septic tanks and cesspits, leachate from rubbish dumps and mine waste tips, agricultural and industrial percolate (too oily or toxic for the sewers) and infiltration from farmland treated with fertilizers. Aquifer contamination by leachate is only a hazard in susceptible geological environments, particularly those with high water-tables in permeable bedrock (limestones) and superficial deposits (sand and gravel). Severe contamination occurs when, with a thin permeable layer on impermeable clay, the rubbish is in direct contact with the groundwater. These sites should be avoided or lined with a heavy-duty impermeable membrane.

Leachate transmission can be slow (which will assist chemical degradation) or can be extremely rapid. For example, the pollution of deep chalk wells in north Kent was recorded about three or four days after sewers overflowed following heavy rain. Furthermore, its persistence meant that the wells had to be pumped for 12 days before the bacterial pollution was eradicated. The depth of infiltration is also important for chemical degradation and bacterial removal. To ensure adequate removal and purification, water in the saturated zone should percolate at least 30 metres of unfissured rock, compared with a mere three metres in the unsaturated zone. It should be noted that rubbish tips are self-purifying over time. For example, biochemical oxygen demand and ammoniacal nitrogen levels of 6000 and 7000 mg l^{-1} respectively, in the first winter after tipping can reduce to 50 and 40 mg l^{-1} after two years. The degradation of leachate by natural processes is mainly associated with ionic exchange with the bedrock.

Another serious aquifer contamination problem is found when nitrates from agricultural areas (with a significant fertilizer base) percolate into the groundwater. For example, in east Suffolk, 41% of the wells monitored had nitrate concentrations in excess of 88 mg l^{-1}, more than twice the universally recognized safe limit. This high level of nitrate pollution in rivers and aquifers is dangerous because it is generally considered that such concentrations cause methaemoglobinaemia (the so-called water-well cyantosis) in babies. Also, carcinogens can be produced in the body from nitrates, and, in parts of Nigeria where nitrate concentrations exceed 90 mg l^{-1}, the death rate from gastric cancer is abnormally high. More research is needed to accurately correlate leachates with pathogens and carcinogens, but there is little doubt that great care must be taken with the abstraction of river water and groundwater.

12.3.3 Ionizing radiation

World-wide publicity and public anxiety followed the nuclear accident at the Chernobyl nuclear power station in the Soviet Union on Saturday 26 April 1986. This highlights the immense significance of ionizing radiation as a contemporary pollution issue. During the first few weeks following the Chernobyl blast, about 30 people died either as a result of the immediate impact or from radiation burns and radiation sickness brought about by close contact with large amounts of radiation from the split core of the reactor. Although such a death toll is itself unacceptable, Chernobyl drew attention to more sinister factors. In assessing the likely damage as a result of the accident there appeared to be a great deal of confusion and speculation, rather than hard facts and information. It was unclear, for example, how far radioactive contamination would spread, how long it would remain in the atmosphere, or what lasting effects this pollution would have. This indeed is the real problem with nuclear radioactivity: there are simply too many unanswered questions. Radioactivity occurs naturally, with the greatest amount of radiation coming from rocks, soil, water and even the human body itself. A great deal of radioactivity is also currently generated artificially through human activities such as food processing and medicine. However, it is primarily from the nuclear power and the nuclear defence and research industries that the bulk of radioactivity is released. Most arises from the processing and reprocessing of nuclear fuel.

The use of radioactive materials by the nuclear power industry leads not only to the generation of power, but also to the creation of a large amount of gaseous, liquid and solid wastes, many of which (such as plutonium) are not found in nature. More significantly, it is extremely difficult to destroy these substances once they have been created. For this reason (among others), nuclear radioactivity, and in particular its disposal, is likely to remain a pressing environmental issue throughout the 1990s. For every year that a one gigawatt nuclear reactor is operational, 5×10^9 curies (or 70×10^6 times more radioactivity than was present in the original fuel) are created. Although much of this radioactivity is lost fairly quickly, some will remain for an extremely long time. For example, 470 curies will still be present after 10 000 years (Bunyard, 1988).

Clearly then, and even ignoring the severe dangers of pollution from accidents such as Chernobyl, the disposal of radioactive wastes is itself a significant pollution problem. At Sellafield in Cumbria, for example, three varieties of waste (high, intermediate and low level) are variously 'disposed of' either by their incarceration (for high level wastes) in stainless-steel and concrete-sided containers which blot and mar the landscape, or simply through their discharge via a pipeline two miles into the Irish Sea. Speculation surrounding the environmental damage incurred by this intentional discharge from the plant is well known, ranging from connections drawn between abnormally high childhood cancer rates (six or more times the national average in the vicinity of the plant) to the contamination of foodstuffs through fish stocks and dairy products.

The story is never as clear-cut as it may at first seem, for the pro-nuclear lobby is keen to emphasize the greater environmental disturbances, and more specifically the air pollution problems, resulting from traditional power generation through coal burning (Section 12.2) to support their cause. Indeed, a number of environmentalists have been known to specifically support nuclear power for its 'environmentally friendly' nature. Despite these protestations, however, it is true to say that the nuclear power industry brings about far more thermal pollution than that found for coal-fired plants (Section 12.2). There is clearly still a long way to go along the road to pollution-free energy generation, and indeed a strategy of conservation may be more appropriate (see next section). However, the logic of supporting one form of environmental damage by denigrating another is clearly outdated.

12.4 STRATEGIES FOR ACTION

This chapter has illustrated the far-reaching and global consequences of continued ignorance of the pollution problem in the 1990s and beyond. It has shown the relationship between economic and social development and pollution levels as well as drawing attention to the main sources and types of environmental pollutants. Case studies of three contemporary pollution issues in the preceding section have pointed to the complexity of the pollution and development question. They also have highlighted the difficulties associated with the development and maintenance of any pollution control legislation or pollution policy.

This issue will be addressed in three main parts. Firstly, the difficulty of developing any form of pollution control strategy when it is all too easy to estimate the costs of such controls but rather more difficult to assess (at least in monetary terms) the benefits of protecting our environment. A secondary issue under this heading relates not only to what the costs of pollution are, but also who should bear these costs. Secondly, the problems of integrating strategies for ameliorating the impact of different pollutants will be addressed. An additional aim here is to highlight the fact that pollution strategies should be closely integrated and should include all geographical scales from local to global. Progress already made along these fronts will be referred to. Finally, and in summary, the chapter will be brought full circle with a discussion of the relationship between pollution and development. This relationship must be broken before progress can really begin in any of the proposed strategic directions.

12.4.1 Costs and benefits of pollution control

It is difficult to place a cost on the limitation of pollution damage and yet it is all too easy to argue that cost is prohibitive in attempting to maintain the status quo. The costs of action are relatively easy to estimate as new forms of waste treatment and disposal and investment in technology to prevent the emission of environmentally damaging substances can be priced. In contrast, the costs of inaction are much less easily calculated. What cost do we place on clean water in our rivers or clean air in our lungs? Equally, 'it is difficult to put a precise monetary value on acidified forests and lakes, dead animals and plants, reduced rural vistas, corroded buildings or impaired human health' (McCormick, 1989). Also important is the fact that the costs of pollution control measures often fall on particular industries, which thus have a strong political interest in maintaining the current situation. Porritt (1984) points to a classic example of this complexity for lead in petrol. After several years of campaigning by various environmental pressure groups, the British Government finally conceded to remove lead from petrol 'before 1990', but as yet very little has been done to implement this decision.

The current privatization frenzy in the UK also provides a case in point. When water as a public utility was privatized, the costs of maintaining water quality fell to the private sector rather than the government. In reality, however, such costs are now being added to the bills of the consumer while the private sector continues to reap the benefits of its financial investment. In addition to the difficulties associated with evaluating and distributing the costs and benefits of pollution control measures, there is also the problem of what Porritt (1984) calls the 'ecological time lag'. In simple terms, pollution problems are often not recognized until it is much too late to do anything about them, let alone decide who should pay for the privilege. The time is right for the development of a sensible economic, social and environmental accounting procedure to be applied to pollution issues. However, the difficulties of deciding who pays and how much, as well as enforcing such decisions, remain immense.

12.4.2 Integrated pollution control strategies

For too long, progress in pollution prevention and control has proceeded by singular, isolated or piecemeal legislation. For example, little is known of the combined effects of different pollutants, and yet while time passes through the development of international agreements on acid rain or nitrates, the combined effects of such damaging materials are probably already doing their worst. Of course, the authors would not wish to suggest that such strategies are not worthwhile. Rather, more time and effort should be spent on evaluating exactly what the combined effects of different pollutants are likely to be and on how society can do its utmost to minimize their damage. Likewise, air, land and water pollution should be tackled simultaneously. A similar integrated strategy is also required in the scale of pollution amelioration and legislation. Many local councils have developed their own 'nuclear-free zones' and environmental groups have produced their own version of local pollution strategies only to find that, within a few miles, public and private industrial establishments may be discharging all manner of dangerous substances into the air or the sea. The authors do not wish to decry these local and grassroots movements; instead they would like to see much greater co-operation and co-ordination on global and local environmental matters.

Happily, there has been some progress along this front. In particular, the EC has, since its inception in 1973, been keen to maintain a consistent environmental stance and has published a series of directives on environmental issues. The 1973 Programme of Action on the Environment proposed action to reduce pollution and nuisances, to deal with environmental problems caused by the depletion of natural resources, to promote awareness of environmental problems and education and to improve the natural and urban environment. The EC has become a prime mover on international co-operation on pollution issues. Taking water pollution as an example, its control in Britain is maintained both through acts of parliament and through English common law. The most important acts relating to water pollution are the 1973 Water Act and the 1974 Control of Pollution Act, both of which contain vague statements on the maintenance of a certain level of cleanliness and 'wholesomeness' of freshwater.

As a component of its environmental policy, the EC has issued a number of directives on water quality. These include legislation on the quality of drinking and bathing water as well as ceilings on the amounts of pollutants that can be discharged by EC members into Europe's seas and rivers. Notwithstanding this legislation, there is once again the thorny issue of enforcement. Britain has continually flouted European legislation on water quality and has even been summoned to the European Court of Human Justice for its non-compliance with EC directives. Despite all this, there is still a long way to go in bringing UK water quality in line with EC standards (Dudley, 1990).

Such initiatives do at least represent a step in the right direction for co-operation and collaboration to develop an integrated pollution policy. Indeed, as pollution has multiplied in its scale and severity, it has been increasingly recognized that a problem of such clear global proportions clearly requires global solutions. The United Nations (UN) and its agencies, such as the United Nations Environmental Programme (UNEP), have been instrumental in the development of such a global programme. The UN has played a significant role in the discussion of environmental problems and in co-operation over environmental difficulties.

UNEP was established in 1973 following the 1972 United Nations Conference on the Human Environment in Stockholm. This conference was significant in that it represented a turning point for global collaboration on issues of environmental importance. In 1990 the UN still had a central role to play in the integration of environmental policy. Indeed, 75 of its member countries recently agreed to work towards a world climate convention which will maintain a global commitment on controlling greenhouse gas emissions. Geographers are advantaged in their ability to see pollution in terms of its massive geographical scale and to appreciate its spatial and temporal problems. It is important, then, to encourage and develop all-encompassing international (and even global) agreements on pollution levels. It is at the same time important not to neglect the potential and far-reaching impact of local and even individual actions.

12.4.3 Pollution and development: a necessary relationship?

This chapter has emphasized the necessary relationship between economic and social development and increasing levels of pollution. However, the increasingly polluted world is a rather narrow view of 'development' (as ultimately the costs of a ruined environment will be borne by everyone) and a number of other issues concerning the pollution and development relationship should be mentioned here. Firstly, it is often stated that 'pollution is the price of progress' (French, 1990) and that in any pollution debate, we must decide in favour of either economics or the environment. This polarized view must at best be regarded as short-sighted, and at worst be seen as fundamentally flawed. Pollution control measures and pollution prevention strategies are often opposed on the grounds of cost. This equation is only relevant in the short term and must then be seen as rather limited financial expediency which will ultimately founder in the face of the economic costs of long term environmental damage. In any case, there are documented examples of environmental strategies having economic as well as social and ecological benefits (Porritt, 1984).

Secondly, despite the conspicuous association between pollution and First World development, it is clear that environmental pollution is by no means a purely capitalist phenomenon. The demolition of the Berlin Wall in 1990 revealed the critical environmental difficulties of the so-called eastern bloc. In particular, East Germany has per capita sulphur dioxide emission rates higher than anywhere else on earth. Clearly, the thawing of the 'cold war' offers tremendous potential for environmental co-operation on a scale never anticipated in the past. Finally, pollution and development problems are not exclusive to the First World. There are obvious political difficulties associated with a global pollution strategy which restricts Third World development on the grounds of pollution control, only a few years after First World multinational corporations have stripped many Third World countries of their natural resources. Furthermore, they have left in their wake a series of pollution problems which, at the time, were viewed as the 'acceptable price of progress'. Such thorny issues clearly demand a rethinking of current strategies of Third World development.

There are some signs of progress along the lines of the strategies discussed here. Global co-operation on issues of environmental protection continues apace, while in the UK the government is currently refining an Environmental Protection Bill, which will include co-ordinated action on all levels and types of pollution. Also, an

Environment White Paper was published in the autumn of 1990. In the meantime, industries and retailers world-wide are keen to display their environmental credentials through 'green' advertising, products and packaging. It would appear, then, that questions of pollution damage and strategies for action have quickly reached the top of the political, social and commercial agenda. However, and in reality of course, all such worthy environmental measures are simply tinkering with the edges. It will take far more than 'ozone-friendly' products, tax concessions on unleaded petrol and even international agreements to get to the root of the global pollution problem. Ultimately, the only way to halt the pollution problem is to generate less pollution by reducing the amounts of resources used. This includes measures such as the promotion of energy efficiency, the use of non-polluting renewable resources, recycling and a minimized dependence on motor vehicles. In reality such actions require a fundamental rethink of patterns of production and consumption, of lifestyles and material aspirations. There are evidently 'costs' associated with such dramatic changes, rendering them steps which politicians, industries and the public may not yet be ready to take. If these are considered seriously, however, they could make a world of difference in every respect!

12.5 REFERENCES

Bunyard, P. (1988) *Health Guide for the Nuclear Age*. London: Macmillan.
Dudley, N. (1990) *Nitrates: The Threat to Food and Water*. London: Merlin Press.
French, H. F. (1990) *Cleaning the Air: A Global Agenda*. Washington: Worldwatch Institute.
Leighton, P. A. (1966) Geographical aspects of air pollution. *Geographical Review*, **56**, 151–174.
Mason, B. J. (1990) Acid rain—cause and consequence. *Weather*, **45**, 70–79.
Maunder, W. J. (1969) *Pollution*. Victoria: University of Victoria, British Columbia, p. 16.
McCormick, J. (1989) *Acid Earth: The Global Threat of Acid Pollution*. London: Earthscan.
Porritt, J. (1984) *Seeing Green: The Politics of Ecology Explained*. Oxford: Basil Blackwell.
Smith, K. (1975) *Principles of Applied Climatology*. New York: McGraw-Hill, p. 67.

12.6 FURTHER READING

Bridgman, H. (1990) *Global Air Pollution Problems for the 1990s*. London: Belhaven Press.
Elsom, D. (1987) *Atmospheric Pollution*. Oxford: Basil Blackwell.
Kemp, D. D. (1990) *Global Environmental Issues: A Climatological Approach*. London: Routledge.
World Commission on Environment and Development (1987) *Our Common Future*. Oxford: Oxford University Press.

CHAPTER 13

Use and Abuse of Wetlands

J. M. R. Hughes

13.1 WHAT IS A WETLAND?

Wetlands include a range of inland, coastal and marine ecosystems that have been formed, and whose characteristics and processes are dominated, by water. A number of overlapping definitions for wetlands have been put forward in the last two decades, many having been generated for legal as well as geographical reasons. The most versatile definition is that used by the 1971 Ramsar Convention on wetlands, which states that wetlands are 'areas of marsh, fen, peatland or water, whether natural or artificial, permanent or temporary, with water that is static or flowing, fresh, brackish or salt, including areas of marine waters, the depth of which at low tide does not exceed six metres'. They may include 'riparian (associated with a river bank) and coastal zones adjacent to the wetlands or islands or bodies of marine water deeper than six metres at low tide lying within'.

Thus the range of wetland types is very large, and a widely used wetland classification is given in Table 13.1. Figure 13.1 summarizes wetlands into seven landscape units: estuaries, open coasts, floodplains, freshwater marshes, lakes, peatlands and swamp forests. These units can be used to define the planning framework for wetland conservation.

Table 13.1 Wetland classification. Adapted from Dugan (1990) by permission of IUCN

Salt water:
  Marine
    (a) Subtidal
        permanent unvegetated waters less than 6 m depth at low tide
        subtidal aquatic vegetation, e.g. sea grasses
        coral reefs
    (b) Intertidal
        rocky marine shores
        shores with mobile stones and shingle
        mobile unvegetated mud, sand or mudflats
        vegetated sediments, e.g. salt-marshes, mangroves

(continued overleaf)

Environmental Issues in the 1990s. Edited by A. M. Mannion and S. R. Bowlby
© 1992 John Wiley & Sons Ltd

Table 13.1 (*continued*)

Salt water (*continued*)
  Estuarine
    (a) Subtidal
      estuarine waters
    (b) Intertidal
      mud, sand or salt flats
      marshes, e.g. salt meadows, raised salt marshes
      forested wetlands, e.g. nipa swamps, tidal freshwater swamp forest
  Lagoonal
    brackish/saline lagoons with narrow connections to the sea
  Salt lake
    permanent or seasonal, brackish/saline/alkaline lakes, flats and marshes

Freshwater:
  Riverine
    (a) Perennial
      rivers and streams, including waterfalls
      inland deltas
    (b) Temporary
      seasonal and irregular rivers and streams
      riverine floodplains
  Lacustrine
    (a) Permanent
      freshwater lakes (>8 ha)
      freshwater ponds (<8 ha)
    (b) Seasonal
      freshwater lakes (>8 ha)
  Palustrine
    (a) Emergent
      permanent freshwater marshes and swamps on inorganic soils
      permanent peat-forming freshwater swamps, e.g. tropical upland valley swamps,
      seasonally freshwater marshes on inorganic soils e.g. pot holes, dambos, sloughs
      peatlands and fens
      alpine and polar wetlands
      freshwater springs and oases
      volcanic fumaroles
    (b) Forested
      shrub swamps
      freshwater swamps forest
      forested peatlands

Artificial wetlands:
  Aquaculture
    fish ponds, shrimp ponds
  Agriculture
    farm ponds
    irrigated land and irrigation channels, e.g. rice fields, canals
  Salt exploitation
    salt pans and salines
  Urban/industrial
    sewage farms, settling ponds, gravel pits
  Water storage
    reservoirs, hydro-dams

E **Peatlands** : these form under conditions of low temperature, low nutrient supply, waterlogging and oxygen deficiency, where decomposition of dead plant matter is retarded and it accumulates as peat. Peatlands cover 500 million hectares in temperate and tropical environments

D **Freshwater marshes** : where groundwater, surface springs, streams or runoff cause frequent flooding or semi-permanent shallow water, freshwater marshes are common. *Typha*, *Phragmites* and *Cyperus papyrus* are well known marsh species

C **Floodplains** : periodically flooded land along a river channel, e.g. flooded forests, marshes, oxbow lakes. Many flood plains are exceptionally productive, e.g. the inner Niger delta in Mali, and support waterbirds and wildlife as well as being economically important

A **Estuaries** : from where river mouth widens into the sea. Salinity is intermediate between salt and fresh ; tides regulate biotic system. These are very productive wetlands which support a range of habitats, e.g. salt marshes, intertidal mud and sandflats, mangroves and nipa palm

F **Lakes** : form under a variety of processes including folding and faulting, e.g. Lake Baikal, glacial action, e.g. cirque lakes, permafrost, e.g. Prairie Potholes in the USA. The highest productivity occurs on lake margins and as far as light penetration supports rooted vegetation. Lakes trap overland flow and stream flow and thus influence water quality, trap sediments and moderate flood peaks

G **Swamp forest** : develop around lakes and floodplains where water is present for long periods. Swamp forests vary with geographical location, e.g. in South-east Asia and Australia many swamp forests are dominated by *Melaleuca*, whereas in the USA species such as red maple, larch and black spruce dominate. In Indonesia, swamp forest covers over 17 million hectares

B **Open coasts** : not influenced by river water or lagoons. Open coastal wetlands support a diversity of habits such as mangroves and mudflats

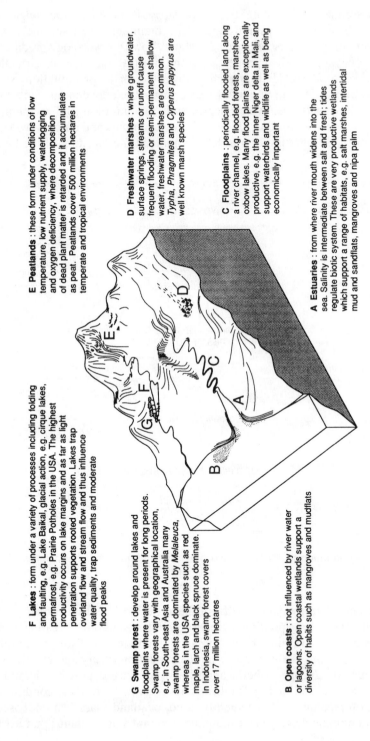

Figure 13.1   Wetland landscape units.   Adapted from Dugan (1990) by permission of IUCN.

13.2   WETLAND HYDROLOGY AND ECOLOGY

Mitsch and Gosselink (1986) state that 'hydrology is probably the single most important determinant for the establishment and maintenance of specific types of wetlands and wetland processes'. Figure 13.2 shows how the hydrology modifies and determines chemical and physical properties of the substratum, which in turn allow a specific ecosystem response. Hydrological pathways such as precipitation, surface run-off, groundwater, tides and flooding rivers transport energy and nutrients to and from wetlands. Flow patterns, water depth, and the duration and frequency of flooding, which are the results of all of the hydrological inputs and outputs, will affect the soil biochemistry and are important determinants in the selection of the biota of wetlands.

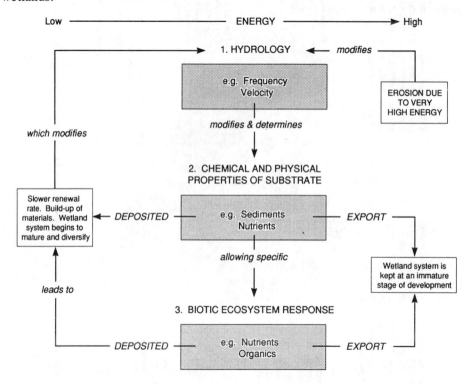

Figure 13.2   Direct and indirect effects of hydrology on wetlands. Adapted from Mitsch and Gosselink (1986).

The mechanisms whereby peatlands form illustrate the importance of hydrology. Peatlands occur where precipitation exceeds evapotranspiration, creating a water surplus. Additionally, there must be a drainage impediment that prevents drainage of the surplus water, and there must be an accumulation of organic matter. Large expanses of northern Europe, Siberia and North America experience humid conditions with flat terrain and high groundwater levels, which are ideal conditions for the formation of peatlands. A frequently used peatland classification based on hydrological and mineral conditions, exemplifies the range of peatlands that exist:

(a) Minerotrophic peatlands receive water that has passed through mineral soil. They have a high groundwater level and occupy a low point of relief in a basin.
(b) Ombrotrophic peatlands (i.e. rain-fed) are true raised bogs that have developed peat layers higher than their surroundings. They receive minerals and nutrients exclusively by precipitation.
(c) Transitional peatlands are intermediate between mineral-nourished (minero-trophic) and precipitation-dominated (ombrotrophic) peatlands.

Temporal variability in wetland hydrology is referred to as the hydroperiod (Mitsch and Gosselink, 1986). This is the seasonal pattern of the water level which integrates inflows and outflows of water, and it is influenced by the physical features of the terrain and proximity to other water bodies. For example, a coastal salt-marsh has a hydroperiod of semidiurnal (twice in 24 hours) flooding and recession superimposed on a twice monthly pattern of spring and ebb tides. The annual fluctuation of a tropical floodplain forest such as that of the Amazon River at Manaus, has an enormous seasonal fluctuation of almost 8 m. In contrast, the below ground water levels of peatlands have hydroperiods with less pronounced seasonal fluctuations than other wetlands.

Hydrology has various specific effects on the ecology of wetlands (Mitsch and Gosselink, 1986). Firstly, hydrology leads to a unique vegetation and animal composition and can limit or enhance species richness. There are many thousands of flowering plants on the earth, but only a relatively small number have adapted to waterlogged soils. Thus wetlands support water-tolerant species in both freshwater and saltwater conditions. For many wetlands, long durations of flooding produce a lower species diversity than do less frequent floods. When rivers flood riparian wetlands, or when tides rise and fall in salt-marshes, erosion, scouring and sediment deposition can allow diverse habitats to develop.

Secondly, the primary productivity (the total amount of organic matter or biomass formed by photosynthesis) of wetlands is enhanced by flowing conditions and a variable hydroperiod, and can be depressed by stagnant conditions. For example, saltwater tidal marshes which are subject to frequent tidal inundations are more productive (have a higher biomass) than those that are only occasionally inundated. Freshwater tidal marshes may have a higher productivity because they avoid the stress of saline soils, and peatlands with flow-through hydrology have much higher productivity than stagnant raised bogs. Thus variable hydrology often leads to a high diversity in wetlands and has given some wetlands the distinction of being among the most productive ecosystems on earth.

Thirdly, the hydrology controls organic accumulation in wetlands through its influence on primary productivity, decomposition and the export of particulate organic matter. A study of an alluvial swamp in North Carolina, USA, found that the decomposition of plant litter was fastest in the river, slower on the wet swamp floor, and slowest on a dry levee (Mitsch and Gosselink, 1986). A higher rate of organic export is to be expected from wetlands that have a considerable flow-through of water. Thus riparian wetlands often supply streams with large amounts of organic detritus such as leaves, branches and bark. Finally, hydrology affects nutrient cycling and availability by controlling the outflow of water.

To summarize some of the ideas just stated, it is useful to examine mangrove

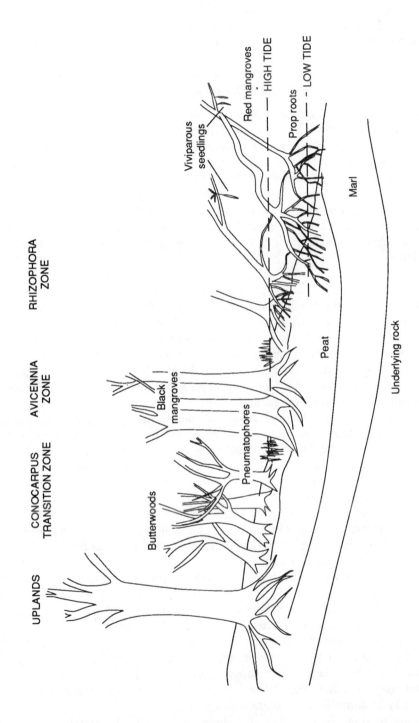

Figure 13.3   Zonation of Florida mangrove wetlands.   Adapted from Mitsch and Gosselink (1986).

ecosystems in terms of their zonation and productivity. Mangrove wetlands are the tropical and subtropical equivalents of temperate salt-marshes. They are forests which tolerate salt and occupy an intertidal, muddy niche. There are about 60 species of mangrove tree and shrub and over 2000 species of fish, invertebrates and epiphytic plants that are dependent on mangroves for their survival (Maltby, 1986). Mangrove wetlands are extremely varied, ranging from giant closed forests of red mangrove (*Rhizophora*) and black mangrove (*Avicennia*) growing to 50 metres high in Brazil, Venezuela and Ecuador, to stunted shrubs less than one metre high at the extremes of their distribution (such as Florida and the Japanese Pacific islands). The greatest concentration of mangroves is in the Indian Ocean–West Pacific region, where there is about one-fifth of the world's mangroves. Within this region, the largest remaining contiguous area is the Sundarbans forest, which covers nearly a million hectares of the Ganges Delta.

Mangrove wetlands have subzones (Fig. 13.3), and red mangroves are often pioneer species growing in the permanently flooded zone below low water. Towards high tide level, red mangroves give way to black mangroves, creating a zonation of vegetation. Such a zonation reflects a plant succession, and, once established, the horizontal prop roots, which anchor the trees to the soft mud, trap more sediment. This process encourages the 'pioneer' species to move seawards, which accelerates shore development and ensures coastal stability.

Table 13.2   Primary productivity and litter fall measurements for two types of mangrove wetlands. Adapted from Mitsch and Gosselink (1986)

	Riverine mangrove wetland	Basin mangrove wetland
Net daily primary productivity (kcal m^{-2})	57	25
Daily litter production (kcal m^{-2})	14±2	9.0±0.4

The primary productivity of mangrove wetlands is dependent on tidal and run-off factors and chemical conditions. Table 13.2 shows that the productivity and litter production is highest for riverine mangrove wetlands and lowest in basin mangrove wetlands. Additionally, many mangrove wetlands are important sources of detritus for adjacent aquatic ecosystems. Mangrove detritus is grazed by crabs, amphipods, insect larvae and shrimps, which are in turn grazed upon by fish and other organisms higher up the food chain. The removal of mangrove wetlands can, therefore, cause a decline in commercial fisheries in adjacent waters.

13.3   DIRECT AND INDIRECT BENEFITS OF WETLANDS

Wetlands can be regarded as having natural functions and economic uses, and both functions and uses have economic values (Pearce and Turner, 1990; Barbier, 1989). An assessment of the economic values of wetlands is very important at the political and decision-making stages of wetland conservation, and to ensure their sustainable utilization (Chapter 2). It is, therefore, useful to examine the benefits of wetlands in terms of their economic values:

(a) Direct use values are those derived from the economic uses made of a wetland's resources (products) and services, e.g. transport, fishing, tourism, water supply and agriculture (such as rice).

(b) Indirect use values are the indirect support and protection provided by the wetland's natural function or 'environmental services', e.g. hydrological uses include groundwater recharge and discharge, flood storage and desynchronization, shoreline anchoring and dissipation of erosive forces and sediment trapping, and ecological uses include nutrient retention, removal and recycling, food chain support, and habitat for fisheries and wildlife.

(c) Non-use or preservation values are those derived neither from current direct nor indirect use of the wetland, e.g. the scenic beauty of wetlands.

It is the combination of functions, products and ecosystem attributes that make wetlands important to society (Table 13.2). The hydrological values of wetlands are worth considering in more detail because of their usefulness to rural communities. Groundwater recharge occurs when water seeps from the wetland down into the underground aquifer. Percolation into the aquifer cleans the water, and it may then be drawn out via wells for water supplies, or it may flow laterally into another wetland as groundwater discharge. Recharge is also useful for flood storage as run-off is stored underground (Dugan, 1990; Mitsch and Gosselink, 1986). In Tunisia, the relatively freshwater of Sebkha Kelbia on the Plain of Kairouan recharges groundwater on the coastal Plain of Enfidaville and supplies wells and agriculture there (Hollis, 1990).

Groundwater discharge occurs when water from underground moves upwards into a wetland and becomes surface water. Studies have shown that groundwater discharge wetlands usually support stable ecological communities because water levels and temperature do not fluctuate as much as those in wetlands dependent upon surface flow. For example, in Amboseli National Park in Kenya, a series of springs originating on Mount Kilimanjaro percolate through the porous lava soils and re-emerge as a series of small swamps. The freshwater attracts wildlife, which is a major attraction of the Park.

Wetlands can control floods by storing precipitation and releasing run-off evenly. Thus, by preserving natural storage, the costly construction of dams and reservoirs is avoided. On the Yellow River of China, the valley area between San-Me and Tokato provides flood peak reduction, and Poyang Lake (China's largest lake) can store one-third of the annual flood waters from Jiangxi Province and reduced the June 1954 flood peak by one-half. Wetlands also stabilize shorelines via vegetation, which reduces the energy of waves and holds the sediment in place by roots. Where the vegetation has been removed, such as in parts of the Norfolk and Suffolk Broads, the cost of maintaining artificial reinforcement can be very high. Sediment retention in wetlands will lengthen the lifespan of downstream reservoirs, and retain toxicants which adhere to sediment particles. Sediment settling is enhanced by the presence of vegetation, but it will also settle in pools within the wetland.

A most important ecological value of wetlands is nutrient retention. Wetlands that remove nutrients will improve the quality of water and prevent eutrophication. Nutrients such as phosphorus and nitrogen are retained in the subsoil or by storage in wetland vegetation. In some circumstances wetlands can be used for treating waste

Table 13.3   Value of wetlands. From Dugan (1990). Reproduced by permission of IUCN

Value of wetlands	Estuaries (without mangroves)	Mangroves	Open coasts	Floodplains	Freshwater marshes	Lakes	Peatlands	Swamp forest
Functions:								
Groundwater recharge	○	○	○	■	■	■	●	●
Groundwater discharge	●	●	●	●	■	●	●	■
Flood control	●	■	○	■	■	■	●	■
Shoreline stabilization erosion control	●	■	●	●	■	○	○	○
Sediment/toxicant retention	●	■	●	■	■	■	■	■
Nutrient retention	●	■	●	■	■	●	■	■
Biomass export	●	■	●	■	●	●	○	●
Storm protection/windbreak	●	■	●	○	○	○	○	●
Micro-climate stabilization	○	●	○	●	●	●	○	●
Water transport	●	●	○	●	○	●	○	○
Recreation/tourism	●	●	■	●	●	●	●	●
Products:								
Forest resources	○	■	○	●	○	○	○	■
Wildlife resources	■	●	●	■	■	●	●	●
Fisheries	■	■	●	■	■	■	○	●
Forage resources	●	●	○	●	●	○	○	○
Agricultural resources	○	○	○	■	●	●	●	○
Water supply	○	○	○	●	●	■	●	●
Attributes:								
Biological diversity	■	●	●	■	●	■	●	●
Uniqueness to culture or heritage	●	●	●	●	●	●	●	●

Key: (○) Absent or exceptional; (●) present; (■) common and important value of that wetland type.

water and effluent. In Uganda, for example, the National Sewerage and Water Corporation is supporting the conservation of papyrus swamps around Kampala because they absorb sewage and purify water supplies.

There are many examples of the hydrological and ecological values of wetlands, some of which have been summarized by Hollis *et al.* (1988) using data from Maltby (1986) (see also Table 13.3):

(a) Firewood: the mangrove Sundarbans of India and Bangladesh provide a commercial crop of firewood and help support 80% of the fish caught in the Ganges–Brahmaputra estuary.

(b) Fisheries: the Inner Delta of the Niger River in Mali produces 100 000 tons of fish per year and supports many villages in the region.

(c) Flood prevention: the value of wetlands preventing serious flooding along the Charles River in the US is $13 500 per hectare per year.
(d) Water treatment: sewage purification by flooded cypress groves in Florida, USA, is 60% cheaper than comparative mechanical and chemical methods.
(e) Energy and carbon dioxide storage: peatlands lock up carbon in dead plant matter, and it has been calculated that organic soils store 500 times the carbon released annually from burning fossil fuels. The destruction of peat areas could reinforce the enhanced greenhouse effect (Chapters 3 and 4).

These data show the multiple values of wetlands, which are also illustrated by mangrove ecosystems. These ecosystems are of major importance to the livelihood of local communities in Asia, Africa, Latin America, the Pacific and the Caribbean Islands. Many villages are sited within the mangrove areas, particularly at river mouths, and the population utilizes the wetlands by harvesting mangrove products, cutting trees for firewood, charcoal and poles, and by fishing. Mangrove fish products in Thailand in 1982 represented 30–100 and 200–2000 US dollars per hectare per year for fish and shrimps, respectively, and mangrove forestry products in Trinidad in 1974 represented 70 US dollars per hectare per year (Maltby, 1986).

13.4 DISAPPEARANCE AND ALTERATION OF WETLANDS

Wetlands cover an area of 6% of the earth's land surface, and are found in all climates and regions. Despite this coverage, wetlands everywhere are under threat from agricultural intensification, pollution, urban development and engineering schemes. Wetlands have been disappearing world-wide at an unprecedented rate: 90% of California's original wetlands have disappeared; 35% of wetlands in Tunisia have been drained or built upon in the last 100 years; 40% of coastal wetlands in Brittany have gone in the last 20 years; and in Ireland 80 000 hectares of bog have been drained since 1946.

Many of the causes of wetland loss are listed in Table 13.4, and examples of threats to protected wetlands are listed below (from Dugan, 1990):

(a) Groundwater abstraction: the Tablas de Daimiel National Park (Spain) has an acute shortage of water caused in large part by the use of groundwater for irrigation.
(b) Dams: the water supply of Lac Ichkeul in Tunisia, a World Heritage and Ramsar site, has been diminished through dam building on the input rivers.
(c) Pollution: in Brazil gold mining in the Paraguay River basin has resulted in mercury poisoning of the waters of the Patanal.
(d) Drainage: one third of the Banados del Este, Uruguay, a Ramsar site and Unesco Biosphere Reserve, has been drained for agriculture.

Mangrove ecosystems have also suffered overexploitation of their natural resources. In South-east Asia, mangroves have been converted to rice paddies and aquaculture at an unprecedented rate, and oil spills have also resulted in large losses (Maltby, 1986). In the Philippines, 24 000 hectares of mangroves were lost per year between 1967 and 1975, and 5000 hectares per year were lost in Malaysia for wood chip production. Many mangrove areas are being lost to agriculture and industry, as

Table 13.4 Causes of wetland loss. From Dugan (1990). Reproduced by permisssion of IUCN

Causes of loss	Estuaries	Open coasts	Floodplains	Freshwater marshes	Lakes	Peatlands	Swamp forest
Direct human actions:							
Drainage for agriculture, forestry and mosquito control	■	■	■	■	●	■	■
Dredging and stream channelization for navigation and flood protection	■	○	○	●	○	○	○
Filling for solid waste disposal, roads and commercial, residential and industrial development	■	■	■	■	●	○	○
Conversion for aquaculture/mariculture	■	●	●	●	●	○	○
Construction of dykes, dams, levees, and seawalls for flood control, water supply, irrigation and storm protection	■	■	■	■	■	○	○
Discharges of pesticides, herbicides, nutrients from domestic sewage and agricultural run-off and sediment	■	■	■	■	■	○	○
Mining of wetland soils for peat, coal, gravel, phosphate and other materials	●	●	●	○	■	■	■
Groundwater abtraction	○	○	●	■	○	○	○
Indirect human actions:							
Sediment diversion by dams, deep channels and other structures	■	■	■	■	○	○	○
Hydrological alterations by canals, roads and other structures	■	■	■	■	■	○	○
Subsidence due to extraction of groundwater, oil, gas and other minerals	■	●	■	■	○	○	○
Natural causes:							
Subsidence	●	●	○	○	●	●	●
Sea-level rise	■	■	○	○	○	○	■
Drought	■	■	■	■	●	●	●
Hurricanes and other storms	■	■	○	○	○	●	●
Erosion	■	■	●	○	○	●	○
Biotic effects	○	○	■	■	■	○	○

Key: (○) Absent or exceptional; (●) present, but not a major cause of loss; (■) common and important cause of wetland degradation and loss

can be seen in Trinidad, Haiti and other West Indian islands. In the Sundarbans of Bangladesh, the increase of plantation forestry is diminishing the natural diversity and ecological quality of the original mangrove forest.

Why have so many wetlands been lost? Wetlands are destroyed because society views their elimination as beneficial, and because the replacement land use is regarded as being more economically productive. Such attitudes are increasingly condemned and the sustainable utilization of wetlands is encouraged. Unacceptably high wetland losses has led to high social costs in many regions, and although many of the losses are deliberate, many are due to ignorance and poor management. Such losses are usually due to limited information on the benefits of wetlands, poor distribution of costs and benefits, deficient planning concepts, policy deficiencies and institutional weaknesses.

13.5   CASE STUDIES

13.5.1   Rio de la Pasion, Guatemala

Seasonally flooded, tropical lowland rivers can provide useful hydrological and ecological services to the people who live along them. This is exemplified by a study of the Rio de la Pasion in northern Guatemala (Hughes, 1989). The river flows westwards across the southern portion of the limestone Yucatan Peninsula in northern Guatemala (Central America), and the floodplain is flooded during the wet season from May to December. Most of the catchment is covered in lowland tropical rain forest, but much of this vegetation is being cleared and converted to ranching or maize and bean cultivation. Additionally, a dam scheme has been planned down-stream of Sayaxche, which is likely to permanently flood the river upstream and to prevent seasonal inundation of the floodplain.

The floodplain wetland of the river is highly productive of fish and rice and is extremely important to the livelihood of people living in Sayaxche (the principal town) and other villages in the area. If the river is not flooded annually, fish productivity will decrease to a minimum and rice cannot be grown. At present, fish are caught by net, smoked or salted for transport to Guatemala City and sold between October and Easter (with very high consumption during Holy Week). The temporary lagoons and flooded areas of Rio de la Pasion are more productive than the river itself, and approximately 200 fishermen work from Sayaxche for six months of the year with a catch of 150 000 pounds of fish during that period. The most important fish species are *Lepisosteus tropicus, Megalops antlanticus* and *Ictiobus meridionalis*. Crocodiles and crayfish are also caught along Rio de la Pasion but both are consumed locally because they cannot be dried or smoked for transportation. Rice grown on the floodplain of Rio de la Pasion is sown in April and harvested in August. Rice production is the most important activity of the inhabitants of Sayaxche: 300 families are involved, each cultivating an average of 8.4 hectares with a yield of 2.6 tonnes per hectare.

Rio de la Pasion also serves as a transport route. Motorized canoes transport local people from village to village and into Mexico (the river flows over an international border), and there is a thriving industry transporting tourists to Mayan archaeological sites (e.g. El Ceibal) and to take advantage of the scenic qualities of the river.

The hydrological values of the river are equally important but are not so obvious. The river flows through limestone and recharges the underlying aquifer. Domestic water supplies in Sayaxche and other villages are obtained from wells which are recharged with river water throughout the year. Additionally, during the wet season, much of the flood water is stored in lagoons on the floodplain, thus desynchronizing flood peaks by releasing the flood water gradually. Flood control is a vitally important function in an area that has numerous habitations along the river bank and where the local population is reliant upon the river for transport and food.

This example shows the sustainable utilization of floodplain wetlands along Rio de la Pasion. If the dam scheme goes ahead, the hydrological and ecological functions of the river will be diminished, and the livelihoods of the local population may be removed.

13.5.2 Lake Baikal, USSR

Lake Baikal is the earth's oldest lake (25 million years old), and also the deepest at 1637 metres. It covers an area of 34 000 square kilometres and contains the world's largest volume of surface freshwater. The lake is situated in Siberia, and to the local native people, the traditionally Buddhist and shamanist Buryats, Baikal is a holy sea. Ecologically, Lake Baikal is unique, being very rich in biomass and number of species. There are 1550 species and variants of animals, and 1085 species of plants, of which 1000 are endemic (Massey Stewart, 1990). An average European lake might have three species of shrimp-like amphipods and eight of flatworms, but Lake Baikal has 255 amphipods and 80 flatworms, as well as 22 species of cottoid fish (bullheads) and an endemic species of seal. The lake has supported a thriving fishing industry, with species such as the endemic omul being a great delicacy. However, Lake Baikal is threatened by pollution from industry and agriculture and many local industries have collapsed as a consequence.

Until the beginning of this century Lake Baikal was largely untouched, but loggers started to clear the forest within the lake's catchment, which resulted in erosion and landslides. The input rivers carried vast amounts of sediment which were deposited in Lake Baikal, as well as carrying timber rafts. Between 1958 and 1968, 1.5 million cubic metres of logs sank to the bottom of the input rivers. As the logs decayed, bacteria depleted the water of oxygen, and many rivers were three to four metres deep in logs. Streams and springs disappeared through excessive felling of timber, and fish, including omul, could no longer spawn in 50 tributaries. The worst problems have been caused by effluent, and in particular from the Selenga River. This river transports partially treated sewage and effluent from 50 factories in Ulan-Ude, and in 1988 it carried 500 tonnes of nitrates into the lake. Two other major sources of pollution are the pulp and cellulose mills at Baikalsk and Selenginsk, which pump toxic waste into Lake Baikal. The plant at Baikalsk has been open for 23 years and since then has discharged 1.5 billion cubic metres of industrial waste into the lake. In 1985, industries in the Irkutsk and Buryat regions emitted 1.2 million tonnes and 204 000 tonnes of air pollutants respectively over Lake Baikal.

The main effects of the pollution have been to alter the chemistry of the lake water as well as the lake's microbial flora. By January 1989 the omul had stopped spawning naturally and its population is now maintained artificially. Since 1987 thousands of the endemic seals have died, because they are at the top of Lake Baikal's food chain and consume toxic pollutants in the fish they eat. Many of these ecological processes are irreversible. Despite public outcry and international pressure, and although some of the industrial plants have been fitted with water purifiers, the pollution still continues. The lake's salvation may lie with its adoption as a Unesco World Heritage Site and the establishment of the Baikal International Centre for Ecological Research, which was inaugurated in October 1989 and has been supported by researchers from around the world. Lake Baikal exemplifies the effects of extreme pollution on a wetland ecosystem.

13.5.3 Halvergate Marshes, UK

The Halvergate Marshes, in the easterly part of the Norfolk Broads in England, are uninterrupted and inaccessible grazing marsh covering an area of 16 square kilometres. During the 1980s this wetland area was the setting for an environmental conflict between small graziers who wanted to conserve the marshes for grazing, and large, wealthy arable farmers who wanted to drain the wetland for cereal growing.

The temptations to drain the marshes by the arable farmers were irresistible because of the financial incentives from the European Community's (EC) Common Agricultural Policy to convert to cereals in preference to livestock, and because of the major funding for drainage offered by the Ministry of Agriculture. For the graziers, the new drainage rates would have been hugely increased by the local Internal Drainage Board (IDB), and consequently many of them would have been bought up by the arable farmers. Between 1979 and 1982, the IDB proposed various pumping and drainage schemes for Halvergate, and in November 1982 the Ministries of Agriculture and the Environment agreed on a compromise solution to avoid a public inquiry. They supported the Tunstall/Acle pumping scheme, but turned down all other schemes for increasing the capacity of the drainage pumps.

The UK Government's decision to turn down part of the IDB scheme was an important precedent towards moves on planning restraint on agricultural practice. In 1984 it was calculated that with 5000 acres of wetland at risk, compensation payments for conservation of the marsh under the 1981 Wildlife and Countryside Act would cost the taxpayer as much as £1 million a year (index linked). This problem of compensation remains, and in June 1984, after various demonstrations by dissatisfied farmers, the Prime Minister was drawn into the debate and she insisted that an exceptional Planning Direction to prevent drainage was imposed on the farmers in question. In March 1985 the Broads Grazing Conservation Scheme was launched, which offered all landowners a fee of £50 per acre to retain livestock farming on the marshes, and in 1987 the Broads was designated an Environmentally Sensitive Area by the EC.

The Halvergate Marshes debate represents a success story for British wetland conservation. The critical issues in the conflict were the expense to the nation, in terms of draining and ploughing, financed by the Ministry of Agriculture, and of compensation to those farmers denied drainage, financed by the Department of the Environment. Furthermore, the issue set a precedent for the workability of the voluntary agreements between farming and conservation under the 1981 Wildlife and Countryside Act. If these had failed, the protection of British wetlands would have been very much at stake.

13.6 THE FUTURE

The ecological, hydrological and economic advantages of wetlands have been highlighted. In most developing nations the rural economy is dependent on the benefits wetlands can provide, and there is an urgent need in these countries to establish an environmentally sound management or wise use of wetlands. The adoption of the Convention on Wetlands of International Importance Especially as

Waterfowl Habitat in Ramsar, Iran, in 1971, and its subsequent ratification by 60 countries, designating 514 wetlands (from January 1991) gives explicit recognition to the shared nature of many wetland benefits. For example, many fish are dependent for breeding upon wetlands in one country, but are harvested offshore by fishing fleets from another. Many migratory birds, especially waterfowl, use wetlands in different countries during the course of their bi-annual migration. The 1987 Conference of the Parties to the Ramsar Convention outlined the principle of the wise use of wetlands by stating that 'it is their sustainable utilization for the benefit of humankind in a way compatible with the maintenance of the natural properties of the ecosystem'. This draws attention not only to internationally important wetlands but also to those that bring benefit to communities, even if only on a local scale.

The major ways for developing such policies should include (from Hollis *et al.*, 1988):

(a) A national inventory of wetlands.
(b) Identification of the benefits and values of these wetlands.
(c) Definition of the priorities for each site.
(d) Proper assessment of environmental impacts.
(e) Use of development funds for projects which allow conservation and sustainable utilization of wetland resources.

This approach to the wise use of wetlands is adopted by the Ramsar Convention, the International Union for the Conservation of Nature and Natural Resources, and other conservation organizations. It is to be hoped that the wise use of wetlands will make them an asset rather than an obstacle for sustainable development.

13.7 REFERENCES

Barbier, E. B. (1989) *The Economic Value of Ecosystems: 1 Tropical Wetlands*. Gatekeeper Series No. LEEC 89–02. London: IIED and Environmental Economics Centre.
Dugan, P. J. (Ed.) (1990) *Wetland Conservation: A Review of Current Issues and Required Action*. Gland, Switzerland: IUCN.
Hollis, G. E., Holland, M. M., Maltby, E. and Larson, J. S. (1988) Wise use of wetlands. *Nature and Resources*, **24**, 2–13.
Hollis, G. E. (1990) The hydrological functions of wetlands and their management. In: G. A. Gerakis (Ed.). *Conservation and Management of Greek Wetlands—Strategies and Action Plan*. Gland, Switzerland: IUCN.
Hughes, J. M. R. (1989) *Methodology for Quantifying the Hydrological Functions of the Rio de la Pasion, El Peten, Guatemala*. Report for IUCN and CATIE under contract no. 9567. Gland, Switzerland: IUCN.
Maltby, E. (1986) *Waterlogged Wealth: Why Waste the World's Wet Places?* London: Earthscan.
Massey Stewart, J. (1990) The Great Lake is in great peril. *New Scientist*, **126**, No. 1723, 58–62.
Mitsch, W. J. and Gosselink, J. G. (1986) *Wetlands*. New York: Van Nostrand Reinhold.
Pearce, D. W. and Turner, R. K. (1990) *Economics of Natural Resources and the Environment*. London: Harvester Wheatsheaf.

13.8 FURTHER READING

Jefferies, M. and Mills, D. (1990) *Freshwater Ecology*. London: Belhaven Press.
Williams, M. (Ed.) (1990) *Wetlands: A Threatened Landscape*. Oxford: Blackwell.

Current wetland news can be obtained from *Ramsar*, the quarterly newsletter of the Convention on Wetlands of International Importance Especially as Waterfowl Habitat, and from *IWRB News*, put out by the International Waterfowl and Wetlands Research Bureau. Both publications can be obtained from the relevant office at: The Wildfowl and Wetlands Centre, Slimbridge, Gloucester GL2 7BX, UK.

CHAPTER 14

Soil Erosion and Conservation

A. C. Millington

14.1 INTRODUCTION

In 1939 Jacks and Whyte published a classic book on soil erosion entitled *The Rape of the Earth*. Rape is a menace to society; this chapter will show that soil erosion is equally menacing to society and poses many problems of which its impact on global food security is the most important. The role of soil erosion in contemporary environmental degradation was first recognized in 1864 by George Perkins Marsh. However, it was not until the Dust Bowl of the 1930s in the American Mid-West that scientists and agriculturalists began to investigate the processes that constitute soil erosion and attempted to find solutions to the problem. Jacks and Whyte's book rode on the wave of concern generated by the Dust Bowl. This environmental disaster also prompted research into soil erosion, including the development of predictive models and soil conservation techniques.

Soil erosion is a ubiquitous, naturally occurring process in which weathered material is removed from the landscape. It encompasses the processes of sediment entrainment and transportation, leading to lowering of the land surface. Ultimately, the eroded material is deposited elsewhere. Natural erosion rates vary globally. Under natural vegetation, erosion rates by water are highest in semi-arid areas and erosion rates by wind are highest where vegetation cover is sparsest, i.e. in arid areas and in some coastal regions. This natural erosion is termed geological erosion. However, the global distribution of erosion is made more complicated where anthropogenic disturbance has occurred. Deforestation, agriculture, mining, and construction all give rise to accelerated erosion (Fig. 14.1). This can create a variety of environmental problems such as a decline in land productivity, excessive siltation of drainage systems and dune encroachment. Although this chapter is concerned with such problems, and ways of reducing accelerated erosion, it is important to realize that there are additional soil degradation processes which lead to environmental damage. These include chemical degradation of soils [e.g. salinization (see Chapter 15) and the build-up of chemicals of toxic levels], physical degradation (e.g. soil compaction) and biological degradation of soils (e.g. declining humus levels).

Environmental Issues in the 1990s. Edited by A. M. Mannion and S. R. Bowlby
© 1992 John Wiley & Sons Ltd

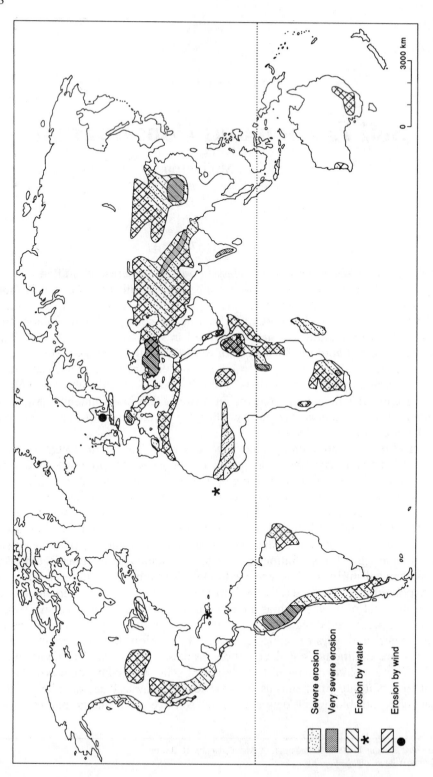

Figure 14.1   World map showing areas of accelerated erosion.

Important linkages exist between the different types of soil degradation, but an investigation into these is beyond the scope of this book.

14.1.1 Rates of soil erosion

Quantitative assessments of soil erosion facilitates spatial and temporal comparisons of the severity of erosion. More importantly, they are vital inputs to soil conservation, and hence agricultural, planning. However, the interpretation of such data is problematic. Most erosion rates are calculated on the basis of data from experimental plots, but errors ensue when such data are extrapolated to wider regional or national contexts. Caution must, therefore, be exercised when interpreting soil erosion rates.

14.2 ENVIRONMENTAL FACTORS AFFECTING SOIL EROSION

Years of scientific experimentation have led to the accumulation of information on soil erosion processes and the factors affecting such processes. This information is important because it has helped our understanding of erosion processes, it is fundamental to the development of soil erosion prediction models, and it can be used to help design conservation techniques.

14.2.1 Environmental factors affecting the erosion of soil by water

The erosion of soil by water is influenced by rainfall erosivity, run-off erosivity, soil erodibility, slope parameters, vegetation cover and land management and cultivation practices.

Erosivity describes the aggressiveness of erosive agents such as rainfall and run-off. Raindrop size, as it is directly related to the kinetic energy released when it strikes the soil aggregate, is an important determinant of rainfall erosivity. As it is not feasible to regularly record raindrop size, rainfall intensity can be used as a surrogate measure. Rainfall intensity can also be directly compared with the soil infiltration capacity and is thus a good predictor of overland flow caused by rainfall in excess to the infiltration capacity. It is therefore also important in predicting inter-rill erosion.

Soil erodibility is the ability of a soil to withstand the detachment and entrainment of particles by rainsplash or water flowing over the soil surface. The factors that affect the infiltration capacity of the soil also affect the generation of overland flow. Of particular importance in the erosion of soil by water are the soil aggregate stability, which is in turn influenced by the soil chemistry, the soil-bonding agents released by decomposition of organic matter (e.g. organic acids and gums) and soil texture. Generally the most erodible soils have high proportions of very fine sand and silt and very little organic matter. Such soils are usually structureless, or have a single grain or fine granular structure.

Early research into the relationships between soil loss and slope angle and length showed clear positive relationships. There is a quantifiable relationship between soil loss, slope angle and slope length. The soil loss increases as the slope angle and length increase. More recent work has suggested that the slope shape also influences splash erosion and run-off generation, which is itself affected by surface roughness.

Vegetation is an important influence on erosion and soil loss. The cover provided by vegetation protects the soil surface from the impact of raindrops. Plant stems and surface litter create rough surfaces which reduce run-off velocities through friction. Decaying litter releases soil-bonding agents which aid soil aggregation, thereby reducing the erodibility. The vegetation also utilizes soil moisture, reducing the likelihood of saturation run-off, and the plant roots bind the topsoil. As a generalization, the removal of vegetation will increase both run-off and soil erosion, although in terms of quantifying the relationship the difficulty lies in which vegetation parameter to use. Vegetation removal and the change from one crop to another are likely to affect soil losses significantly. Many experimental measurements have been made which show an increase in soil loss when the land is disturbed (see Section 14.6).

Land management and cultivation practices can also reduce or exacerbate erosion rates. Soil conservation measures, if introduced correctly, will reduce soil losses, but other cultivation practices may not. For instance, repeated up-and-down the slope cultivation leads to increased erosion, particularly along vehicle wheel tracks.

14.2.2   Environmental factors affecting the erosion of soil by wind

The concepts of erosivity and erodibility can also be applied to wind erosion. In this context, wind erosivity defines the ability of the wind to erode material. A critical parameter here is wind velocity. Particle entrainment by the wind depends on drag and lift, which are functions of wind velocity and particle size. Other parameters are also important in defining wind erosivity, particularly wind turbulence and gustiness.

The nature of the surface materials and the condition of the surface combine to determine the erodibility. The main material parameters are soil particle size, the 0.1–0.15 mm fraction being the most prone to erosion, and the extent, nature and strength of the aggregates. The mechanisms for binding particles together are no different from those discussed earlier, but their relative importance differs due to the nature of wind erosion. Four types of aggregation structure which reduce wind erosion are recognized: primary (water-stable) aggregates; larger, secondary aggregates (soil clods); fine-textured material trapped in and between clods; and surface crusts formed by the action of rain. The breakdown of aggregates is achieved by wind abrasion and/or mechanical weathering, or, more commonly, by splash erosion and cultivation.

Weather conditions are of vital importance, not only for determining wind velocity and direction, but also because for wind erosion to be effective the soils must be dry. This is because soil moisture promotes cohesion among soil particles, thereby reducing soil erodibility in relation to wind erosion (although wet soils often have a lower strength and are vulnerable to water erosion). Thus weather conditions such as droughts and high temperatures, leading to high rates of evapotranspiration, cause drying of the soil surface, which promotes wind erosion.

The wind fetch, defined as the length of uninterrupted wind flow, is important. If the fetch is long, wind erosion is more likely to reach its maximum efficiency. Wind fetch can be considered as analogous to slope length in erosion by water.

Surface roughness also contributes to erodibility. It is a function of the size of soil aggregates, the presence of plant debris on the soil surface and the effects of cultivation. Wind velocities, and hence wind erosion, are reduced as the friction increases with surface roughness. The vegetation and plant cover also determines the proportion of the soil surface exposed to wind action.

14.2.3 Short-term temporal variations in the factors affecting erosion of soil by water and wind

Although the relationships between the factors discussed in the preceding section and soil loss are relatively straightforward, a level of complexity is introduced because seasonal variations in rainfall, run-off and vegetation growth occur. Rainfall erosivity varies with the type of rainfall. In semi-arid climates in west and southern Africa, for example, the rains early in the season are much more intense than those later in the season. In Britain much of the soil erosion in south and east England is linked to the high intensity rainfall associated with convective summer thunderstorms rather than with lower intensity winter frontal rainfall.

Vegetation growth also varies throughout the year. The natural vegetation may not be dense all year round and crop cover is also low in the early part of the crop-growing cycle. When this is the case and heavy rainfall occurs, erosion rates are extremely high.

Like the factors which influence erosion by water, many of the parameters defining erosion by wind vary with time. The dominant wind direction and strength can also vary throughout the year. The pattern of wind erosion is complicated by the fact that some of the dominant winds may be accompanied by precipitation, thereby reducing the likelihood of erosion.

14.3 PREDICTING SOIL EROSION

Models of soil loss, based on extensive field measurements of the parameters discussed in Section 14.2, have been formulated for two reasons. First, they can be used to assess soil erosion risks in areas where measurements have not been made and, second, they can be used to determine the design criteria for soil conservation measures.

The question of scale must first be considered before examining the efficiency of any model to predict either soil loss or sediment yield. Soil loss prediction models generally operate at the scale of a hillslope or field, while sediment yield models operate at the scale of entire catchments. Predictive models based on physical parameters have their uses, but accelerated erosion is a complex phenomenon which is affected not only by environmental factors, but also socio-economic factors. These are often interrelated in a complex manner which creates problems for modellers. Explanations of soil loss based on socio-economic factors (see Section 14.9) are useful at the catchment level, as are sediment yield models, but are not applicable to soil loss prediction at the smaller scale of an individual field or hillslope.

The most commonly used models for the prediction of soil loss are parametric models. Parametric models rely on statistical relationships between soil loss and various parameters which are derived from large sets of field data. For predicting soil erosion by water, the Universal Soil Loss Equation is the most widely applied model, whereas the Wind Erosion Equation is used for predicting wind erosion (Cooke and Doornkamp, 1990). Although widely used, their accuracy declines significantly beyond the geographical region in which the original data used to develop the model were collected. The more accurate prediction of soil losses lies not in the development of parametric models based on data sets from other regions, but on deterministic models based on an understanding of erosion processes. Such models have been developed by geomorphologists but are not yet routinely used in soil conservation planning.

14.4   IMPACTS OF SOIL EROSION

Various impacts of soil erosion were listed in the introduction. In this section on-site and off-site factors are considered in more detail.

14.4.1   Soil erosion and land productivity

Although rates of soil loss are important for comparing erosion over time or in different environmental situations, in agriculture it is often not the actual rate of soil loss that is important but the resultant decline in land productivity. In particular, the loss of topsoil is important because it contains almost all the available crop nutrients and organic matter, and it is often fairly thin when compared to the rest of the soil profile.

The impact of soil erosion on crop productivity has recently received much attention, e.g. Follett and Stewart (1985). Even small losses of topsoil can cause drastic reductions in crop productivity because of the nutrient concentrations in the topsoil. This is particularly problematic with monocultures with high fertilizer inputs. In these systems it is assumed that chemical fertilizers will maintain the land productivity, but physical parameters which are equally important for crop growth and erosion control (e.g. the water-holding capacity and soil aggregation) are ignored (Table 14.1).

Table 14.1   Soil erosion induced declines in crop production in Kenya as a percentage of production potential under a non-eroding situation. Adapted by permission from El-Ashry and Ram (1987)

Crop	High input regime	Medium input regime	Low input regime
Maize	74	61	52
Beans	68	56	49
Cassava	64	52	41
Wet rice	8	6	8
Irish potato	73	55	51

14.4.2   Effects of transported sediment

Soil loss from fields and the resultant decrease in land productivity are on-site effects of soil erosion. Off-site effects occur due to the transport and subsequent deposition of eroded sediment. The routing of eroded sediment along watercourses is particularly important. Deforestation in the catchments of many rivers has led to changes in run-off and soil erosion. The impact is felt along entire river courses as fertile or infertile sediment is deposited on agricultural land and as increased wet season flooding or

Table 14.2   Sedimentation rates (10^6 m^3 yr^{-1}) in selected reservoirs. Data reproduced by permission from El-Ashry and Ram (1987), Patanik (1972)

Country	Reservoir	Sedimentation rate
Kenya	Kindaruma	4.1
Sudan	Hasm-el-Girba	10.3
India	Bhakar	41.6
India	Panchet	11.8
India	Nizam Sugar	10.8
India	Ukai	26.8

lower dry season flows occur. The siltation of reservoirs and irrigation works is a major concern. It is a serious threat to economic development because it reduces the productive or economic life of reservoirs (Table 14.2), silts up canals and adversely affects aquatic flora and fauna.

For example, if sediment loads in the Tana and Thiba Rivers in Kenya show similar increases to those found elsewhere in East Africa, there will be serious consequences for the recently built Masinga Dam. Its capacity to generate hydroelectric power will not last more than a decade, and it will be completely choked with sediment in 30 to 35 years, causing the destruction of aquatic ecosystems, upstream flooding and sedimentation elsewhere along the rivers (El-Ashry and Ram, 1987).

14.5   CASE STUDIES

Three case studies have been chosen to reflect the impact of soil erosion in different climatic zones, to compare the situation in the developing and developed world, and to emphasize the importance of the global nature of the problem.

14.5.1   Humid tropics

Accelerated soil erosion and its environmental impacts are increasingly becoming a major problem throughout the humid tropics as forests are cleared (see Chapter 5). The developing nations of the tropics with their burgeoning populations, ever-increasing demand for land, food and fibre resources, and the need for food security can ill-afford the detrimental impacts of soil erosion. Although the problem is widespread, the amount of research on soil erosion and conservation in the tropics does not exceed that emanating from the USA, and quantitative studies are still lacking in many areas.

Inter-rill erosion, the combination of splash erosion and sheetwash, is common throughout the humid tropics, although all other types of water erosion occur. Acknowledging the significance of inter-rill erosion, especially as it is visually less spectacular than landscapes dissected by gullies, is important because hitherto its lack of recognition has held back erosion and conservation research in the tropics. The extent of accelerated erosion in the humid tropics can be seen in Fig. 14.1

The reasons for the extent of rainfall erosion in the humid tropics are the frequent high intensity and long duration rainfall events combined with high rates of overland

234

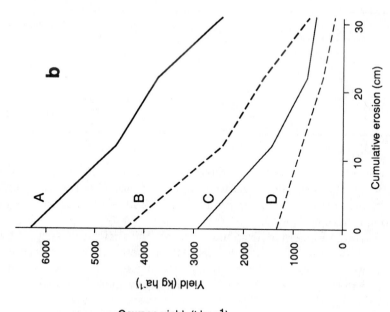

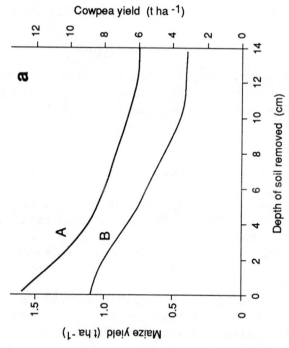

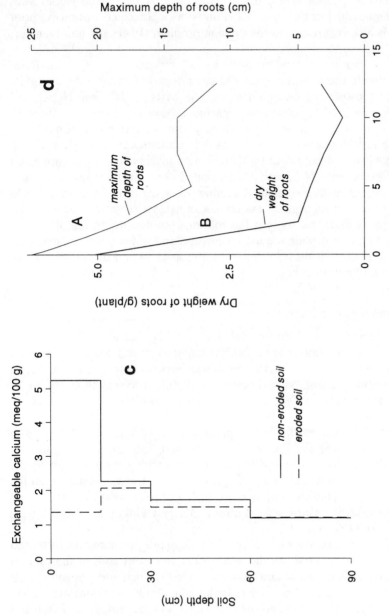

Figure 14.2   (a) Effects of soil removal on (A) maize and (B) cowpea yields on Nigerian alfisols. Adapted from Lal (1976). (b) Simulated crop yield versus accumulated erosion relationships for central Sierra Leone: (A) maize, improved cultivation system; (B) maize, traditional cultivation system; (C) cowpea, improved cultivation system; (D) cowpea, traditional cultivation system. Adapted by permission from Biot *et al.* (1989). (c) Variations in exchangeable calcium levels between eroded and non-eroded ferralsols. Adapted from Moberg (1972). (d) Influence of soil removal (simultaneous erosion) on root development in maize: (A) maximum depth of roots; (B) dry weight of roots. Adapted from Lal (1976).

flow due to the low infiltration capacities of many soils. Many humid tropical soils are highly erodible soils with poor structures and low amounts of organic matter; in addition, there is often a lack of coherence between the topsoil and subsoil. Moreover, such soils are exposed because of land clearance on steep slopes and, particularly in commercial farming systems, long slopes. In some instances, poor cultivation practices also contribute to the erosion problem (El-Swaify and Dangler, 1982).

The impacts of soil erosion on land productivity are important to both subsistence and cash-crop farming in the humid tropics. On-site effects can involve the complete loss of nutrient-rich topsoil from steep hillslopes, e.g. parts of Haiti and Nepal, and effectively stops farming. More commonly, partial topsoil loss occurs, leading to suppressed crop yields. It is difficult to generalize about the latter situation as crop losses are sensitive to individual site conditions, but examples from Nigeria, Sierra Leone and Tanzania involving simulated soil erosion on maize and cowpeas are given in Fig. 14.2. This shows (Fig. 14.2a and 14.2b) that, as erosion rates increase, crop growth and yield decline, with the greatest decline occurring in the first year. Yield decline is caused by a loss of organic matter and nutrients from the topsoil (Fig. 14.2c), as well as physical degradation in the rooting zone leading to inhibited root development (Fig. 14.2d) and poor water transmission and water-holding capacities. In addition to these factors, it must be noted that crop yield and decline in growth vary between different types of soil.

14.5.2 Western and southern Europe

Rates of erosion in many parts of Europe are naturally high because of the climate, rugged topography in mountainous areas, unfavourable rock lithology and erodible soils. Some areas, as in the Mediterranean zone, have a long history of forest clearance and agriculture dating back to Greek and Roman civilizations. Many of these areas are now severely eroded; changes in agricultural systems in the last 50 years have had important repercussions for erosion rates in other parts of Europe.

Erosion by wind is a localized hazard in many coastal regions, e.g. Gascony in France, where sand dune stabilization has been practised since the 16th century. In northern Europe the sandy soils of the Baltic Plain stretching from Poland to the eastern Netherlands suffer serious erosion (Fig. 14.3). After the deforestation on the Prussian Haff (Germany), blowing sand became such a problem that inhabitants had to move and three villages—Karwaiten, Lattenwalde and Pillkopen—were buried under encroaching sand (Zachar, 1982).

Erosion by water is commonplace in Europe, but the major problems are restricted to four situations (Fig. 14.3). First, on the sandy loam and stony soils of the Iberian plateau inter-rill erosion is common and there are extensive badlands. Second, soils in the steeply sloping vineyards of many European wine-growing regions are poorly protected from the effects of rainfall and run-off. Third, soils developed on flysch and shale in the Alps and the Mediterranean countries are highly erodible and show extensive erosion by water. Finally, the sand-rich and loam soils developed on loess in northern Europe show areas of inter-rill and rill erosion due to the combined action of soil compaction and the presence of unstable soil aggregates as a result of the low amounts of organic matter (Table 14.3). In these latter soils the rates of

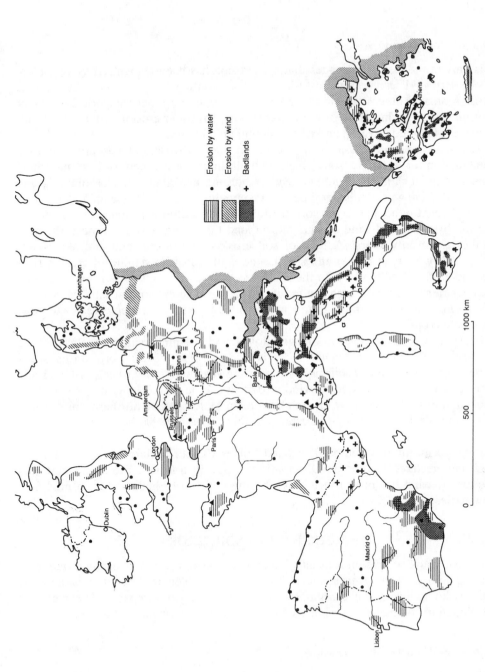

Figure 14.3   Soil erosion in the European Community and Switzerland (compiled from maps of soil erosion in the European Community and Switzerland prepared by the European Association of Soil Conservation).

erosion are high enough to significantly affect crop yield and the prognosis for sustained agriculture through to the next century in these regions is poor (Morgan, 1986)

14.5.3   England and Wales

It is only in the last two decades that soil erosion has been recognized as a problem in the British Isles, despite historical evidence dating back to the 18th century. Erosion on agricultural land in Britain can be grouped into four categories: (a) water erosion caused by high intensity summer storms; (b) water erosion on saturated soils in winter; (c) wind erosion; and (d) erosion in upland areas.

The high rates of erosion that occur on some British uplands are due to a combination of erodible soils, high rainfall intensities, steep slopes, moorland management techniques (such as heather burning and grazing), tree planting and recreational impacts. Gullying of peat is a common problem in these upland areas.

Reverting to the theme of soil erosion and agricultural productivity, it is soil erosion by water and wind on agricultural land that has received the most attention in the last twenty years. A map of soil erosion risk in England and Wales was compiled by Morgan (1985) and six categories of land with significant erosion risks were identified (Table 14.4). In the south and east of England there is evidence that summer rainstorms with high rainfall intensities lead to erosion on sandy and sandy loam soils (similar to those found in north-west Europe) under cereal and row-vegetable crops.

Erosion by wind is prevalent on agricultural land in Humberside, Lincolnshire, the East Midlands and East Anglia. There is evidence that the wind erosion hazard is increasing on light, peaty soils used for growing vegetables, sugar beet, potatoes and cereals. The costs of erosion by wind in Britain can be high. The cost of clearing up choked drains after extensive erosion by wind in March 1968 in northern Lincolnshire was over £4000, and it has been estimated that at least 10 000 hectares of farmland was affected (Cooke and Doornkamp, 1990). It has been estimated that the economic risk of erosion by wind on easily eroded soils is £85 per hectare for sugar beet and £600 per hectare for onions. Observations made in northern France suggest that erosion by wind has been exacerbated by the removal of hedges that formed field boundaries.

14.6   COMBATING SOIL EROSION

The basic cause of erosion is usually inadequate soil and land management on farms, grazing land and other cleared areas. Tackling the problem in these areas is therefore the key to reducing soil erosion rates and alleviating the impacts. The measures developed to combat soil erosion are known as soil conservation measures.

14.6.1   Soil conservation measures

Soil conservation measures aim to reduce the rate of soil erosion to an acceptable level. Ideally, this level should be equivalent to the rate of topsoil formation to enable a state of environmental sustainability to be achieved. However, in the

Table 14.3 Erosion rates on sandy and loamy soils in north-west Europe. All data are from Belgium and England and include data from single storms through to mean annual rates, which are all felt to be representative of annual rates. Adapted from Morgan (1986)

Type of cultivation	Soil erosion rate (kg m^{-2})
Bare soil	0.7 –13.0
Cereals	0.06– 3.0
Sugar beet	0.3 – 2.7

Table 14.4 Areas with high erosion risks in England and Wales. Adapted from Morgan (1985) by permission of Blackwell Scientific Publications

Type of erosion	Area affected (km^2)	Main areas	Main soil types and crops suffering from erosion
Water and wind erosion on arable land	3400	North Norfolk, Nottinghamshire and Vale of York	Mainly on sands and sandy loams under cereals, sugar beet and potatoes
Wind erosion on arable land	3600	East Suffolk, North Norfolk, parts of Nottinghamshire, Lincolnshire and Yorkshire	Mainly on sands and sandy loams under sugar beet
Water erosion on arable land	10 800	Midlands, Bedfordshire, Cambridgeshire, Suffolk, South Devon, Isle of Wight, Kent and Sussex	Sand, sandy loam, loam and silty sand soils under cereals, potatoes, sugar beet, field vegetables and horticulture
Wind erosion on arable low lands	2700	Somerset, Fens, West Lancashire	Peaty soils under cereals, sugar beet, field vegetables and horticulture
Water erosion on upland peats	3300	Pennines, Lake District, Welsh Hills, Exmoor and Dartmoor	Blanket peat at risk from grazing, afforestation and recreation
Wind erosion in coastal environments	700	Most coastal areas	Sand dunes and exposed sands

absence of this information for most soils, the guidelines given for the construction of conservation works are little more than arbitrary figures.

Soil conservation measures to tackle erosion by water consist of two types (Table 14.5). Mechanical measures can reduce the slope angle and/or length. Biological measures can improve soil properties to reduce soil erodibility; the maintenance of a

Table 14.5   Techniques for controlling erosion by water. Adapted from Cooke and Doornkamp (1990)

Mechanical measures	Biological measures
Bench terraces	Cover cropping
Contour bunds	Mulching
Tie-ridging	Afforestation
Strip cropping	Contour cultivation
	Minimum and no-till cultivation

Table 14.6   Techniques for controlling erosion by wind. Adapted from Cooke and Doornkamp
(1990)

Reduction of wind velocity	Reduction of soil erodibility
Vegetative measures:	*Moisture conservation*:
Cover cropping	Mulching
Close-growing crops	Tillage
Sand dune stabilization (grasses and	Timing seedbed preparation
afforestation)	Irrigation
	Terracing
Cultivation measures:	Contour cultivation
Mulching	Strip cropping
Rotation grazing	
Crop rotation	*Topsoil conditioning*:
Planting crops normal to prevailing winds	Correct timing of tillage
Field and strip cropping	Minimum tillage
Primary and secondary tillage	Crop rotation
	Manuring
Mechanical measures:	Chemical stabilizers
Windbreaks	
Shelter belts	
Dune stabilization by brush matting or stones	

better vegetation cover can also reduce the impact of rainfall. Alternatively, land-use
practices which promote less erosion can be introduced. Similarly, conservation
methods used to combat erosion by wind can be divided into two on the basis of their
mode of operation (Table 14.6). One group of methods aims at a reduction in the
surface wind veolocity by increasing the surface roughness. A second group reduces
soil erodibility by altering the soil characteristics.

14.6.2   Application of conservation measures

Although many methods of combating soil erosion have been devised, it is apparent
that the adoption of soil conservation measures among farmers the world over is
proving a more difficult problem than the development of new soil conservation
techniques. Agricultural economists in the USA have conducted extensive research
into the levels of uptake of soil conservation measures in an attempt to identify the

causes of their non-adoption by farmers. Land tenure appears to be the most important factor influencing the adoption of soil conservation techniques. Significant variations in the uptake of soil conservation measures between different types of land tenure have been ascribed to the varying distance between landowners and the decision makers on the farms, differences in conservation attitudes and ethics, variations in farm economies and farm size, and the different lengths of economic planning horizons.

Although many techniques have proved effective under the prevailing environmental conditions in North America and Australia, where most techniques have been developed, their application to developing world agriculture remains a problem. A review of the adoption of conservation measures in Africa (Millington *et al.*, 1989) suggests that three factors dominate farmers' decisions to adopt conservation strategies. First, factors such as land availability, farm size and type of land tenure system. Second, capital resources, access to credit, the substitution of labour for capital and seasonal labour availability. Finally, access to government and political institutions and their functioning with respect to individual farmers. These factors are often interrelated and it is impossible to highlight any one as the most important.

Farm size is an important consideration because land losses are associated with the construction of many conservation measures (Table 14.7). Consequently short-term declines in farm productivity may occur even though there may be a long-term increase in productivity. The loss of land due to the introduction of soil conservation on a small peasant farm of one to two hectares may be more than the farmer can withstand economically, causing the farmer not to adopt soil conservation measures. Alternatively, differences in land loss between conservation techniques may determine which technique to adopt. It has been shown in Latin America and the Caribbean that, as a consequence of the land loss and decreased soil fertility, only larger farms with land surpluses readily accept certain conservation methods. There

Table 14.7   Land losses associated with commonly adopted soil conservation techniques in Africa. Adapted by permission from Millington *et al.* (1989)

Conservation technique	Land loss per hectare as a proportion of total area (%)
Bench terracing*	7.5–52.5
Grass strips†	16.7
Contour bunding‡	9.1
Stick and Stone bunding§	2.2
Broad based banks¶	0
Contour cultivation‖	0
Conservation tillage	0

* According to the Food and Agriculture Organization guidelines for slopes ranging from 5 to 35%.
† Based on field data from Swaziland, assuming 3 m wide strips.
‡ Based on field data from Machakos District, Kenya.
§ Based on field data from Sierra Leone.
¶ Based on an assumed grazing land use.
‖ Based on field data from Malawi.

is also evidence to show that as populations increase, and more arable land is required, farmers find land to cultivate, sometimes to the detriment of effective soil conservation (Millington *et al.*, 1989).

14.6.3   Adoption of soil conservation measures in developing countries

The adoption of soil conservation in developing countries is the most important current problem facing conservationists. This is because individual farmers, as well as the economies of developing countries, are least able to withstand the effects of declining land productivity and the economic costs of the off-site impacts of erosion. Developing countries abound with examples of failed soil conservation schemes where inappropriate methods were introduced into an environment with little understanding of the socio-economic situation. How then can the record of soil conservation in such areas be improved?

If these failures are analysed it is apparent that the underlying cause in most instances is a gap between scientific knowledge and farmer perception. Expatriate experts with formal scientific and economic training formulate conservation strategies and policy according to their perceptions of the erosion problem. Peasant farmers are equally concerned about soil erosion, but at the same time have their own set of resource management priorities and their own understanding of local environmental mechanisms. The gap between the 'expert' and the farmer is often immense and can lead to serious mismatches between the two groups in the perception of the importance of soil erosion as a problem. For example, in Sierra Leone 16 chiefdoms (administrative units) were ranked according to the severity of the soil erosion hazard perceived by 750 farmers from the chiefdoms, and according to erosion hazards measured using 'standard' environmental and socio-economic parameters suggested by previous studies. The two rankings showed very little correspondence (Millington *et al.*, 1989).

These fundamental differences arise because of different perceptions of the importance of environmental problems between the two groups. Peasant farmers have limited risk capital, seasonal labour constraints, little machinery and restricted access to institutions. Therefore they must formulate indigenous, or adopt external, soil conservation techniques within their local set of constraints. Conservation activity promoted by governments and development agencies is strongly influenced by external factors, especially the global political economy and global environmental issues, rather than fluctuations in the severity of environmental problems in the particular countries.

The recognition of this gap in knowledge and preception casts serious doubts on the effectiveness of government directed conservation planning. Not surprisingly it has been suggested that local knowledge of the farming community must be utilized in soil management and this must be the way forward for soil conservation planning in developing countries. If this approach is taken, the main criteria for the introduction of soil conservation planning must be the perception of soil erosion problems by farming communities and individual farmers rather than by the 'experts'.

14.7 CONCLUSIONS

Accelerated soil erosion is widespread, although until recently it was thought to be a problem restricted mainly to arid and semi-arid areas. However, the increased awareness of land degradation processes has shown that soil erosion is present in areas originally thought to be free from such problems, e.g. the humid tropics and north-west Europe.

Along with an increased knowledge of erosion processes has come increasing awareness of the on-site and off-site impacts of erosion. The socio-economic impacts of decreased land productivity and the environmental and economic costs of off-site impacts are widespread, persistent and costly.

Despite the recognition of erosion problems in countries such as Britain, Belgium and the USA, it is the developing countries that are least able to cope with the effects of erosion, both at the state and individual farm levels. It is now recognized that socio-economic factors, as well as environmental factors, cause soil erosion. This has led to an increasing realization that, in developing countries, trained scientists, economists and policy makers are often at odds with farmers when it comes to their perceptions of the relative importance of soil erosion compared with other environmental problems, the causes of erosion and solutions to the problem. It has been suggested that the local knowledge of the farming community should be utilized in conservation planning. This suggests a reorientation in soil erosion research and conservation planning, away from top-down scientific remedies and towards a more balanced use of the scientific knowledge of erosion processes and the socio-economic problems which hinder their adoption.

Soil erosion first became a global environmental issue in the 1930s. Since then much has been learnt about the processes. Models to predict erosion have been formulated and conservation measures devised. Nevertheless, soil erosion continues to be a major global environmental problem, and one that is growing. The global community must be even more concerned about 'the rape of the earth' in the 1990s.

14.8 REFERENCES

Biot, Y., Sessay, M. and Stocking, M. (1989) Sustainability of Agricultural Land. *Land Degradation and Rehabilitation*, **1**, 263–278.

Cooke, R. U. and Doornkamp, J. (1990) *Geomorphology in Environmental Management*, 2nd edn. Oxford: Oxford University Press.

El-Ashry, M. T. and Ram, B. J. (1987) Sustaining Africa's natural resources. *Journal of Soil and Water Conservation*, **42**, 224–227.

El-Swaify, S. A. and Dangler, E. W. (1982) Rainfall erosion in the tropics: a state-of-the-art. In: El-Swaify, S. A. and Dangler, E. W. (Eds). *Soil Erosion and Conservation in the Tropics*. Madison: American Society of Agronomy and Soil Science Society of America, pp. 1–25.

Follett, R. F. and Stewart, B. A. (Eds) (1985) *Soil Erosion and Crop Productivity*. Madison: American Society of Agronomy and Soil Conservation Society of America.

Jacks, G. V. and Whyte, R. O. (1939) *The Rape of the Earth*. London: Faber & Faber.

Lal, R. (1976) *Soil Erosion Problems on an Alfisol in Western Nigeria and Their Control*. Ibadan: International Institute of Tropical Agriculture.

Millington, A. C., Mutiso, S. K., Kirkby, J. and O'Keefe, P. (1989) African soil erosion— nature undone and the limitation of technology. *Land Degradation and Rehabilitation*, **1**, 279–290.

Moberg, J. P. (1972) Some soil fertility problems in the West Lake region of Tanzania, including the effects of different forms of cultivar on the fertility of some ferralsols. *East African Agricultural and Forestry Journal*, **37**, 35–46.

Morgan, R. P. C. (1985) Assessment of soil erosion risk in England and Wales. *Soil Use and Management*, **1**, 127–131.

Morgan, R. P. C. (1986) Soil degradation and soil erosion in the loamy belt of northern Europe. In: Chisci, G. and Morgan, R. P. C. (Eds). *Soil Erosion in the European Community*. Amsterdam: Balkema, pp. 165–172.

Patanik, N. (1972) Soil erosion: a menace to the nation. *Indian Farming*, **24**, 7–10.

Zachar, D. (1982) *Soil Erosion*. Amsterdam: Elsevier.

14.9 FURTHER READING

Blaikie, P. (1986) *The Political Economy of Soil Erosion in Developing Countries*. London: Longman.

Blaikie, P. and Brookfield, H. (1987) *Land Degradation and Society*. London: Methuen.

Morgan, R. P. C. (1986) *Soil Erosion and Conservation*. London: Longman.

CHAPTER 15

Desertification

J. Wellens and A. C. Millington

15.1 INTRODUCTION

Desertification is an evocative and a misleading term. It is evocative in that it conjures up a picture of encroaching desert—of sand dunes advancing over agricultural land around the edges of deserts. It is misleading in that it implies a single process rather than a number of different processes and that the desert edge does not advance intact. It is the combined effect of the following processes in drylands: accelerated wind and water erosion, woodland destruction, waterlogging and the salinization of irrigated land.

Confusion exists between the concepts of desertification, drought and progressive desiccation (Goudie, 1990). The idea of progressive desiccation, albeit with fluctuations between wetter and drier periods, of dryland climates since the last glacial period is supported by much evidence, but it is not the sole cause of desert expansion. Anthropogenic influences on environmental degradation in drylands were first postulated by George Perkins Marsh in 1864, and disasters such as the Dust Bowl in the American Mid-West in the 1930s and the successive African droughts of the 1970s and 1980s have focused our attention on human-induced degradation. There is now little disagreement that the processes leading to desertification are largely due to an overexploitation of dryland resources. Short-term climatic fluctuations, such as one or two year droughts, exacerbate the problem, but do not cause desertification.

There are many definitions of desertification which relate to the processes involved and/or the significance of climate and people. This chapter promotes the view that desertification comprises well defined processes which operate singly or in combination in dryland regions to cause environmental degradation. These are naturally occurring processes which are aggravated under adverse climatic conditions and where population pressure is high.

15.2 CAUSES OF DESERTIFICATION

Overgrazing, overcultivation, woodland destruction and poor irrigation practices all contribute to desertification. These lead to an acceleration of processes already

Environmental Issues in the 1990s. Edited by A. M. Mannion and S. R. Bowlby
© 1992 John Wiley & Sons Ltd

common in drylands, e.g. the physical and biological degradation of soils, wind and water erosion, and soil salinization. The combination and intensity of causes and processes varies under different land uses. For example, on rangelands overgrazing leads to a diminution of the sparse vegetation in addition to soil compaction, which then often lead to soil erosion. Soil erosion is also the greatest threat in areas of rain fed cultivation. In irrigated areas the dominant processes are soil waterlogging and salinization.

15.2.1 Overgrazing

Increases in herd sizes have led to increased grazing pressure on many arid and semi-arid rangelands. The expansion in livestock populations is due to the increased availability of veterinary services, well-drilling programmes and the investment in livestock by urban-based entrepreneurs. At the same time there has been a reduction in rangeland areas due to the expansion of rain fed cultivation into regions traditionally set aside for stock rearing.

Range management practices among nomadic pastoralists were, and still are, ecologically sustainable given the levels of grazing resources available. However, recent changes in pastoral communities have led to an acceleration of desertification processes in many instances. Traditional methods of rotational pasture use based on the seasonality and productivity of individual pastures such as nomadism and transhumance, have been curtailed by settlement programmes. Under these programmes grazing is often restricted, leading to year round use of some pastures and soil compaction by the concentrations of livestock at water holes. This makes vegetation recolonization difficult and increases the risk of water and wind erosion. Heavily grazed land around villages and along grazing routes are similarly affected. The management of rangeland vegetation by fire can also lead to accelerated erosion. Late season, high temperature fires are used to kill off shrub and tree seedlings, without damaging dormant grass species which then thrive and provide good pasture for grazing animals. However, such fires leave the ground bare prior to the start of the rains in many semi-arid areas, leading to high levels of erosion in the early wet season.

The extent of overgrazing varies in relation to the type of animals. Firstly, grazing animals are selective in the plants that they eat, with palatable young foliage being preferred. As grazing pressure increases, more foliage is consumed, impairing the reproductive potential. Ultimately, this results in a reduction in the amount of vegetation and increases the extent of bare soil. Secondly, gregarious animals, such as sheep, graze in concentrated groups, thus increasing the potential to severely degrade small areas. Thirdly, the resource demands of animals vary. For instance, camels are better adapted to arid rangelands than cattle, and produce more milk for each unit of biomass consumed. They graze a wider range of plants and need fewer water holes than cattle. Finally, it should be noted that not all overgrazing is attributable to domestic livestock. Domestic animals can cause severe degradation problems when they become feral, for example, rabbits, cattle, donkeys, horses and goats cause problems in Australia, and ungulates and elephants are known to overgraze and cause extensive vegetation destruction in eastern and southern Africa.

15.2.2 Overcultivation

Rain fed agriculture in drylands is mainly restricted to semi-arid and subhumid areas where rainfall and soil moisture levels are adequate for annual crops to grow. However, the traditional areas of rain fed agriculture are expanding to more arid areas where such types of agriculture are less sustainable because the rainy season is shorter and less reliable. For example, Goudie (1990) noted that rain fed dryland cultivation occurs in areas with a mean annual rainfall as low as 150 mm in North Africa and the Middle East.

The restrictions placed on crop growth by the soil moisture content results in extensive cultivation methods with a restricted choice of crops and poor yields. Despite these problems the increasing food requirements of expanding populations and the need for foreign exchange earnings from cash crops has led to an expansion of rain fed cultivation in many dryland areas. This expansion has been effected in two ways, firstly by extending cultivation into less productive, marginal areas traditionally reserved for stock rearing and secondly by reducing the fallow period between cultivation. For example, in the Oglat Merteba region of southern Tunisia, the conversion of rangeland to cereal farming in the late 1970s could have resulted in profits of 10–18 dinar per hectare and the cultivation of almonds could realize 30 dinar per hectare. This compared with the gross profit of 2–3 dinar per hectare resulting from grazing sheep (Unesco, 1980). As a result of reducing the fallow period between cultivation, there has been a reduction in average crop yields, whilst those rangelands still available to pastoralists have witnessed declining productivity due to overgrazing.

To reduce competition for water and nutrients, most of the natural vegetation of an area is removed prior to cultivation. This interferes with vegetation–soil interactions, which are important for maintaining environmental stability. Crops provide little cover during the early stages of growth, and once harvested there is no cover until the next season, unless the stubble is left in the ground. In rain fed areas with short growing seasons this may result in the ground being left unprotected for many months. The low yields obtained from such land mean that there is little short-term economic advantage to be gained by investment in fertilizers, pesticides, soil conservation measures or other land management techniques.

The overall reduction in vegetation cover, the timing of farming operations and the disruption of critical vegetation–soil interactions means that water and wind erosion (see Chapter 14) are commonplace in overcultivated areas. The type and rate of erosion is intricately linked to medium- and short-term climatic fluctuations. For example, a drought lasting for several years will lead to soil desiccation and a poor vegetation cover. Such conditions promote wind erosion. Alternatively, runs of wet years can give pastoralists and cultivators false expectations, encouraging them to increase their herds or extend into marginal zones. These can have equally disastrous consequences, leading to accelerated soil erosion.

15.2.3 Woodland destruction

Although woody biomass resources are scarce in drylands, woodland destruction continues to occur to meet timber, fuelwood and charcoal needs as well as to provide agricultural land. The impact of fuelwood collection on woodlands is important in most dryland areas. Fuelwood collected in such areas is destined for either the local rural community or, more likely, urban areas. The effect of this urban demand has been to create 'rings of deforestation' around many cities. Fuelwood and charcoal destined for major cities in drylands such as Dakar, Karachi and Khartoum is transported over 800 kilometres in some instances, and cities such as Kano, Niamey and Ouagadougou have extensively cleared areas of over 10 000 square kilometres. A study of deforestation between 1972 and 1982 in a 30 000 square kilometre area around 41 Indian cities found that between 48 and 92% of the forest had been lost around dryland cities such as Jaipur and Bhopal (Bowonder *et al.*, 1987).

Woodland clearance to provide land for rain fed agriculture is common in Africa, south Asia and Latin America. Although it provides a short-term increase in woody biomass resources because large amounts of wood become available over a short period, the medium-term result is usually environmental degradation. For example, Grainger (1990) reports that 88 000 hectares of woodland are cleared annually in the Kordofan and Darfur regions of Sudan to make way for mechanized sorghum cultivation. After being cropped for three to four years about 45% of the land is degraded to such an extent that it has to be abandoned.

15.2.4 Waterlogging and salinization

Throughout history cultivators trying to extend croplands through irrigation have encountered soil waterlogging and salinization. A prime example is that of the ancient civilization of Mesopotamia about 4000 years ago. Irrigation is carried out to maintain soil moisture within the limits for optimum plant growth and provides a method by which agricultural productivity in arid and semi-arid areas can be increased. Water-logging and salinization can dramatically reduce land productivity and ultimately take land out of production. This is a serious and widespread problem in drylands (Table 15.1).

Waterlogging occurs when the water-table lies close to the soil surface. High water

Table 15.1 Proportion of irrigated land affected by salinization for selected dryland nations (proportions as percentages). Adapted from International Institute for Economic Development (IIED) and World Resources Institute (WRI) (1987)

Asia		*Africa*	
India	27	Algeria	10–15
Iran	<30	Egypt	30–40
Iraq	50	Senegal	10–15
Israel	13	Sudan	<20
Jordan	16		
Pakistan	<40	*North and South America*	
Sri Lanka	13	Colombia	20
Syria	30–35	Peru	12
		USA	5–20
Others			
Australia	15–20		
Cyprus	25		

levels in the soil created by irrigation restrict the plants' rooting depth by reducing soil aeration. Increases in water-table height may be due to the lateral movement of groundwater from an area with a higher water-table to an area of lower lying land. Additionally, the water-table may be high due to the accumulation of irrigation water above an impermeable or slowly permeable horizon. Most major crops, notably wheat and cotton, which are grown in irrigated areas cannot tolerate waterlogging. Rice is the main exception and thrives in such conditions. Within the fields, the distribution of irrigation water may not be regulated, often due to the notion that if some is good then more must be better. This can lead to large surpluses of water being applied, which increases the likelihood of waterlogging.

Salinization is the increase in the concentration of soluble salts in a soil. In drylands, high temperatures can cause high rates of evaporation, which draw water and salts up through the soil by capillary action. As the water evaporates from the soil surface, salts precipitate out and accumulate. This can happen even when the initial salt concentration of the water is fairly low, and is a particular problem in clay soils. Salinization also occurs when saline irrigation water is applied to slowly permeable soils. For example, the application of 10 000 cubic metres of mildly saline irrigation water (50 mg/l of salts) per hectare of land deposits two to five tonnes per hectare of salt in the soil (Kovda, 1980). The use of artesian groundwater for irrigation can cause such problems because it is often more saline than water from other sources. The Saharan oases in Tunisia and Algeria have been affected in this way. The salt content of the soil varies depending on the season. During wetter periods salts may be leached, while during the dry season salts move upwards. Some degree of control over salinity levels can take place if the water is applied during the dry season to leach the salts accumulating near the surface.

Salinization is detrimental when the salt concentration is high enough to impair plant development. Crops vary considerably in their salt tolerance, with date palms and barley being two of the most salt-tolerant crops (Table 15.2). Generally, soil is considered to be toxic if the salt concentration is greater than 0.5–1.0% (Kovda,

Table 15.2 Salt tolerance of various crops

Tolerance to ESP (range at which affected)*	Crops	Growth response
Extremely sensitive (ESP 2–10)	Deciduous fruits, nuts, citrus fruits, avocados	Sodium toxicity even at low ESP values
Sensitive (ESP 10–20)	Beans	Stunted growth at low ESP values, even if good soil conditions
Moderately tolerant (ESP 20–40)	Clover, oats, rice	Stunted growth
Tolerant (ESP 40–60)	Wheat, cotton, alfalfa, barley, tomatoes, beet	Stunted growth due to poor soil conditions
Most tolerant (ESP >60)	Wheat grasses, Rhodes grass	Stunted growth due to poor soil conditions

* Exchangeable sodium percentage (ESP) is calculated as the content of sodium ions as a percentage of the total exchangeable bases in milliequivalents.

1980). The concentration of sodium in the soil is also important because of the effect it has on the uptake of soil moisture by roots. Sodium concentrations, expressed as exchangeable sodium percentage (see Table 15.2), greater than 2% are detrimental to most crops.

Salinization is also a serious problem in the wheat growing areas of North America and Western Australia. Saline seeps occur on slopes with soils derived from naturally saline substrates. When deep-rooting trees and shrubs are replaced by shallower rooting crops the amount of water lost through evapotranspiration is reduced while percolation increases. This results in salts being transported in solution through the soil, seeping out on to adjacent lowlands and leading to salinization and reduced land productivity.

Although salinization affects relatively little land compared with the other desertification processes, the economic loss per hectare of salinized, irrigated land is three times as great as for areas of desertified rain fed cropland and 100 times as great as for desertified rangeland (Dregne, 1983). An approximate cost for this can be estimated from the fact that in 1980 a third of the World Bank's lending for agricultural development (about $2.6 billion) was for irrigation projects (Grainger, 1990).

15.2.5   Other causes

Goudie (1990) lists a number of specific causes of desertification. These are more localized or less well known than those outlined in the preceding sections:

(a) dung is often used as a fuel in southern Asia, at the expense of its use as a manure; as a consequence soil fertility may decline while susceptibility to erosion may increase

(b) eucalyptus trees are extensively planted for fuelwood, shelter belts and for shade in drylands, but there is evidence that they use such large amounts of soil moisture that they desiccate soils

(c) off-road vehicles can seriously disturb fragile desert ecosystems and affect the soil

(d) the overexploitation of groundwater resources is common and can lead to loss of habitats, drying up of oases, ground subsidence, reduced aquifer storage capacity and increased pumping costs

(e) the diversion of water from catchments for irrigation can lead to lake shrinkage, for example, the Aral Sea has contracted by 67% in volume since 1960 due to the diversion of water from the Amu Darya and Syr Darya to irrigated areas in Turkmenistan

(f) changes in farming systems can have environmental consequences; for example, in Yemen and Greece labour shortages have resulted in the lack of maintenance of soil conservation structures, which eventually fall into disrepair causing rates of erosion to increase

(g) soil compaction and crusting occurs as a result of rainsplash erosion (see Chapter 14); sediment-rich irrigation water infills soil pores and agricultural machinery causes compaction; crusts form a barrier on the soil surface which reduces the infiltration rate—this reduces the water available for plants, and increases run-off and erosion by water, restricting the emergence of new seedlings and further reducing vegetation cover.

15.2.6 Destruction of vegetation

A key factor in many of the processes that cause desertification is the destruction of vegetation. Vegetation is an important resource in arid and semi-arid regions because of the generally low levels of biomass and its irregular distribution. Grazing and browsing animals depend on foliage and fruits for food, and many plants provide food, fibre, dyes and medicines for humans. Woody plants also provide fuel and timber resources.

Vegetation also has an equally important role in environmental protection. It provides a supply of organic matter to the soil, which increases nutrient levels and promotes soil aggregation. Wind speeds are reduced by the friction that occurs between the vegetation and near-surface air movements, and wind-borne soil and dust particles are trapped by vegetation. Vegetation intercepts rainfall, evaporating some of it back into the atmosphere and releasing some to the ground as stemflow and leaf drip. Consequently, run-off is reduced and infiltration rates increase, which lead to a reduction in erosion by water. The removal of vegetation not only exposes the soil to erosion processes but leads to a higher proportion of incoming solar radiation being reflected back to the atmosphere. This may inhibit further rainfall, leading to further desiccation.

15.3 DESERTIFICATION: SPATIAL AND TEMPORAL DIMENSIONS

Many surveys have been undertaken to determine the spatial extent of desertification (see discussion in Grainger, 1990). The estimates quoted here are from Mabbutt (1984), a survey carried out for the United Nations Development Programme. These have been selected because they include information on subhumid regions as well as semi-arid, arid and hyper-arid regions, and because they are related to population distribution. Nevertheless, it should be noted that the estimates are based on questionnaire surveys sent to national governments and not on field measurements.

15.3.1 Spatial patterns of desertification

Mabbutt (1984) classified desertified land into four categories (Table 15.3). The total area suffering from at least moderate desertification was, after adjustments, 34 750 km² (Table 15.3), the distribution of which is shown in Fig. 15.1. These statistics show distinct regional differences in the proportions and in the types of land desertified. A number of trends are apparent from Table 15.3:

(a) most desertified land is rangeland; cultivated land accounts for a smaller area, of which rain fed cropland is areally more extensive than irrigated cropland

(b) the relative importance of different desertification processes varies between regions mainly because of differences in the dominant types of land use between regions: soil compaction and erosion of rain fed cropland is a problem over much of Africa, in south Asia and in South America; salinization and waterlogging of

Table 15.3 Extent of areas at least moderately desertified. Adapted from Mabbutt (1984) by permission of the Foundation for Environmental Conservation, Geneva. Moderately desertified land was defined as a <25% loss in cropland productivity and a <25% loss in the livestock carrying capacity of rangelands

Region	Rain fed cropland		Irrigated cropland		Rangeland		Total	
	Area (10³ km²)	(%)	Area (10³ km²)	(%)	Area (10³ km²)	(%)	Area (10³ km²)	(%)
Africa	1290	37	19	5	6100	38	7409	37
North Africa	150	4	5	1	680	4	835	4
Sudano-Sahel	720	21	8	2	3420	21	4148	21
South of Sudano-Sahel	420	12	6	2	2000	12	2426	12
Asia	1350	39	280	70	5850	36	7480	37
West Asia	150	4	30	8	980	6	1160	6
South Asia	1050	30	200	50	1270	8	2520	13
Soviet Asia	120	3	20	5	1500	9	1640	8
China/Mongolia	30	1	30	8	2100	13	2160	11
Australia	120	3	3	1	1000	6	1123	6
Europe	130	4	16	4	150	1	296	1
North America*	240	7	40	10	1800	11	2080	10
South America†	330	10	40	10	1250	8	1620	8
Total‡	3350		400		31 000		34 750	

* Excludes Mexico.
† Includes Mexico.
‡ Adjusted to include remote, non-productive rangelands.

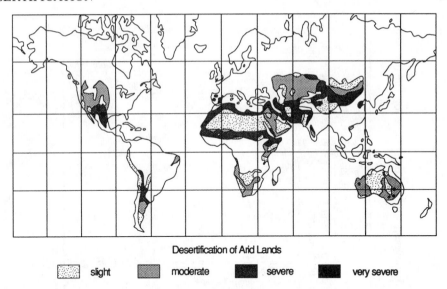

Desertification of Arid Lands

slight   moderate   severe   very severe

Figure 15.1   World map of areas at risk from desertification. Adapted from Grainger (1990) by permission of Earthscan Publications Ltd.

irrigated agricultural land is a serious problem in south Asia and, to a lesser extent, in the Americas; rangeland degradation, and the resultant acceleration of erosion processes, is serious over much of Africa, Central Asia (China, Mongolia and Soviet Asia) and North America

(c) overall the regions with the most extensive desertification are Africa and West Asia (Middle East); in these areas at least 80% of all drylands are moderately desertified and in terms of the amount of productive dryland affected in five regions at least 10% of the land is at least moderately desertified, i.e. Sudano-Sahelian Africa (21%), south Asia (13%), Africa south of the Sudano-Sahel region (12%), China and Mongolia (11%) and North America (10%)

(d) desertification is a more significant problem in developing, rather than developed, nations.

15.3.2   Populations affected by desertification

Mabbutt (1984) also provided data about the numbers of people exposed to desertification (Table 15.4). He estimated that about 280 million people living in rural areas were affected by at least moderate levels of desertification in the early 1980s. If the urban populations of these regions are included the population at risk increases to about 470 million, i.e. approximately 10% of the world's population. Most of the people affected by desertification live in areas of rain fed cropland. Irrigated areas are also densely populated, exposing their populations to desertification risks. Rangelands, although areally extensive, have much lower population densities. The extent of desertification is not, however, only restricted to the people living in the regions at risk. Many of these regions export products such as food and industrial crops, meat and fuelwood to urban areas outside the area at risk. This inevitably increases the number of people at risk.

Table 15.4 Populations affected by desertification. Adapted from Mabbutt (1984) by permission of the Foundation for Environmental Conservation, Geneva

Region	People (in millions) affected by at least moderate levels of desertification			People (in millions) affected by	
	Rain fed cropland	Irrigated croplands	Rangelands	At least moderate desertification on all types of land	Severe levels of desertification on all types of land
Africa	79	3.5	25.5	108	61
North Africa	11	1	4	16	8.5
Sudano-Sahel	36	1.5	13.5	51	27.5
South of Sudano-Sahel	32	1	8	41	25
Asia	56	50	17	123	55.5
West Asia	16	12	4	32	8.5
South Asia	34.5	23	9	66.5	16
Soviet Asia	1.5	4.5	1	7	29
China/Mongolia	4	10.5	3	17.5	2
Australia	0.1	0.1	0.03	0.23	6.5
Europe	13	1.5	2	16.5	0.03
North America*	2	1	1.5	4.5	6
South America†	22.5	2.5	4	29	13.5
Total	172.6	58.6	50.03	281.23	142.53

*Excludes Mexico.
† Includes Mexico.

15.3.3   Rates of desertification

Only one estimate of the global rate of desertification has been made and this should be treated with caution. Dregne (1983) calculated that each year 177 000 km² of rangeland, 20 000 km² of rain fed cropland and 5460 km² of irrigated cropland decline to the point of reaching a negative economic return.

15.4   CASE STUDIES

Three case studies are presented to illustrate the main points made in Section 15.2 on processes and causes. The first study examines woodland destruction in Yemen, whereas the second study concentrates on waterlogging and salinization in Pakistan. The final study examines overgrazing in southern Tunisia.

15.4.1   Woodland destruction in Yemen

Yemen, on the south-western tip of the Arabian Peninsula, once supported large tracts of subtropical woodlands. Much of this woodland has been cleared to provide agricultural land and to meet the demand for fuelwood, charcoal and construction timber.

Recent analysis (Millington, 1991) suggests that the woody biomass stocks will be severely depleted by the end of the century, and only one area of woodland, Jebel Bura, now remains. Between 1973 and 1987 in the Jebel Bura area there has been a large-scale reduction in woodland and shrubland areas. A little over half the area mapped (51.57%) showed a decline in tree crown density and 8.23% of the land had been converted from woodland and shrubland to agricultural land. In total the overall reduction in tree volume over the 15 year period was a little over 6000 cubic metres.

The major phase of woodland destruction for agricultural land clearance in Yemen occurred centuries ago and currently little wooded land remains. Current woodland destruction is mainly due to demand for fuelwood and charcoal and, to a much lesser extent, construction timber. Increased demand for wood products has occurred contemporaneously with the establishment of the Yemeni road network since the mid-1960s. In addition to the construction of the road network the introduction of four-wheel drive vehicles with their ability to reach far beyond the trunk road network has been equally important in recent woodland exploitation. This has meant that wood-supply areas have been linked to the main cities by roads, facilitating the movement of wood and promoting increased exploitation of the remaining wood-lands. Therefore the pattern of woodland destruction in areas such as Jebel Bura in Yemen can readily be explained in terms of increased accessibility to the area and the increased demand for fuelwood in the last two decades.

Woodland destruction in Yemen has socio-economic and environmental impacts. Taking into account the current and projected supply and demand estimates, the World Bank (Millington, 1991) has estimated that fuelwood stocks in Yemen will be depleted by the end of this century (Fig. 15.2). This will cause severe economic disruption for the government and individual households. The relationships between current woodland management practices, the projected levels of woody biomass exploitation and the Yemeni environment are more complex. However, evidence

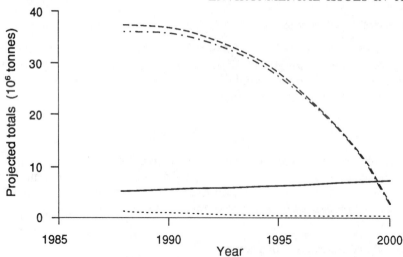

Figure 15.2   Estimates of fuelwood stocks and household energy demand in Yemen. Adapted from Millington (1989). (——) Consumption; (—·—) recently cut wood; (· · · ·) collected dead wood; (– – –) all wood.

suggests that the effects of woodland destruction at the rates projected will lead to increased soil erosion and terrace degradation, decreased recharge of aquifers, increased erosion by wind and encroachment by sand, microclimate degradation, and, through these processes, reduced agricultural productivity.

15.4.2   Waterlogging and salinization in Pakistan

Irrigation in Pakistan probably began with the earliest Indus civilizations, dating back to the 8th century AD. A number of types of irrigation are practised in the country and most of the irrigated land in Pakistan is on the floodplains of the Indus river and the rivers of the Punjab (Fig. 15.3). The present system of barrages and irrigation canals, which was begun in the mid-1800s and is still expanding, constitutes one of the most complex irrigation systems in the world. The area within the entire irrigation system is 13.4 million hectares, of which 13.3 million can be cultivated. The irrigated areas provide most of Pakistan's food and industrial crops, including wheat, cotton, oil seeds, rice, sugar cane and tobacco. Consequently, environmental problems in the irrigated areas that cause a decline in productivity are of critical importance to the Pakistani economy (Biswas, 1987).

Drainage canals, which channel excess water and receive surplus irrigation water, suffer from flow obstruction by weed and fungus growth and sedimentation, which significantly lower their effectiveness. Many of these canals are unlined and seepage losses are high; in fact, they account for 80% of aquifer recharge! (Biswas, 1987). This has led to a gradual rise in the depth of the water-table on irrigated land. Prior to these extensive irrigation developments water-table depths in the Indus Valley and the Punjab ranged from about 24 to 28 metres, but a dramatic rise in the overall levels has been evident throughout this century. This has led to extensive waterlogging and salinization (Table 15.5). Salinization has also been exacerbated by the use of

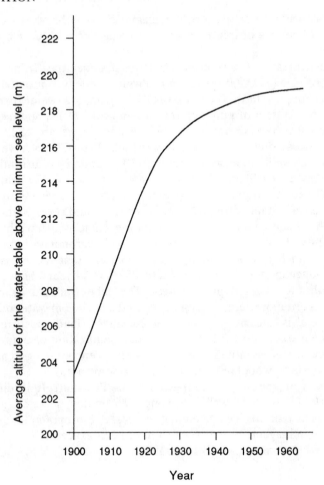

Figure 15.3   Increase in level of water-table in southern Punjab, Pakistan. Adapted from Unesco (1980).

Table 15.5   Waterlogging and salinization in Pakistan. Adapted from Unesco (1980). Salinity (percentage soluble salts) categories: normal (<0.2%), slight (approximately 0.2%), moderate (0.2–2.5%), high (>2.5%)

Province	Waterlogged area (as percentage of area under irrigation)			Salinized area (as percentage of surveyed area)			
	Severe (0–1.5 m)†	Moderate (1.5–3 m)	Total	Slightly saline	Moderately saline	Highly saline	Total saline
Punjab	6.05	25.18	31.23	14.80	4.30	6.50	25.63
Sind	11.73	37.67	49.40	20.64	27.12	50.49	98.26
NWFP*	10.00	1.00	11.00	—	—	—	9.00
Baluchistan	—	—	—	—	—	—	6.88

* North West Frontier Province.
† Values refer to depth of water-table.

saline irrigation water due to aquifer contamination. It has been estimated that currently 40 000 hectares of irrigated land are being lost each year by waterlogging and salinization.

Most remedial measures have been carried out under the Salinity Control and Reclamation Projects (SCARP) programmes which were first initiated in 1958. Two reclamation techniques have been used. First, pumping water from freshwater aquifers through a system of pumping stations and tube wells, and, second, vertical and horizontal subsurface drainage of saline water. There are eight SCARP programmes in the worst affected area, the Lower Indus Plains. They have been planned to cover some 3.6 million hectares but only 0.7 million is currently undergoing reclamation. This area includes 1381 freshwater tube wells, 379 saline water tube wells, 1437 km of surface drains and 10 pumping stations. The results are encouraging; in SCARP 1 the water-table dropped to between three and seven metres in 14 years. Waterlogging has been eliminated and 45% of the saline soils have been reclaimed. More importantly, land reclamation combined with agronomic improvements has improved crop and livestock yields to such an extent that the gross value of the agricultural produce in SCARP 1 increased by 250% in 14 years. However, problems do remain with the SCARP programmes. The tube wells suffer corrosion and encrustation and groundwater pumping has caused a decline in water quality. A more perverse problem is the assumption that reclaimed land must revert back to agriculture, which would probably lead to further salinization of unfavourable land. A change of land use is essential in such areas and it is encouraging to note the switch from wheat to on-farm woodlands and orchards in some areas.

Is salinity control and reclamation winning in Pakistan? It is difficult to say. Although nearly 15 000 tube wells, draining 3.86 million hectares, and over 5000 kilometres of surface drains have been dug, the SCARP programmes are still dogged by technical and management errors which have been exacerbated by the rising cost of energy needed to pump water (Biswas, 1987).

15.4.3 Overgrazing in Tunisia

The population of the Tunisian arid and semi-arid zones has increased seven-fold since the late 19th century, due mainly to advances in health care. Environmental pressures are therefore severe, particularly in regions on the edge of the Sahara where rainfall is only between 100 and 200 mm per year.

It was realized at the time of independence in 1958 that traditional land use practices would not be able to provide enough agricultural output or employment to support the increasing population. Government measures to encourage the settlement of nomads were introduced, including the construction of schools and dispensaries, borehole provision to provide year-round water and in some instances the construction of whole villages of houses with low rents.

The increase in the human population has been mirrored by an increase in stock numbers. However, whereas large flocks of sheep and goats used to be collectively managed, the settlement schemes have resulted in the proliferation of small, privately owned flocks which remain close to the village for most of the year. This has resulted in overgrazing and environmental degradation of the land around settlements. The proportion of goats to sheep is increasing, appearing to indicate a decline in the quality of rangelands because goats are better adapted to browsing poor rangeland than sheep. During years with low rainfall, the flocks, finding no new forage, rely on the surface leaf litter for food and this increases erosion risks. By the end of the summer of 1976–77 (a very dry year), litter made up 89% of the daily diet of sheep and 61% of the diet of goats (Novikoff and Skouri, 1981).

Experimental trials indicated that by deferring the grazing on some areas of the rangeland, a 22% increase in vegetation productivity could be achieved. In addition, there was a more even distribution of vegetation supply throughout the year, allowing supplementary feeding to stop and increasing the soil stability.

15.5 REMEDIAL MEASURES

Reducing the effects of desertification involves more than finding technical solutions to specific problems. The socio-economic problems facing their implementation and widespread adoption must be considered. Scientific and technical research has concentrated on solving particular problems, for example, soil conservation measures to reduce specific types of soil erosion and pasture improvements to improve grazing. It is beyond the scope of this book to review these individual measures. Soil conservation techniques have been discussed in Chapter 14, and methods to combat salinization, waterlogging and rangeland degradation are briefly discussed in Section 15.5.1. A comprehensive review of measures to combat desertification can be found in Goudie (1990). In Chapter 14 the importance of socio-economic factors in determining the rates of adoption of soil conservation measures were discussed. Socio-economic factors are important in the adoption of all remedial measures and, therefore, in combating desertification.

15.5.1 Rangeland improvements

Rangeland improvements are necessary to enhance productivity and to increase the amount and quality of forage. This facilitates grazing and the maximization of stock production. Such improvements allow the multiple use of rangelands consistent with sound ecological principles. The main rangeland improvement measures are:

(a) the introduction of indigenous and exotic forage plants
(b) reseeding
(c) planting fodder trees and shrubs
(d) soil and water conservation
(e) natural regeneration of vegetation by grazing and browsing control
(f) fencing to regulate use
(h) dune stabilization
(i) eradication or control of undesirable plants.

15.5.2 Controlling salinization and waterlogging

Generally, there are three ways in which salinization is controlled. Firstly, through the improvement or reclamation of poorly managed soils. This is usually achieved by drainage improvement allowing soluble salts to be leached out of the soils. In sodic soils the sodium has to be replaced by calcium or magnesium, usually in the form of gypsum. Secondly, the prevention or control of waterlogging and salinization in unsalinized or slightly salinized soils. Thirdly, the introduction of land and water management techniques to minimize inputs of saline water. It can be seen that all of these remedial measures rely heavily on the provision of adequate drainage systems and efficient leaching.

15.6 CONCLUSIONS

It has been shown that desertification is the combined effect of accelerated erosion by wind and water, woodland destruction, soil waterlogging and salinization in dryland environments. Although these are all natural processes in drylands, they are exacerbated by long-term or short-term adverse climatic conditions and anthropogenic activity. The initiating factor, except for salinization and waterlogging, is usually the destruction of vegetation.

The problem is widespread, affecting rain fed and irrigated cropland, rangeland and woodlands. The problem is most prevalent in the developing countries, although significant problems exist in Australia, Canada, Spain and the USA. Well over 300 million people living in rural areas are affected by desertification. The problem is, however, more widespread than this as many large dryland cities, and some cities outside dryland areas, are dependent on these degrading areas for food, fuel and industrial crops. The total number of people affected by desertification was estimated in the early 1980s to be about 10% of the world's population.

Technical solutions to many of the problems faced in drylands exist. Implementing them is another matter, and must involve the recognition at an early stage of the socio-economic factors which lead to the acceleration of the processes that cause desertification. However, remedial measures may prove inadequate. Trying to modify deserts through reclamation and irrigation schemes which do not take into account the capability of dryland ecosystems to recover from disturbance may always be doomed to failure. As Cloudsley-Thompson (1988) puts it 'Instead of trying to change the land to make it conform to present economic and political expectations, development should be adapted to exploit the potentialities of the environment as it exists.'

15.7 REFERENCES

Biswas, A. K. (1987) Environmental concerns in Pakistan, with special reference to water and forests. *Environmental Conservation*, **14**, 319–328.
Bowonder, B., Prasad, S. S. R. and Unni, N. V. M. (1987) Deforestation around urban centres in India. *Environmental Conservation*, **14**, 23–28.
Cloudsley-Thompson, J. L. (1988) Desertification or sustainable yields from arid environments. *Environmental Conservation*, **15**, 197–204.
Dregne, H. (1983) *Desertification of Arid Lands*. Chur: Hardwood Academic.
Floret, C. and Hadjej, M. (1977) An attempt to combat desertification in Tunisia. *Ambio*, **6**, 366–368.
Goudie, A. S. (Ed.) (1990) *Techniques for Desert Reclamation*. Chichester: Wiley.
Grainger, A. (1990) *The Threatening Desert*. London: Earthscan.
IIED and WRI (1987) *World Resources Report*.
Kovda, V. A. (1980) *Land Aridization and Drought Control*. Boulder: Westview.

Novikoff, G. and Skouri, M. (1981) Balancing development and conservation in pre-Saharan Tunisia. *Ambio*, **10**, 135–141.

Mabbutt, J. A. (1984) A new global assessment of the status and trends of desertification. *Environmental Conservation*, **11**, 103–113.

Millington, A. C. (1992) Rapid appraisal of biomass resources: A case study of northern Yemen. *Biomass and Bioenergy*, in press.

Parry, M. (1990) *Climatic Change and World Agriculture*. London: Earthscan.

Unesco (1980) *Case Studies on Desertification*. J. A Mabbutt and C. Floret (eds.) Paris: Unesco.

15.8   FURTHER READING

Glantz, M. N. (Ed.) (1987) *Desertification*. Boulder: Westview.

Mortimore, M. (1989) *Adapting to Drought*. Cambridge: Cambridge University Press.

SECTION III

LOCAL IMPACTS AND REACTIONS
(b) Change in Societies

CHAPTER 16

Transport: Maker and Breaker of Cities

P. Hall

16.1 INTRODUCTION

The title of this chapter commemorates a classic paper by the economist Colin Clark (Clark, 1957). This was written as long ago as 1957, but remains one of the most important contributions ever made to analysing the problems of transport in cities. Clark argued that, at least since the first Industrial Revolution two hundred years ago, the growth of cities had been shaped by the development of their transportation facilities. These in turn were dependent on the evolution of transportation technologies. For each successive development of the technology, there was a corresponding kind of city. However, the relationship was more complex than that: it was a mutual one. The transportation system shaped the growth of the city, but on the other hand the previous growth of the city shaped and, in particular, constrained the transportation alternatives that were available. The pattern of activities and land uses in the city, and the transportation system, therefore existed in a kind of symbiotic relationship.

However, Clark stressed, the two could get out of step, and indeed very often do so. This is particularly true because cities change more slowly than the available technologies change. Almost all cities, except the very newest ones, reflect the slow accretion of buildings and functions over time. This is very obvious in any European city: London and Manchester both go back 2000 years to Roman times, and most other British towns and cities can count at least 1000 years of history. Even in North America, cities are older than they first seem; New York dates back to 1633, while on the west coast San Francisco has two hundred years of history since the establishment of the first Franciscan mission, and nearly 140 years of continuous development since the effective start of the city during the hectic gold rush days of 1849.

In all these cities, the constraints of the past hang heavy. In London, Cologne or Rome the street pattern goes back to the Middle Ages or beyond. In Boston,

Environmental Issues in the 1990s. Edited by A. M. Mannion and S. R. Bowlby
© 1992 John Wiley & Sons Ltd

Massachusetts, the main shopping street is still the one that the Puritan fathers marked out as a trail in 1630. Of course, these past inheritances may be changed by rebuilding or new streets, but they are expensive, and are often politically unpopular. That is why such cities have trouble in accommodating modern traffic, and why many of them, in Europe especially, have given up trying: they have simply excluded the automobile from their downtown areas, returning by choice to the pedestrian world of the Middle Ages.

This chapter traces some of the consequences of the Clark paradox. In particular, it asks how cities and their transportation systems have evolved historically, and how they can get out of step. It will show that there have been four such crises in urban transport planning: around 1850, around 1880, in the 1950s and now again in the 1990s. It concludes by asking what can be done about the problem.

16.2   PAST CRISES IN URBAN TRANSPORT PLANNING

About 150 years ago, most cities, even the biggest, virtually did without what would today be regarded as a transportation system. True, horse-drawn carts brought agricultural produce into the city and took industrial produce out. The biggest cities tended to be ports, such as London, Bristol, Liverpool and Glasgow. This reflects the fact that until this time, the only efficient way of carrying bulk goods any distance was by water, and navigable water often did not extend very far inland. Early canals and the first railways were just beginning to change this state of affairs. People in this *pre-public transport city* tended to get around on their own two feet. The BBC's classic production of *Little Dorrit* gives a realistic feel of that early Victorian city: it is full of people walking, walking, walking. This meant that they had to crowd together close to their work.

The result was cities that by modern standards were extraordinarily small and dense. London extended barely beyond today's City and West End, although it housed two million people. In virtually every other British city, no matter how big, it was possible to walk from the centre to green fields in twenty minutes or less. There was plenty of open land beyond, land that was to become some of the most valuable urban real estate in the world, but it then had no value beyond agricultural value, because it was too far for daily travel on foot. The result, as Clark showed in another classic paper (Clark, 1951), was an extraordinarily steep population density gradient: within the city population densities were very high, but they soon descended to rural levels. Indeed, commuting in its modern sense did not exist.

At that time, when Dickens was writing, this kind of city faced a crisis. It could not grow much further as long as the jobs remained in or near the centre and as long as there was no personal transport. Something had to change. It changed between 1850 and 1870. Steel wheels on steel rails were a great technological development, which reduced the frictional resistance of travel and made it possible to shift not merely bulk goods, but also bulk loads of people, over land. The result was the horse-drawn tramcar, and then (in some cities) a steam-powered version. And duly, from around 1870, there was what Sam Bass Warner has called the development of streetcar suburbs around American cities, and equivalent tramcar suburbs in Birmingham, Manchester, Liverpool, Leeds and other British cities. In the greatest cities of the world, there was another development: commuter rail services. The very word 'commuter', which came into the language in this sense at that time, meant a person who 'commuted' his commitment—it was invariably his commitment, in those days when middle-class women minded the hearth—to buy a daily railroad ticket into one bought by the week or month.

Cities that developed on this pattern, characterized by London, Manchester, Paris or New York around 1890, can be described as *early public transport cities*. Horse-drawn tramcars allowed only a very limited suburban extension of the city—limited because they were very slow. Steam-powered commuter railways allowed a much bigger spread, but it took the form of more or less discrete blobs around the stations, because steam trains are slow to accelerate or decelerate. The classic early suburbs such as Bedford Park in London, Edgbaston in Birmingham, Didsbury in Manchester, Bearsden in Glasgow and their American equivalents, were of this kind.

As these technologies were relatively unsophisticated, again a crisis was in the making. London in the 1880s had four and a half million people. It could not grow much further on the basis of the existing technologies. Although the middle classes could escape to gracious suburbs, the great mass of blue collar workers could not; they were crammed together in horrendous slums, the conditions in which shocked the conscience of the late Victorians. There was something like a social crisis, with mass popular demonstrations and the threat of revolution.

Then, around 1890–1900, came a distinctly new phase. Electricity was applied to propelling trams and commuter trains. Underground electric railways were constructed in the biggest cities, and soon they were extended above ground to serve new suburban rings, as in London. The result, by the 1920s and 1930s, was a new kind of city, which can be called the *late public transport city*. It came in two models. In the biggest cities such as London, Paris, Berlin or New York, there would be fully-fledged underground or metro systems extending outwards perhaps 10 or 12 miles. In smaller cities, there would be tram systems extending perhaps four or five miles. In both, commuter railways would extend the zone of feasible commuting as far as 20 miles or more. As electrified commuter trains could stop more often, the result was a more even spread of development: the beginnings of suburban sprawl. The classic case was Los Angeles. Most people think of Los Angeles sprawl as caused by the car, but that is simply not so: it was a creation of the Big Red Cars of the Pacific Electric Railway of Henry Huntingdon, once the most extensive commuter rail system in the world, which disappeared. In 1990, perhaps ironically, one was reopened as the new Blue Line of the Los Angeles Transit District.

However, it was not sprawl of the modern kind. It was radial sprawl. The lines converged on a traditional central business district. Pictures of Los Angeles from the 1920s show city centre streets which look like those of New York or Chicago, or indeed like London, with the same congestion. There were, however, very many more cars than in those other cities, and 1920s pictures of suburbs such as Glendale show many more: rows and rows of Model Ts parked along the kerbs. In fact, Los Angeles already had the same car ownership in 1924 that it took most of America until the 1950s and Europe the 1970s to achieve, i.e. about one car to every two households.

Thus, before any other great city, Los Angeles began to experience that clash between urban morphology and the urban transportation system which every other place would experience only much later, in the 1950s and 1960s. This was the third great crisis in urban transport. The Angeleno answer was typically innovative; it was to let the public transport system decline and to consciously become an automobile-oriented city. Duly, from the late 1920s, the Pacific Electric system began to lose passengers. A few years after that, trams began to lose patrons everywhere. The downturn in British cities actually occurred from the same time as in Los Angeles, that is from the late 1920s, although the passengers were turning not to cars, which they did not yet own, but to buses.

There has been a strong suggestion, in a celebrated testimony to Congress by Bradford Snell, that the Los Angeles situation was engineered by General Motors, which arranged for its subsidiaries to buy up the lines and convert them as a device for selling its buses (Snell, 1974). This is probably true, but it does not explain the failure of American public transport. More recent work by Jones (1985) shows that public transit did indeed decrease from the 1920s onwards because it was fighting in an increasingly competitive market as though it still had an effective monopoly; it failed to trim costs, in particular labour costs, and failed to adapt its services to the loss of off-peak and weekend revenue that mass automobile ownership brought. Although nothing like this happened in Britain, the trams also began to be converted, a process that turned into a flood after the Second World War, effectively removing trams from British city streets by the early 1960s.

The late public transport city was essentially a city of the period 1900 to 1940. After the war, with the development of the new car-oriented suburbs, it could not survive in its classic form. What developed as a result, first in most American cities and then in European ones, was a curious hybrid. A few great cities, all of them in the American west (Los Angeles, Phoenix, Salt Lake City, Dallas), effectively opted for full motorization. To achieve this, they had either to reconstruct themselves radically (as did Los Angeles) or develop from the start in a new way, as in Phoenix. The key is low density dispersal, not only of homes but also of jobs. It is the latter that is the distinctive feature: most places have dispersed homes, but only these cities have dispersed jobs. In contrast with New York City with its two million jobs on Manhattan south of 60th Street, Los Angeles in the 1960s had only just over 100 000 downtown jobs, a figure more appropriate for a traditional city one-tenth its size. Now it has more, but that is another story which is discussed below.

In New York and in the major European cities, as well as in Tokyo, the story was different. These cities kept strong central business districts, and in the great redevelopment boom of the 1960s they often became stronger still. The result was that they could not adopt full motorization on the Angeleno model. The very biggest, such as New York, Paris, Tokyo and London, are 'strong centre cities', as the transport consultant Michael Thomson calls them; they have one million and more downtown office workers crowded into an area typically ten square miles or less, and it is this feature that makes them rail dependent. They depend largely on their early 20th century rail systems to get people to work, because only rail systems can handle those kinds of flows along corridors: up to 40 000 people per hour on one lane or track (Thomson, 1977) (pedantry compels the remark that there is some evidence that buses can do that too, but only under special conditions).

The next rung of cities, which includes most of the great regional centres of Europe and the United States, as well as some smaller European capitals (Birmingham, Manchester, Brussels, Copenhagen, Frankfurt, Milan, Atlanta and San Francisco), are what Thomson calls 'weak centre' cities: they typically have between a quarter of a million and half a million workers in their downtown office cores, and they bring them in by a mixture of buses, trams and light rail (generally developed out of trams), with a minority using commuter rail (Thomson, 1977).

These are hybrid cities, because outside these dense cores they tend to be as car dependent as the Los Angeles type of city. Everywhere, even in the strong centre cities, only a minority of the total metropolitan area workforce works downtown, and very commonly that workforce is shrinking. There are suburban clusters of offices and shops, and these are growing. These clusters are highly car dependent and often have weakly developed public transport systems (although this varies greatly from one city to another, and European cities in particular have better suburban public transport). Further, the men, and some women, who live in these suburbs use the car almost exclusively in all their non-work journeys—for shopping, entertainment, recreation and weekend social life. So, in effect, such urban areas have two transportation systems, one used by a minority of commuters and by carless people (principally the elderly, the poor, many women and teenagers), and one used by everyone else, and indeed by the commuters outside commuting hours.

This is, of course, very extravagant in resource costs. The buses and trams are tucked away in the depots on weekends just as the cars come out of the garages. This arrangement has some built-in contradictions, which apply to both strong centre and weak centre cities, and also to some degree to full motorization cities.

16.3 THE CURRENT CRISIS IN URBAN TRANSPORT

The first contradiction is that between private and social cost and benefit. Everyone would like to use the car if at all possible, and everyone does, even to get to central jobs. As the builders and managers of central garage space do not consider the cost of the congestion they help to cause, and as the oil companies do not charge any more for a tank of petrol in congested city areas than in the open countryside, the result is a massive social cost which all car drivers impose on all other users at congested times. This social cost includes the time wasted in traffic jams and the environmental impacts not only of damage to the quality of the urban environment (Chapters 17 and 18) but also of pollution from car exhausts (Chapter 12). Such pollution is worse in slow moving urban traffic.

There are ways of handling this clash between social and private costs and benefits. The most effective is to set the marginal private cost of driving equal to the marginal social cost (MSC) insofar as the latter can be identified correctly. Singapore has tried to do this for over ten years by a special licence fee to enter the downtown area in the morning peak; Hong Kong has experimented with a full electronic road pricing system, but has so far not dared to implement it on an everyday basis; Oslo now has 18 toll gates round its city centre, but still does not charge the full marginal cost, imposing the tolls simply to help build more roads. The fact is that MSC pricing, as the economists call it, is politically unpopular. Every city that has considered it, apart from Singapore, has balked at it. So cities stagger on, inefficiently.

A second problem, which is really a special case of the first, is the energy consumption and environmental impact implications of increasing dependence on the car as the only available means of moving around. Two Australian researchers have produced detailed estimates of energy consumption in 32 different urban areas worldwide (Newman and Kenworthy 1989a, 1989b). They found that the average petrol consumption in American cities was nearly twice as high as in Australian cities, four times higher than in European cities and ten times higher than in Asian cities. Even within the United States, per capita petrol consumption was as much as 40% higher in 'full motorization' cities such as Houston than in 'strong centre' cities such as New York or even 'weak centre' cities such as Boston.

Differences in petrol prices, income and vehicle efficiency explained only about half of these variations. What was significant was the urban structure: cities with strong concentrations of central jobs, and accordingly a better developed public transport system, had much lower energy use than cities where the jobs were scattered. Cities such as Houston or Phoenix, Newman and Kenworthy conclude, could achieve fuel savings of 20–30% if they became more like Boston or Washington in their urban structure. Looking at cities world-wide, they also found a strong correlation between energy use and overall density. Toronto in Canada records only just over half the fuel consumption of American cities, although petrol prices are actually lower and vehicle efficiency even poorer. The reason, they conclude, is Toronto's more compact land use patterns, grouped around access to public transport, which are a result of conscious land-use planning—a difference evident even to a first-time visitor flying into Toronto airport from any American city.

Comparing European and American cities, they found another important factor. Hamburg in Germany has about as many central jobs as Houston, but it makes commuting by rail easier and cheaper than commuting by car, whereas Houston provides plenty of cheap parking. Overall, there is a strong relationship between energy use and the use of public transport, especially rail, and provision for the car. Newman and Kenworthy conclude that 'physical planning agencies have a major contribution to make in the conservation of transportation energy in cities' and that 'there is a large potential for conserving gasoline in US cities by shifting to land use and transportation patterns that are evident throughout the world and also within some US cities'; specifically, 'The policies of reurbanization and reorientation of transportation policies . . . should reduce gasoline use, and may also provide economic, social, and environmental benefits' (Newman and Kenworthy, 1989b, p. 35). Their findings have an obvious relationship to the issue of creating a sustainable urban economy through energy efficient land-use planning discussed in Chapter 17 and the problems of counterurbanization discussed in Chapter 18.

There is yet a third problem. In many large urban areas the central workforce, on which the traditional radial pattern of public transport was based, is no longer growing. It is at best static and at worst declining, as homes and jobs decentralize to ever more far-flung suburbia. Employment is everywhere decentralizing in a relative sense, because the growth of jobs in the suburbs is far faster than that in the central business district. This is true not only in American cities, but, ominously for the Newman–Kenworthy recommendations, in Europe also (Cheshire and Hay, 1989). Far from America adopting energy-conserving European patterns, it appears that Europe is going down the profligate American road.

City politicians do not like this. In city after city, in Europe and in North America, they have planned very expensive public transport systems to try to improve accessibility to central areas, thus hoping to attract private capital to invest in ever more office space that they hope will generate ever more jobs. There is not much evidence that they are succeeding at any level. Such systems almost never pay their keep; not only the construction, but also the operation, must be massively subsidized, sometimes by as much as two-thirds of total cost. There have been some spectacular planning disasters, the most notorious being the public transport system of Miami, which is carrying only 15% of its forecast riders.

The fact is that if people have a choice, they continue to find the car hard to beat for comfort and convenience at a bearable price. The only way around that is to raise the price of motoring or lower the price of public transport, or both. Numerous studies world-wide show that commuters are price-sensitive to some degree, and that if offered a sufficiently attractive package they will respond. British experience, above all in London, suggests that travelcards, which for a fixed weekly or monthly price give unlimited rides, are extremely effective. Since London Transport intro-duced them, ridership on the underground has increased by 60%, a far greater increase than could be explained by the relatively modest upswing in central London employment in the same period. Still, all this may be expensive to the public purse, especially if it is combined with new construction. In Germany, a country which has built more new public transport rail systems than any other in the last 15 years, recent research indicates that these new public transport networks have had little measurable effect on the volume of economic activity in the city centres. Those places that invested a lot seem to have results not very different from those that invested little. It may be that a follow-up study after a few years would show something different, but the only verdict at present is the old Scottish one, not proven. The real nightmare, for policy makers, is that the city invests a great deal of money on new or refurbished public transport systems and then finds that economic activity has declined so badly that the system is massively underused. This has happened in Liverpool and Glasgow, two British cities badly affected by the 1980s recession and high unemployment. The new systems opened but the customers were not there.

Meanwhile, the new jobs are mainly in the suburbs. That indeed is one reason why public transport systems are showing poor financial results: they are still serving patterns of travel, predominantly radial, that are declining. In the United States during the 1980s, one of the most remarkable geographical changes has been the appearance of new so-called 'back office complexes' (large record handling and administrative offices with little face-to-face public contact), as along the I-680 corridor in the San Francisco Bay Area, or the Zip Strip around Princeton in New Jersey. There are similar campus office complexes at the periphery of major metropolitan areas elsewhere in the United States and in Europe, for example, Stamford in Connecticut, 43 miles from downtown New York City, Reading, 40 miles west of London, and the Parisian new towns, such as St Quentin-en-Yvelines. However, these three are different from San Francisco and Princeton in having good access to public transport, including feeder buses and commuter rail.

In these instances companies are moving out of large cities for a number of good reasons. Rents are much lower. The right kind of middle-class female clerical labour is available, looking for employment that can be combined with short commuter trips. So the new office developments tend to skip over the inner-city cores such as East London, or the Bronx in New York City, and locate 20, 30 or 40 miles out. The problem is that the move often puts them outside the sphere of effective public transport service. Tyson's Corner, which is in the middle of the Washington suburbs about 15 miles from the White House, was a filling station and grocery store in the mid-1960s; it is now the biggest commercial downtown area in the state of Virginia, but it is a downtown without effective public transport access, as Washington's excellent new Metro system terminates three miles away. The result, here and in many other similar American towns, is what the Berkeley transport expert Robert Cervero has called suburban gridlock—traffic congealing along a 10 or 20 year old freeway system never built with these kinds of journeys in mind; indeed, ironically, built as by-passes around the urban congestion.

In fact, public transport can seldom extend fast enough or far enough to keep up with the suburbanization wave, which layers homes and jobs like a marble cake, all sunk in a dough of car dependency. To grapple with this problem, starting in the 1960s, metro areas have created new commuter networks on the model of the San Francisco BART, the Parisian REF or the Frankfurt S-Bahn: express systems which link the downtown areas directly with distant suburbs and satellite towns. In Paris especially, this strategy was linked with the creation of strong subcentres in the satellites, thus distributing and reconcentrating employment, and creating balancing sets of commuter flows; St Quentin-en-Yvelines, mentioned earlier, is one of these. Such commuter networks may extend 40 or 50 miles from the urban cores, thus serving most demands for radial transport (although in London and Tokyo, a small minority of commuters are now beginning to use high speed trains to commute up to 100 miles each way each day). The main problem with this approach is that it still cannot cater adequately for the cross-flows, which are left on an overloaded highway system designed for the very different purpose of carrying inter-city or by-pass traffic. Frankfurt offers a dramatic example.

16.4 WHAT SOLUTIONS?

There is a tendency to go on throwing more expensive capital investment at the resulting problem; to build more motorways, develop light rail schemes, extend the underground. It is almost certain that such solutions will prove self-defeating. Neither the policy of pouring more concrete to build roads, or of welding more steel rail to extend the underground or build new light rail lines, is likely to solve the suburban gridlock problem. Some of the investments do not even serve the main patterns of demand, whereas others will probably prove white elephants because of underuse. There is very little evidence that commuters in low density suburban areas will abandon their cars for less convenient public transport, and that will be true even if the quality of car commuting notably declines. The reason, first noted with San Francisco's BART in the 1970s, is that people care not about the speed on the line haul but on the total door-to-door elapsed journey time, and on this criterion the car has a big built-in advantage.

One reason is that as jobs and homes deconcentrate, they will seldom, if ever, reconcentrate in ways that give the same advantage to public transport as the old central areas did. This is clear from Europe, as well as from one or two North American metropolitan centres such as Toronto. A regional public transport authority may work in conjunction with a regional planning agency to group jobs and apartment homes around public transport nodes, as in Stockholm, Paris and Toronto over the last 30 years, to give public transport the best possible chance to compete with the private car for the commuters, including that very elusive but important breed, the reverse commuter. The Toronto example is particularly interesting, because, as the Newman–Kenworthy evidence proves, it shows that North Americans will readily accept ways of living and working that are radically different from those in the typical North American city.

Equally interesting are the experiments now taking place in the Californian cities of Sacramento and San Jose, where architects such as Peter Calthorpe are joining forces with city planning offices to develop single-family home areas very different from the traditional tract development. They are based on maximizing pedestrian access to public transport spines and on restricting car use. The point is that these designs are not, as sometimes occurs in Western Europe, forced on reluctant public housing tenants who have no choice. They are being sold in the open market, admittedly to a public that is reeling under the impact of some of the highest housing prices in the United States, but constituting a market nonetheless.

The other, closely related problem is that conventional public transport cannot readily scoop up the increasing number of cross-commuters who use corridors that are insufficiently dense to support good public transport. It might do so if the money were available to create networks that would cater for these flows. The right approach may be to start from this fact, and consider how to provide a system that will compete with the car on its own terms. It would need to offer door-to-door service in the same or less time, in conditions of equal personal comfort and convenience, and it would need to compete on price. Only some kind of demand-responsive service is likely to achieve this, in the form of dial-a-bus or frequently circulating minibuses (the model used in many cities in developing countries) or van or car pools. These should be encouraged by fiscal incentives as well as by physical priorities such as diamond lanes for high occupancy vehicles. These might indeed be physically separated from ordinary freeway traffic and thereby allowed to operate at higher speeds. Experience shows that buses operating in this way can deliver the same numbers of downtown commuters, per hour and per lane, as traditional commuter rail systems. They may therefore provide a good solution for downtown trips as well.

Very belatedly, some cities have begun to think of adapting their systems to give a more even pattern of grid-type access between subcentres. This was explicit in the original plan for the San Francisco BART; the Frankfurt S-Bahn, which is a net connecting the cities of Frankfurt, Mainz, Wiesbaden and Darmstadt, has some of the same characteristics; Paris is now pondering an outer circular REF connecting the satellites and other destinations such as the airport.

The problem is that, mapped on to metros of this size, this kind of system still leaves a very large number of traffic desires inadequately catered for. This brings the alternative answer based on buses and so-called paratransit vehicles, which are small buses or vans intermediate in size between cars and conventional buses. In North America in the last decade, a few cities such as Ottawa and Houston have made remarkable progress in developing many-to-many bus transit systems based on an idea borrowed from the airlines, hubbing through nodal points, with guaranteed fast transfers. These systems use dedicated rights of way, either exclusive busways or so-called HOV (high occupancy vehicle) lanes on freeways. Adelaide has its O-Bahn guided busway system on which buses travel at 60 miles per hour between the central area and the suburbs, then leaving to travel as conventional buses on ordinary streets, thus giving a flexibility that light rail systems can never offer. Nancy in France has pioneered dual-mode vehicles which operate as trolley buses in the centre of town and in the inner city, but change over to diesel for the ride to the suburbs. Essen in Germany has combined such buses with the guideway principle, allowing buses converted to trolley buses to travel rather like rubber-tyred light rail vehicles through tunnels into the city centre. Curitiba in Brazil has a very extensive network of exclusive, segregated bus lanes in the central reservations of dual carriageways.

Such schemes are well adapted to suburban, low density, many-to-many type journeys from dispersed origins to dispersed destinations. However, they can also be used, indeed have been used, in radial journeys to city centres, such as in Adelaide, Nancy and Curitiba. They may provide low cost alternatives to expensive rail transit schemes, especially where the future of downtown employment is uncertain. The point is that once a high occupancy exclusive right of way comes into existence, then it can be utilized by many different kinds of vehicles as an exclusive rail right of way cannot. Moreover, these vehicles can fan out to serve dispersed suburban origins and destinations in a way that rail vehicles cannot.

All these experiments offer a way of developing public transport systems that specifically cater for the new journey patterns typical of the dispersed metropolis. The problem is that the basic unit, which is the conventional bus, may be too large to cater for such dispersed demand patterns. A system based on small vans, like those that are being used so successfully for service to and from American airports, may be the answer. These systems essentially provide a way of moving small numbers of people at a premium fare between their own homes and traffic hubs. They could be as suitable for commuting as for airport access, as the manager of one of the most successful among them suggested in *The Wall Street Journal*. California has made extensive use of van pools and ride sharing generally, and such schemes have proved highly successful for the daily commuter trip across the bridge from Oakland into San Francisco, which ironically has taken passengers away from the local bus system.

This, of course, is not to say that rail schemes are entirely irrelevant to a solution of the urban transport problem. In Canada, Calgary's light rail system, which uses old rail right of way and existing downtown streets, provides an effective radial network to a concentrated downtown that has had a lot of growth in the last decade, and it seems to be carrying good traffic. Washington's Metro seems to be justifying itself, because it similarly serves traffic good for rail. However, in many contexts, road based solutions seem to be the only practicable answer.

The paradox is that these systems do not use notably new technologies. As the thesis of this chapter is that cities change under the influence of their transportation systems and that these change under the impact of technological change, it would be expected that some dramatic shifts in technology might occur. The odd fact is that no new urban transportation technology has been acquired for nearly 90 years. Electric trains, subways, light rail, even the private car were all available by 1890. It is surely about time for a change.

The new phenomenon of superconductivity, about which there has been so much publicity, might have a dramatic impact here. First, most experts seem to think that it will at last make feasible, on a widespread scale, the development of magnetic levitation (maglev), which has been the subject of repeated experiment during the last 20 years but which is being used so far only on a small-scale experimental basis. Most people seem to be thinking in terms of trains, but this technology might well represent a way of revolutionizing travel by car using special guideways. After all, once levitated a whole bunch of disparate vehicles could share the same controlled airspace. If this is combined with dramatic falls in the cost of transmitting electricity, the use of electrically powered vehicles on a large scale might be feasible. Maglev will surely be applied to revolutionize inter-urban transportation over high traffic corridors such as Tokyo to Osaka, where the Japanese confidently propose a new system by the year 2000, but its major use may well be in urban areas. If it is, it will once again, as with the underground lines of the 1900s and the motorways of the 1960s, require a vast programme of new investment to make it possible.

These examples suggest a possible prototype for a future metropolitan transportation system which would give many-to-many accessibility, partly over ordinary streets and partly over an automated guideway system. It would be based on small van-like vehicles similar to the airport shuttles. They would be either electrically powered, a real possibility given the impetus for the development of electrical vehicles which comes from the recent Californian air quality regulations, or perhaps dual mode, capable of switching between petrol and electricity. At nodal points they would be coupled together, either physically or electronically, to run as automated trains along special guideways. At yet other nodes they would split again to serve local destinations. Of course, no metropolitan area in the world currently offers anything of this kind. Nevertheless, elements of the system exist—in the guided busways in use in Adelaide and Essen, in the dual-mode vehicles already in service in Essen and Nancy, and in automated public transport systems in Lille, Vancouver, London Docklands, Osaka and Kobe.

There is an alternative, perhaps only a partial alternative, which is to try to cater for these flows by transportation systems management techniques designed to divert demand from the conventional, single-driver no-passenger automobile into more collective or shared modes such as vans and car pools. This is achieved by a variety of carrots and sticks ranging from access to preferential HOV lanes and parking slots, through direct cash incentives, to physical limitations and charging for access to employment nodes. Washington DC has gone the whole way, restricting one freeway to HOVs only at peak hours. At least one suburban downtown area, Bellevue in the state of Washington, has been equally audacious in restricting parking spaces while exploiting its position as a transit hub. Road pricing, such as has been in force in central Oslo since early 1990 and may be introduced soon in central Stockholm, could be used in the same way, not only in congested downtown areas, but also in the suburban nodes. All these experiments deserve intensive study because they are far more likely to succeed than the expensive rail systems, which have invariably proved to be catastrophic failures in terms of ridership and revenue.

This is only a partial alternative because it could evolve into the first system. There is already a blurring of traditional distinctions between one mode of transportation and another; the difference between a rubber-tyred metro and a guided trolley bus is extremely subtle, as is the distinction between a small van and a taxi. Also, as more and more information technology is injected into the traditional highway system, it increasingly takes on the characteristics of a public transport operation. So, within the next decade, there may be a convergence of the different modes, wherein cars are electronically locked together on transportways while public transport systems increasingly offer on-demand service from any place to any other place.

16.5 REFERENCES

Cheshire, P. C. and Hay, D. G. (1989) *Urban Problems in Western Europe: An Economic Analysis*. London: Unwin Hyman.

Clark, C. (1951) Urban population densities. *Journal of the Royal Statistical Society, A*, **114**, 490–496.

Clark, C. (1957) Transport: maker and breaker of cities. *Town Planning Review*, **28**, 237–250.

Jones, D. W., Jr. (1985) *Urban Transit Policy: An Economic and Political History*. Englewood Cliffs, NJ: Prentice-Hall.

Newman, P. W. G. and Kenworthy, J. R. (1989a) *Cities and Automobile Dependence: A Sourcebook*. Aldershot and Brookfield, VT: Gower.

Newman, P. W. G. and Kenworthy, J. R. (1989b) Gasoline consumption and cities: a comparison of U.S. cities with a global survey. *Journal of the American Planning Association*, **55**, 24–37.

Snell, B. C. (1974) *American Ground Transportation: A Proposal for Restructuring the Automobile, Truck, Bus, and Rail Industries. (Subcommittee on Antitrust and Monopoly, Committee on the Judiciary, US Senate)*. Washington, DC: Government Printing Office.

Thomson, J. M. (1977) *Great Cities and their Traffic*. London: Gollancz.

16.6 FURTHER READING

Newman, P. W. G. and Kenworthy, J. R. (1989) Gasoline consumption and cities: a comparison of US cities with a global survey. *Journal of the American Planning Association*, **55**, 24–37.

Richards, B. (1990) *Transport in Cities*. London: Architecture Design and Technology Press.

Thomson, J. M. (1977) *Great Cities and their Traffic*. London: Gollancz.

CHAPTER 17

Towards Sustainable Urban Development

M. J. Breheny

17.1 INTRODUCTION

At the beginning of the 1990s there appears to be a genuine popular and political concern about the state of the environment. Bouts of such concern have occurred at intervals before, but sadly have been short-lived. The current interest in the environment does, however, seem to be sufficiently heartfelt to suggest that it will be maintained. One of the reasons for this is that some common world-wide understanding of both the problem and the solution has been reached. An example of this is the goal of 'sustainable development' (see Chapter 2), which was emphasized in the Brundtland Report (World Commission on Environment and Development, 1987). As Chapter 2 explains, the concept has been only vaguely defined, but nevertheless conveys the readily understood message that there must be a balance between levels of development and the stock of natural resources; that is, development must be at a level that can be sustained without prejudice to the natural environment or to future generations.

Although 'sustainable development' has merit as a slogan, for it to be of practical value a clearer view is needed of what is meant by the term. The environmental debate has tended to focus on the natural environment, despite the fact that cities—the core of the human-made environment—are obviously a major consumer and degrader of that natural environment, as Fig. 1.1. shows. To counter this focus, the emphasis here is on the urban environment: as an important user of the natural environment, but also as an important resource itself. Consideration of the role that cities play in affecting the natural environment produces an appreciation that maybe cities are themselves a resource that needs to be protected as development activities are sustained: hence 'sustainable *urban* development'.

After some clarification of terminology, this chapter considers the city as a resource. It then examines the possibilities for the planning of cities for greater sustainability. The effect of future urban form on energy consumption is considered;

Environmental Issues in the 1990s. Edited by A. M. Mannion and S. R. Bowlby

in particular, the debate on concentrated versus dispersed forms is assessed briefly. When energy objectives are put together with quality of life aspirations, the picture becomes more complex. This chapter is very much a tentative introduction to issues that are now regarded as crucial, but which have been neglected hitherto. In this sense it is consistent with the UK White Paper on the Environment, *Our Common Inheritance* (Department of the Environment, 1990). The chapter concentrates on urban sustainability as it relates to developed nations.

17.2   DEFINING SUSTAINABLE DEVELOPMENT

As a result of the general confusion over definitions of sustainable development, Pearce *et al.* (1990) present a step by step explanation of their understanding of the term.

They begin by defining 'development' as simply the achievement of a set of aspirations of a group in society:

> What constitutes development depends on what social goals are being advocated by the development agency, government, analyst or adviser. We take development to be [a set of] desirable social objectives; that is a list of attributes which society seeks to achieve or maximise. (p. 2, parentheses added)

'Sustainable development' is then defined, again simply, as the desire to maintain the achievement of such development aspirations over time. At this stage, the definition does not relate specifically to the physical environment, nor does it refer to any conditions.

The question of time is important here. An assumption that development goals are to be met very positively over a long time period—what Pearce *et al.* (1990) call 'strong sustainability'—is likely to be inconsistent with resource preservation. More modest benefits from development, 'weak sustainability', will be more consistent with resource preservation. When faced with the question of future benefits, economists usually attempt to discount these, that is, give them a lower weighting, relative to present benefits. Pearce *et al.* (1990) argue that economists will have to change their thinking and grant greater weight to future benefits if development is to be sustained.

Next, the minimum conditions for sustainable development are introduced. Here, the assumption is that the natural capital stock, that is, the stock of natural resources, should not be allowed to decrease over time.

Although Pearce *et al.* (1990) do allow that non-natural capital stock might also be preserved, they argue it is not of such fundamental importance as natural capital stock, which has the particular characteristics of irreversibility—once destroyed it is lost forever—and diversity—when available in abundance it is resilient to shocks and stress. Owing to the irreversibility characteristic of natural resources it is argued that an approach is required that avoids risks. Although technological advancements may provide substitutes for some natural resources, it would be very unwise to rely on this. It is better to play safe and assume that society cannot afford to lose any type of resource.

Although different groups might have different objectives in mind when pressing for sustainable development, Pearce *et al.* (1990) argue in favour of two fundamental considerations: intragenerational and intergenerational equity. They argue that the idea of maintaining a constant or increasing natural capital stock is likely to best serve the development objective of intragenerational equity; that is, of favouring the disadvantaged in society at any one time. This is illustrated with reference to the benefits to be gained in poor, developing countries from a constant supply of natural resources. Intergenerational equity, which, for many people, is the crux of the sustainable development issue, would ensure that 'the next generation should have access to at least the same resource base as the previous generation' [Pearce *et al.* (1990) p. 14].

A point that is implicit in the explanation of Pearce *et al.* (1990), but warrants specific attention, is the idea that in achieving sustainable development subject to conditions, a balance may have to be struck between development aspirations and appropriate levels of resource use. Aspirations may have to be reduced for the conditions to be met. This is consistent with their idea that 'weak sustainability', where the present benefits are consistently positive but modest, is the most likely outcome.

Basically, sustainable development requires that development activities are planned to minimize long-term resource use. A constant test of success in doing this must be whether the next and subsequent generations will have at least the same natural resource base that is available now.

17.3 DEFINING SUSTAINABLE URBAN DEVELOPMENT

Given, then, some understanding of the meaning of sustainable development, how can this be related to urban areas? What constitutes 'urban sustainability'? Here there is little prior thinking to draw upon, as most of the sustainable development debate has concerned the natural environment.

There is, of course, a long history of concern about the economic, social and environmental problems of cities, and in particular of the inner cities. There is also a long tradition of anti-urbanism, stretching back to the 18th century, and a more recent concern developed from the 1930s onwards about urban pressures on the countryside. Despite this history, debates on urban issues have never seriously considered questions of sustainability. Even in the current sustainability debate, discussion of ways of planning cities for sustainability has been very limited. This is surprising, given that cities are obviously both great consumers and degraders of the natural environment.

To move from questions of general sustainability to urban sustainability, it is useful to consider how each one of the definitions and qualifications of Pearce *et al.* (1990) might be adapted.

Sustainable urban development is easy to relate to their basic starting point in defining sustainable devlopment. A set of development aspirations translates as a set of urban development aspirations. Sustainable development translates as sustainable urban development, in which the achievement of aspirations is to be sustained over time.

The notion of sustainable development, subject to constant or improving natural capital stock, can be translated easily to sustainable urban development subject to constant or improving natural capital stock. There are, of course, many urban development activities—for example, industry, housing, transport—which consume vast amounts of natural resources and which, in turn, contribute to environmental degradation. The bigger question is whether these activities can be 'sustainably developed' without jeopardizing the resources of the wider environment (see Fig. 1.1).

The translation can be extended further by considering what urban intragenerational and intergenerational equity might mean. Social and economic equity among urban citizens may be hard to achieve, but is easier to envisage. Likewise, the idea of making sure that future generations of urban dwellers are not fundamentally constrained by actions taken now is clear in principle.

The translation from sustainable development to sustainable urban development seems straightforward enough. However, further consideration reveals three basic problems. The first problem concerns scale. The assumption behind the attempt of Pearce *et al.* (1990) to link development to resource constraints is that there is a clear relationship between the nature of development and resource consumption and environmental degradation. At the global or general economic scales, to which most published work refers, this may be so. At this scale it is possible to conceive of a balance between development and resource use. If more specific development aspirations are considered, urban or otherwise, the link is less clear. It is unreasonable to expect a balance between development and resource use to be struck within a local area. Local initiatives must be seen as contributions to the achievement of sustainability at larger, national and global scales. Hence the exhortation to 'think globally, act locally'. It is not possible to measure the contribution of these local initiatives to sustainability. All that can be said is that progress is in the right direction, and possibly that it is moving slowly or quickly.

This uncertainty about local contributions supports another view of sustainability, which argues that judgements should err on the side of caution. It may be that future technological change will provide substitutes for natural resources. Nevertheless, given that in general the natural world is irreversible, such assumptions should not be relied upon; if in doubt, be cautious.

A second problem that arises in considering sustainable urban development concerns the degree to which urban activities are resource depleting. In urban areas, there are many development aspirations, the achievement of which may be hindered by environmental problems, but which themselves do not tend to drain the natural physical stock. For example, the maintenance of the architectural heritage, the creation of cultural activities and the desire to improve skill levels would seem to have little to do with natural resource depletion.

It may be necessary to make a distinction between urban development activities which create resource threatening effects (e.g. fuel consumption, toxic emissions, water pollution; see Chapters 4, 6, 11 and 12) and those that do not. This latter group covers a whole variety of activities which range from the totally benign to those that have deleterious effects on the quality of life, and, hence on the accumulated human-made resource that the city represents, but not necessarily on the natural environment (e.g., noise, visual intrusion, crime). There exists a continuum between these extremes, but for present purposes it is useful to make a simple distinction.

The relationship between the first, the natural resource consuming group and the second, the relatively benign group of urban activities is complex. For example, the consumption and degrading effects of apparently benign activities might in fact be indirect and hence difficult to trace. Also, there may be activities that are mainly consumers, such as office buildings, or degraders, such as sewage works, and not both.

This leads on to a third, and possibly the most profound, problem that arises in trying to translate from sustainable development to sustainable urban development. This is the possibility that the city itself should be regarded as a precious resource.

17.4   THE CITY AS A RESOURCE

Having reviewed the principles behind the sustainable development debate, and translated these to the urban context, there appears to be two findings—one expected and the other less so. The first is that much more careful thought needs to be given to the role that cities play in consuming and degrading natural resources. This is profound, but unsurprising in principle. As to the practice, this will be considered later. The second finding, and something of a bonus, is that the sustainable development debate provokes the idea that the city itself is a resource, and should be subjected to the rigorous 'development–sustainable development–conditions–intergenerational equity' examination. In effect, the conditions stage of Pearce *et al.* (1990) can be divided into natural and human-made resource constraints. The suggestion is that the human-made constraints warrant as much attention as the more familiar natural constraints.

Cities pose a double threat to the natural environment: as consumers of precious resources and as destroyers at source of those very same resources. However, the city is much more than simply a gluttonous devourer of precious resources and a crude disgorger of effluents. At its best, it is the productive base for economic growth, of rising living standards, of innovation, of education, of culture; indeed, of civilization itself. The most appropriate form of human habitation is debatable, of course. There is a long tradition of anti-urbanism, which finds its current expression in calls for a return to decentralized, small-scale living and the adoption of less materialistic, more 'rural' values (Chapter 10). However, there is little doubt that existing cities, particularly at their best, are an enormous resource. In the clamour surrounding the new found environmentalism, focusing as it has on the natural world, this has been rather forgotten. Just as the city is seen as a major threat to the natural environment, it should also be regarded as a resource base that sustains economic, social and cultural life. It is, in effect, a secondary resource. The natural world supports the city, but the city's human-made resources, in turn, give the city its distinctive, dynamic character.

These human-made resources are less precious than those from the natural physical stock; they can be replaced, at a price. Indeed, the abilities of the city to adapt is one of its fundamental characteristics. The history of cities shows clearly this ability to change as economic, technological, cultural and political pressures have varied. However, although this adaptability is, in principle, a positive feature of cities, specific responses are not always for the good. For example, economic change has often blighted particular cities while favouring others; decentralization has consumed more land and blighted inner city areas; increased affluence has created road congestion and intensified pollution. Thus, much of the accumulated capital stock of buildings, history, culture and local environment—built up at high cost and consuming large amounts of the natural resource stock—is lost. Every time the human-made stock is replaced, the old stock is not only wasted, but more natural resources are consumed.

Thus, it would be wise to look upon urban sustainability as involving the achievement of urban development aspirations, subject to the condition that the natural and human-made stocks of resources are not so depleted that the long-term future is jeopardized. Just as some consumption of natural resources is appropriate under this approach, so some loss of the human-made stock is inevitable. Indeed, an important difference between the natural and human-made stock is that the former does not (unless damaged by humans) deteriorate. The human-made stock, be it buildings, institutions, technologies, or ideas, can deteriorate, and should be replaced. Much human inventiveness is directed at doing just that, hence the inability of the systems approach (Section 1.2.1) to predict social change.

What the sustainability debate has achieved is to make people think more clearly about the city as a resource and about defining development aspirations more precisely, and about the appropriate degree of preservation and consumption of the existing urban resource. The consumption and/or replacement of the human-made stock of urban resources has now to be undertaken with greater care than in the past. It has to be consistent with sustainable urban development aspirations, which in turn have to be consistent with reasonable calls on both natural and human-made resources.

17.5   MOVING TOWARDS THE SUSTAINABLE CITY

What can be done, then, in policy terms, to achieve the sustainable city? Here, there is a sudden switch in the logic, from general discussions about sustainability to detailed urban policy issues. However, if the argument is accepted that cities are major contributors to the environmental problem and that local initiatives must be the means to global sustainability—'think global, act local'—then the switch may not seem so strange.

The limited debate so far about the contribution of urban policy to sustainability has focused on two sets of ideas. These coincide neatly with the division of conditions discussed earlier; one set concerned with urban development and the natural resource stock, and hence focusing on urban energy consumption and pollution; and the second set concerned with urban development and the maintenance of the human-made resources, particularly quality of life, of the city. As will be clear, these two debates overlap, but for the moment they will be reviewed separately.

17.5.1   Urban sustainability and natural resources

The two major areas in which the energy efficiency of cities might be improved are by reducing travel and hence fuel consumption, and by developing the built environment to reduce domestic and industrial energy consumption. In the former case, at least, carbon dioxide emissions would also be reduced. Most of the research aimed at devising more energy efficient urban forms has focused on these two issues.

Owens (1987), in reviewing the relationship between energy use and urban form, makes a distinction between changes brought about by the 'market' if energy costs rise, or are raised by taxation, and changes that public measures, including planning policy, might introduce. On market responses, Owens is wary of attempts at prediction because so little is known of how individuals and households might respond in practice. In general she expects short-term responses to increased energy costs to consist of 'coping' strategies, in which marginal adjustments in lifestyles, such as buying a smaller car or installing secondary double glazing, suffice. She notes in particular that there is likely to be resistance to reductions in mobility; the demand for mobility is likely to be relatively price inelastic, even if that for petrol is relatively elastic. Short-term adjustments are likely to have little effect on urban form, given the range of coping strategies and the massive inertia built into large urban structures. Owens (1987) does not discount, however, the possibility of long-term market responses that will affect urban form. Just as increased mobility, partly resulting from the falling real cost of petrol, has contributed to urban decentralization, so in the long term might reduced mobility induce concentration.

Although Owens (1987) is doubtful about the effects on energy consumption and pollution of price rises, the alternative of redesigning urban form must be put into perspective. There is a strong body of opinion that the price mechanism, via taxation, has to be used to reflect now the long-term 'true' costs of environmental damage. High taxes on fuel and on polluters, plus regulation over technology such as catalytic converters in cars, are likely to have profound effects on consumption patterns and ultimately on urban form. These effects are likely to be much more powerful than any planning policy designed to rearrange urban form. This is not to say that planning cannot play a role, clearly it can, particularly if politicians fail to grasp the taxation option, but simply to put its contribution into perspective.

In view of the difficulty of predicting market responses to changes in energy costs, Owens (1987) concludes that the most effective short-term approach to energy concerns is for planners and architects to devise ways of designing more energy-efficient built environments. At the intra-urban scale Owens suggests that there is some loose consensus on the appropriate characteristics of efficient forms. Compact and mixed land uses facilitate public transport systems and allow the introduction of communal heat and power arrangements.

One apparently obvious solution to the energy crisis is to promote the high density, compact city. This seems to have the merits of short private journey lengths, least land consumption and the greatest public transport and communal heat and power possibilities. One very strong advocacy of this approach comes from the European Commission's *Green Paper on the Urban Environment* (CEC, 1990). This paper is intended as a device to further the Commission's work on urban areas, which have been neglected relative to regions, and to address the role of urban areas in creating and, ultimately, reducing environmental problems.

The urban environment is defined, implicitly, as the physical environment and also the general 'quality of life' for residents and workers. Thus, the urban aspirations of the Commission, i.e. the development it wishes to see sustained, are broad and varied. Also, it definitely sees the city as a resource, which, as has been argued above, must be necessary for addressing the question of urban sustainability.

The report provides both an analysis of the causes of urban environmental problems and proposals towards a Community strategy to overcome these problems.

The analysis concludes that urban environmental problems can be traced to two factors: 'The first of these is the uncontrolled pressure placed on the environment by many of the activities which are concentrated in the cities. The second—and not unrelated—factor is the spatial arrangement of our urban areas.' (CEC, 1990, p. 31.)

Planning practices which have allowed the spatial separation of functions and large-scale suburbanization are identified as the source of this second factor. Suburbanization, pejoratively referred to as 'sprawl', is regarded as an evil, whereas high density urban cores are the ideal.

This provocative analysis is, however, based on a number of contentions which can be challenged. On the question of energy efficiency, is it certain that the high density city is the most efficient form available? Public transport patronage is highest in large urban cores, but so is traffic congestion. Slow moving, dense traffic creates very high fuel consumption and hence concentrated carbon dioxide emission levels.

There is also an implicit assumption in the Green Paper that commuting patterns are still largely from peripheries to existing cores and, hence, high density urban areas will produce lower average journey lengths. This may be true in some countries, but not in most. Commuting patterns have become much more diverse in recent years, often from suburb to suburb, because of increasing mobility.

Owens (1986) comes to a very definite conclusion in her review of the merits of the compact city, at least in its extreme form:

> The compact city idea has generally, and not surprisingly, been greeted with some scepticism . . . Quite apart from the serious questions that could be raised about the flexibility and sociological implications of such a form, its apparent energy advantages do not stand up to detailed scrutiny. (p. 62)

The CEC Green Paper also comes out very strongly against the idea of new towns or new settlements. It argues that these are unacceptable on two grounds: (1) by detracting from the compact existing city, they are energy inefficient, and (2) their failure to create the quality of life of the urban core. However, again there is little evidence presented to support this view.

The Green Paper argues that further urban growth should be accommodated within the boundaries of existing urban areas. This view can be challenged on two grounds: is this possible, even if it were desirable, and is it desirable? In Britain, it must be concluded that it is not possible. In the south-east region, for example, it is estimated that one million houses will have to be built in the next 20 years. Most of these will have to be built either as suburban extensions to existing cities, towns and villages or as new settlements. This is the real challenge of urban sustainable development.

On the question of desirability, the Green Paper does not produce evidence to show that the building of new settlements will be any less energy efficient than other forms of urban development. Obviously, the building of a new settlement provides more opportunities for energy efficient forms to be tried than does, say, suburban extension.

There has been a number of studies of the relative efficiency of new towns. Cresswell and Thomas (1972) carried out an assessment of the journey to work closure of a wide range of new towns relative to 'natural' towns, using an 'independence index' (internal work trips, divided by in-commuting plus out-commuting). Their conclusion was that the planned new towns were considerably more efficient in this respect than the other towns. However, if the Cresswell and Thomas work is updated, the situation has changed.

Table 17.1 shows Cresswell and Thomas's results for new towns for the years 1951, 1961 and 1966. Equivalent results have been calculated for 1971 and 1981, and later new towns, not covered by the original analysis, have been added. High values indicate a high degree of self-containment and low values the contrary. It is immediately obvious that average levels of self-containment peaked in the new towns in the period 1961–66, at a time when many of the early new towns were approaching their planned capacity, but before mass car ownership significantly increased personal mobility. Since that peak, the values of the index have fallen, suggesting declining self-containment. The values for the 'natural towns' chosen by Cresswell and Thomas (1972) have remained relatively constant over time. They have declined, but gradually. In 1981 the independence index for both types of town was similar.

The most likely explanation for the changes, leaving aside definitional problems, is that rapidly increasing car ownership since 1966 has increased mobility and hence

Table 17.1 Independence index values for new towns, 1951–81. The 1951, 1961 and 1966 figures are taken from Cresswell and Thomas (1972) by permission of the Town and Country Planning Association and the Office of Population Censuses and Surveys. 1971 and 1981 figures have been calculated from tabulations in the respective Census Journey to Work Tables

New towns	Independence index				
	1951	1961	1966	1971	1981
Aycliffe	0.08	0.52	0.57	0.44	0.74
Basildon	0.36	0.96	0.96	0.87	0.76
Bracknell	0.90	1.13	1.02	0.87	0.82
Central Lancs	—*	—*	—*	1.88	1.88
Corby	1.41	1.91	2.51	0.69	1.79
Crawley	0.98	1.59	1.58	1.69	1.15
Cwmbran	0.72	0.74	0.88	0.75	0.80
Harlow	1.42	2.04	2.05	1.92	1.44
Hatfield	0.65	0.63	0.66	0.32	0.45
Hemel Hempstead	1.31	1.82	1.72	1.43	1.00
Milton Keynes	—*	—*	—*	1.36	1.44
Newtown	—*	—*	—*	1.03	1.32
Northampton	—*	—*	—*	2.88	2.43
Peterborough	—*	—*	—*	1.84	1.99
Peterlee	0.34	0.20	0.36	0.41	0.39
Redditch	—*	—*	—*	1.30	1.12
Runcorn	—*	—*	—*	0.73	0.94
Skelmersdale	—*	—*	—*	0.67	0.87
Stevenage	0.92	2.29	2.03	1.63	1.14
Telford	—*	—*	—*	2.61	2.41
Warrington	—*	—*	—*	1.74	1.32
Washington	—*	—*	—*	0.56	0.67
Welwyn	1.12	1.09	1.12	0.97	0.68
All new towns†	0.85	1.24	1.29	1.00	0.95
All 'natural towns'	—	—	1.04	0.89	0.98

* Figures not presented in Cresswell and Thomas (1972).
† Average for those 12 new towns for which values are available for all years.

generated more diverse types of journey to work trips. Thus, for both new towns and other towns, self-containment has declined. The decline has been greatest in new towns. This may be a consequence of 'artificially high' rates of self-containment in the early years of new towns, when the policy was to link housing with jobs for new residents.

Rickaby (1987) provides an analysis of the energy efficiency of six urban forms. His conclusion is that the most efficient forms are likely to be 'centralisation' and 'decentralised concentration'; that is, the compact city favoured by the CEC, and separate concentrated subcentres around existing large cities. This latter finding, which will be familiar to advocates of new towns, suggests that the European Commission's assumptions about the superiority of the high density, compact city need to be rigorously examined.

17.5.2 Urban sustainability and quality of life

The European Commission's Green Paper (CEC, 1990) is interesting because it discusses the two urban concerns—effects on the natural environment and the quality of the human-made environment—together. It is concerned about the effects of cities, as consumers and degraders, on the natural environment; hence the proposed solution of the compact city. It also regards the city as a precious resource which must be protected and enhanced:

> The recreating of the diverse, multifunctional city of the citizen's Europe is thus a social and economic project for which the 'quality of life' is not a luxury but an essential. (p. 10)

Thus, the one solution neatly solves two problems: the desire to protect the natural environment and to preserve the quality of life that the healthy city provides.

The report argues that the recent rediscovery of the value of urban living:

> . . . reflects the failure of the periphery: the absence of public life, the paucity of culture, the visual monotony, the time wasted in commuting. By contrast, the city offers density and variety; the efficient, time and energy-saving combination of social and economic functions; the chance to restore the rich architecture inherited from the past. (p. 7)

On the question of the quality of life in urban areas, the Green Paper extols the virtues of living in a compact, culturally diverse, exciting urban core, and denigrates suburban living. This logic seems strange to many people who would come to exactly opposite conclusions. Whereas some European cities do show the features that the Green Paper prizes, most do not. Indeed, many urban cores have serious problems. In many British cities, people have been leaving core areas in large numbers, in part because of the desire for more space, but also because of the deteriorating quality of life. Also, although most people argue against uncontrolled sprawl, planned suburban development, arguably, has been successful, at least in Britain.

The Green Paper's ideal compact, diverse and culturally-rich city—which society might try to preserve where it still exists, or indeed create afresh in a limited number of cases—ignores most of the space, people, and hence accumulated urban resource that actually exist within cities. Most of the citizens of Europe live in suburbs, many of them happily.

There is a great danger in equating urban cores with urban areas. These cores have a deep history; many have a wealth of cultural and architectural assets and they still have the diversity and richness of activities, albeit under attack, that is so valued. These cores should be preserved and enhanced and, in extreme examples, recreated. However, they constitute a small part of the city. How does the European Commission suggest that existing suburbs are converted into the high density, diverse, culturally-rich environment desired? The reality of much modern urban living is ignored in the Green Paper.

This does not deny that the European Commission's advocacy of the compact city has merit. Indeed, it has been argued in Section 17.4 that the city's accumulated human-made resources must be regarded as precious cultural, economic and social assets. The point is that the large city is a complex phenomenon, which in many instances now consists of extensive suburbs as well as historic cores. As the virtues of the compact city are debated, this must not be forgotten.

17.5.3 Contradictions of the compact city

Although the European Commission's Green Paper is to be welcomed in raising the debate about the future form of cities, there are many contradictions that arise from the advocacy of the compact city that it fails to consider. Table 17.2 lists the more obvious of these.

Table 17.2   Contradictions of the high density, compact city

High density versus suburban quality of life
High density versus access to urban green areas
High density versus renewable sources of energy
High density versus vigorous rural economy
High density versus telecommunication-rich dispersal

The question of the relative merits of high density cores versus what, in some places at least, might be regarded as the superior suburban lifestyle, has already been rehearsed. The real challenge for the planners may be in what to do with what already exists, most of it various forms of suburbia, rather than with marginal opportunities for new development.

The environmentally prompted call for high density cities tends to clash with the demand for more urban green space that comes from the same source. Some extreme protagonists of the compact city movement actually regard urban open space as undesirable because it detracts from the intensive built environment that they favour. The European Commission does not go this far, but nevertheless the trade-off between density and open space has to be resolved.

Another clash of environmental principles that arises with the advocacy of the compact city concerns the development of renewable energy sources. Owens (1986) suggests that for any large-scale development of solar heating or wind power in cities, lower density development is required. She suggests that a linear grid form of development, with buildings focused at high density on a grid of roads and public transport routes and open areas in the centre of each grid, might facilitate both high density living and opportunities for renewable energy sources. Likewise, it might also accommodate green open space.

The concentration of future development into compact cities would also tend to exacerbate what is seen by some researchers to be a major problem; that is, declining rural economies. Newby (1990), for example, has argued that the success of urban containment policies in Britain has deprived rural areas of economic activity that could have sustained them. Again, this is a contradiction that the European Commission has not considered.

Interestingly, the current vogue for high density cities has come along just as another diametrically opposed view of the urban future has been established. Over the last few years a number of writers have stressed the virtues of an anti-urban lifestyle. Although there is a long tradition of anti-urbanism, the ideas have been resurrected recently and have been given a new technological imperative. The basic philosophy envisages a future that is decentralized, in both the institutional and land-use sense. Some advocates stress the need for a change of values: from urban values, which are 'concerned with the possession of things' to rural values, which 'are to do with ways of life, with wholes and what is qualitative, and hence with where we belong' (Ash, 1987, quoted in Robertson, 1990). Other authors stress the emancipatory effects of new technologies on working and living lifestyles. Through new telecommunications technology, they argue, the centuries-old need for face to face contact is disappearing. This holds out the prospect of the 'electronic cottage' notion of future living and working. Robertson (1990), for example, promotes the idea of continued urban dispersal, a reaffirmation of rural values and the exploitation of technologies to bring this about.

Although Robertson (1990) sees this as being accompanied by greater energy efficiency, this view is clearly inconsistent with that of the European Commission. It implies low density, land-consuming development. Also, it is debatable as to whether this geographically decentralized, telecommunications-rich lifestyle would increase or decrease conventional travel. It could reduce the need for conventional travel and hence be more energy efficient, particularly if new growth is focused into a number of compact (polynucleated) settlements. In this instance, the reduction in travel resulting from telecommunications would have to more than offset the increase in travel caused by a move from concentrated cities to polynucleated settlements. In recent years, particularly in southern Britain, there has been the development of polynucleated settlements, as cities have decentralized, but without any significant evidence that the increased use of telecommunications affects lifestyles and hence travel patterns.

One obvious problem with the vision of a new decentralized lifestyle is that, if taken up by large numbers of people, the idea becomes undermined. Here, the 'tragedy of the commons' would be the literal outcome. Lots of people individually choosing to live in the countryside could collectively undermine the very lifestyle that they seek. The prospect of this outcome is what has fuelled the NIMBY ('not in my backyard') protectionism in many rural areas of Britain (see Chapter 18).

None of these contradictions of the compact city deny that the proposal has merit. What they do suggest is that the issue is complex and needs careful analysis and extensive debate. Many of the issues have been raised before, but now the questions of the quality of urban life and the effect of cities on the natural environment, which together make up the issue of 'sustainable urban development', have to be considered together.

17.6 CONCLUSIONS

This chapter has attempted to develop the idea of 'sustainable urban development' by distinguishing between conditions relating to the natural environment and those relating to the human-made environment. It is concluded that the city must be considered as a resource, just as the natural environment is a resource. Both are precious. With this distinction, ideas about balancing all urban development aspirations with resource use can be developed. Some such aspirations make demands on the natural environment, others make demands upon the human-made environment of the city. It is this linking of the natural and human-made environments that constitutes the idea of 'sustainable urban development'.

This conceptual framework matches neatly with some current thinking on the future of cities. In particular, the European Commission's Green Paper (CEC, 1990) tries to relate the environmental sustainability of the city with the question of quality of life. The task is to find solutions that perform well on both grounds. The Commission tentatively suggests that the solution is to promote the high density, compact city, in the European tradition, as both energy efficient and socially desirable.

Although instantly appealing, this proposal does in fact warrant careful examination. Is it energy efficient and is it socially desirable? Some research on the energy efficiency of different urban forms has been reviewed. This offers some support for the CEC view, but also suggests that other forms might be equally appropriate. These forms include the polynucleated settlements, possibly akin to new towns, that the Green Paper opposes. Obviously, this issue of urban form and energy efficiency is very important and deserves much more research and debate.

It is necessary, however, to be wary of assumptions that there can be a direct relationship between changes in urban form and environmental improvement. The forces that determine energy consumption, urban change and environmental degradation are complex. They are political, social and economic, as well as physical. Just as market responses to changes in energy prices are difficult to predict, so too are responses to planned physical change. This should not prevent research into appropriate urban forms; research that should inform planned change to existing urban stock as well as to new settlements. Planned change needs to be integrated with political and economic initiatives if urban areas are to become more sustainable. In the circumstances, the best strategy, as Owens (1986) and Rickaby (1987) suggest, might be to find urban forms that are robust; that is, that are energy efficient in a variety of circumstances.

The sustainable urban development debate must link, as does the CEC Green Paper, environmental sustainability and the quality of urban life. Questions of closing the environmental 'circle', of desirable urban lifestyles, of the economies of our cities, of the future form and function of cities, and more, need to be related. It is necessary to determine development aspirations, decide on the degree to which they are to be sustained, and then to relate them to resource contraints, both natural and human-made. These are complex questions. As shown here they throw up contradictions: some people prefer suburban lifestyles, despite the apparent merits of concentration; some argue for greater, not less decentralization, based on improved telecommunications; some people are worried about the effects of concentration on rural economies. The more these issues of sustainability are debated, the more complications and contradictions seem to arise. In the short term the problem may seem to become more, not less intractable. However, with so much at stake, despite the obvious urgent need for action, time spent on clarifying concepts and ideas is essential.

17.7 REFERENCES

Ash, M. (1987) *New Renaissance*. Bideford: Green Books.
Commission of the European Communities (CEC) (1990) *Green Paper on the Urban Environment*. Brussels: EEC.
Cresswell, P. and Thomas, R. (1972) Employment and population balance. In: Evans, H. (Ed.). *New Towns: The British Experience*. London: Charles Knight for the Town and Country Planning Association, pp. 66–79.
Department of the Environment (1990) *Our Common Inheritance*. London: HMSO. (Cmnd 1200.)
Newby, H. (1990) Revitalizing the countryside: the opportunities and pitfalls of counter-urban trends. *Journal of the Royal Society of Arts*, **CXXXVIII** (5409), 630–636.
Owens, S. (1986) *Energy Planning and Urban Form*. London: Pion.
Owens, S. (1987) The urban future: does energy really matter? In Hawkes, D., Owers, J., Rickaby, P. and Steadman, P. (Eds). *Energy and Urban Built Form*. London: Butterworths, pp. 169–186.
Pearce, D., Barbier, E. and Markandya, A. (1990) *Sustainable Development: Economics and Environment in the Third World*. Aldershot: Edward Elgar Publishing.
Rickaby, P. (1987). Six settlement patterns compared. *Environment and Planning B, Planning and Design*, **14**, 193–223.
Robertson, J. (1990) Alternative futures for cities. In: Cadman, D. and Payne, G. (Eds). *The Living City: Towards a Sustainable Future*. London: Routledge, pp. 127–135.
World Commission on Environment and Development (1987) *Our Common Future*. Oxford: Oxford University Press.

17.8 FURTHER READING

Cadman, D. and Payne, G. (Eds) (1990) *The Living City: Towards a Sustainable Future*. London: Routledge.
Cullingworth, B. (Ed.) (1990) *Energy, Land and Public Policy*. New Brunswick and London: Transactions Publications.
O'Riordan, T. (1988) The politics of sustainability. In: Turner, R. K. (Ed.). *Sustainable Environmental Management*. London: Belhaven Press, pp. 29–50.
Redclift, M. (1987) *Sustainable Development: Exploring the Contradictions*. London: Methuen.

CHAPTER 18

Counterurbanization and Environmental Quality

D. Spencer and B. Goodall

18.1 INTRODUCTION: THE NATURE OF COUNTERURBANIZATION

Counterurbanization is a term applied to the fundamental reshaping of the geography of population and economic activity which has taken place over the last 30 years in most developed societies. It has been used generally and specifically, resulting in confusion over its precise meaning (Champion, 1989). It is defined here as the reorganization of urban living over a much enlarged spatial scale. Counterurbanization involves the dispersal of population and economic activity to free-standing settlements in 'rural' localities, and deconcentration, whereby the smallest settlements grow the most. A more even population distribution gradually emerges. Clearly, 'counterurbanization' is not synonymous with suburbanization because population redistribution occurs beyond the physically built up town or city. Through this process, both accessible and relatively remote rural regions have become the destination for new and relocated jobs and homes.

Counterurbanization may simply be recognized as metropolitan decline paralleled by rural growth. Such a 'broad brush' perspective can easily overlook the finer spatial outcome of urban–rural shifts which have brought expansion to medium and small sized towns in particular. What passes for counterurbanization at the regional level is, therefore, urbanization at a smaller scale, characterized by peripheral suburban growth and internal reorganization. Many issues relating to environmental quality addressed later in this chapter stem essentially from the ways in which rural settlements are urbanizing as a result of widely operating counterurbanization forces.

The aim of this chapter is to examine the impact of counterurbanization on environmental quality in rural subregions. The underlying argument is that there are broad reciprocal relationships between socio-demographic changes, the spatial reorganization of settlement patterns, and the quality of the environment. This will be considered in general terms, and then illustrated by a case study. Ultimately, this chapter will discuss how far the environmental implications of counterurbanization

Environmental Issues in the 1990s. Edited by A. M. Mannion and S. R. Bowlby

are likely to become matters of urgent social concern in the 1990s, and what the policy responses might be.

It is tempting to view counterurbanization as 'top down', thereby stressing societal forces which have conferred advantages on 'rural' regions for living and working. Inevitably, this has local spatial and environmental implications: there will be an upsurge in demand for land for development in and around specific settlements so that their character is altered. Such environmental change can subsequently affect people's well-being, both positively and negatively. The advantages of growth may include more living space and better services and amenities. Conversely, pressures and conflicts may threaten to erode visual environmental quality in more fragile settlements and the surrounding countryside. However, the nature of local environments themselves can also affect how far societal changes reach down to them. Localities are unique: some areas are thus more receptive to change than others, depending on their inherited human and physical characteristics, political attitudes towards further development, and the extent to which local decision making is autonomous (see Chapter 1). Moreover, the experience of living, working and doing business within an environment structured in a particular manner may affect the very way that processes subsequently operate. Dispersal, for example, may give way to centralization if moving people and goods between relatively scattered settlements becomes too costly and threatens environmental damage.

18.2 ENVIRONMENTAL QUALITY OF SETTLEMENTS

All settlements, and the neighbourhoods within them, can be viewed as sets of resources or environments for people to use. Indeed, all sections of a particular community will, within constraints, search out an environment which affords the best practical conditions for living, working and recreation. The quality of the environment in a given place will influence the quality of life for its residents and workers, and the built environment plays an important role here.

It is debatable which of a multiplicity of factors should be taken into account when assessing environmental quality. Nevertheless, four broad areas stand out. First, there is the question of shelter, which clearly depends upon the quality and amount of housing. Second, there is public health and safety, concerned with the prevention of diseases, accidents and pollution. Third, there is the provision of an efficient environment which facilitates human activities and movement. Fourth is the creation of an environment which maximizes people's comfort and enjoyment of living: here more intangible amenity variables, such as privacy or visual and aesthetic beauty, are important. All in all, the quality of the environment in any settlement will amount to an amalgam of factors covering quality of location, neighbourhood and buildings, together with a wide range of access and amenity variables (Burke and Taylor, 1990).

Environmental quality is a subjective concept. This is largely due to the fact that the objective, physical and socio-economic environment is filtered by people's perceptions, feelings and attitudes. Thus, environmental quality means different things to different people. It may reflect their roles as residents, entrepreneurs, planners, or councillors. In addition, the perception of environmental quality may vary according to social class and lifestyle. Even within a given social group, individuals may view the same objective environment differently because of their circumstances, such as gender, age, or disability. Female residents caring for pre-school children in dormitory suburbs may experience a certain isolation imposed on them by their environment during the working week when the family car is used by their spouse for journeys to work. Likewise, the physically disabled and elderly with mobility problems, if public transport is infrequent and distances to bus stops are excessive.

Environmental quality varies spatially between settlements and within them at the neighbourhood scale. Differences in quality between settlements often reflect size and regional location, and the legacy of previous phases of development. In country towns, for example, there is frequently considerable diversity over a small scale, often amounting to an amalgamation of residential, commercial and industrial functions derived from piecemeal development through time. Quality here is there-fore closely related to the juxtaposition of past and present built forms.

Popular perceptions now centre on the belief that environmental quality rises as settlement size diminishes; indeed, this has been one explanation of the urban–rural shift of industry and population. In reality, however, the issue is more complex, as the environmental advantages of small size (proximity to countryside, shorter journeys to work, reduced traffic congestion, less pollution and crime) may be offset by disadvantages (such as the narrower range of public services, leisure and cultural opportunities). The growing trend towards privatized lifestyles, leisure activities and consumption patterns centred on house and garden may mean that less value is now placed on public facilities requiring larger threshold populations.

At the intra-urban or neighbourhood scale, there are also significant variations in environmental quality. One of the key factors at work here is the juxtaposition of land uses. Quality depends, for example, on the extent to which non-conforming uses (such as noxious industries and scrap yards) penetrate residential environments. In addition, efficient operation within some industrial areas may be hampered by inappropriate adjacent activities (such as where access needed by a particular business is impeded by unorganized car parking caused by another firm).

Environmental quality is not static. In the short term, residential environments may be enhanced as landscaping matures and building materials weather. Over a longer time span, however, such neighbourhoods deteriorate unless regularly main-tained. There may be little incentive to initiate environmental improvements due to pressures stemming from two main sources.

First, the societal population-related changes discussed in Chapter 7 (such as the decline in average family size, the emergence of more smaller households and rising personal aspirations for more living space) mean that buildings may not match current demand and may need to be replaced or modified. Allied to this, natural increase and in-migration lead to pressures for urban peripheral growth (if densities are not to be increased) or redevelopment (if urban containment is advocated). Second, technological advances can render many buildings and much infrastructure economi-cally obsolete so that their physical replacement may be a viable alternative to continued maintenance. This may present further difficulties when the buildings have intrinsic historic and/or architectural merit, as their replacement is then clearly not simply a matter of economics. One possible solution here is to adopt building conservation policies, where renovation is frequently combined with limited internal and/or external change to the physical fabric to give a new lease of life to older structures. Such policies may also be carried out on an area basis to sustain and enhance the quality of the townscape as a whole. In either case, conservation is clearly not synonymous with preservation, as sensitively induced change, as opposed to a reluctance to sanction even the slightest alteration, is acceptable.

Hence the exact form environmental quality takes in a given place and at a particular time will depend upon how far societal forces have induced urban change. Taking settlements as a whole, variation in quality is often a reflection of age (particularly of the degree of 19th century legacy) and of the subsequent fortunes of the regional economy: settlements in stagnating or declining regions tend to be of lower environmental quality. At the neighbourhood level, quality tends to decline when locational advantages have been lost or densities increased, and improves where there has been urban renewal, rehabilitation, or conservation.

18.3   SPATIAL VARIATIONS IN ENVIRONMENTAL QUALITY: A CAUSE AND CONSEQUENCE OF COUNTERURBANIZATION

This section examines the reciprocal relationship between social processes, spatial form and environmental quality. It is argued that migration by businesses and people (the tangible evidence for counterurbanization) can, in part, be seen as a search for a higher quality environment. It is stressed that this search does not, however, occur in a vacuum. The choices of people and firms are usually made within constraints attributable to forces at work at the societal level in the spheres of production and consumption. Any comprehensive explanation should therefore commence at this macro-scale, but not be restricted to it.

One cause of counterurbanization lies in the shift of investment from metropolitan regions to peripheral and other 'rural' localities. The driving force behind this has been industrial re-structuring, where manufacturing firms in particular have reorganized their activities in the face of international competition to maintain or enhance profitability. Tasks have become increasingly spatially separated, and as part of this routine production has been decentralized to 'rural' settlements.

Rural localities have thus become enmeshed into this 'new spatial division of labour' (Massey, 1984). From the entrepreneur's point of view, they have possessed the advantages of supplies of inexperienced labour (particularly non-unionized women unaccustomed to the disciplines of factory work), and operating costs lower than in urban agglomerations (wages, rents, rates, land values). The decline of traditional agriculture and rural industries and the lack of alternative employment has not only meant that a potential labour force has been at hand, but also that key decision-makers in rural communities naturally welcomed inward job-creating investment. A better environment in which to produce (in terms of sufficient space for efficient operation, room for expansion, lack of congestion and the residential attractions of the countryside) is also an integral part of the explanation for rural industrial development.

This trend cannot, however, be entirely attributed to inward investment in manufacturing. In Britain, for example, service sector growth has become relatively more important. Some of the larger settlements in East Anglia have, for example, participated in the boom in producer services employment. In scenically attractive areas, seasonal tourism has become a major source of income (albeit low) for many residents. In other rural areas, country towns have experienced growth in public services and the administration, retailing and distributive trades as the centralization of private and public sector activities into prominent central places has occurred to take advantage of economies of scale (Hudson and Williams, 1986). Indeed, this has been promoted by planning policies aiming to strengthen the economic base of 'key settlements'. Consequently, accusations of territorial injustice have been heard: while retailing and personal services in many villages has declined (if not disappeared), stability or growth has been the outcome of a combination of market forces and planning policies in the larger settlements. Through this, some attributes of environmental quality may be enhanced, but at the expense of others.

Such far-reaching changes in the realm of production and the rendering of services have had demographic consequences. Typically, selected rural communities have experienced expansion, primarily through a positive migration balance. Rural economic growth has also stimulated repopulation when transfers of key personnel needed to set up 'branch plants' or 'back offices' has occurred. This has fuelled in-migration *per se* and meant that the newcomers have been predominantly middle class or skilled working class, which has distorted rural social structures (Fielding, 1989). Economic growth in the countryside can also halt depopulation by offsetting the loss of agricultural employment and jobs in ancillary industries. In addition, as the number of women in employment increases, so do the problems of securing employment for two-job households in alternative locations, which retains some rural populations.

Nevertheless, it would be a mistake to interpret rural demographic growth solely in terms of residential inertia or involuntary inward moves. Rural areas offer clear gains, many of them environmental, for those whose favourable economic circumstances enables them to conduct spatial searches and exercise spatial choices.

In peripheral rural regions, therefore, the bulk of population increases can be attributed to spontaneous in-migration by a range of subgroups anxious to maximize environmental satisfaction by acquiring a rural residence. Long distance migration has brought in economically active people who have taken up local job opportunities and others who have started businesses. Their anti-urban value system is reflected in the way they have traded off lucrative urban employment for lower pay plus perceived gains in the quality of their lives found in a more rural locality. In some parts of Britain, such as Devon and Cornwall, most newcomers are engaged in 'conventional' employment; there appears to be only a limited revival of a romantic 'back to the land' subculture (Perry *et al.*, 1986; Bolton and Chalkley, 1989).

In the more accessible rural areas, counterurbanization shows rather different characteristics. Economically active people feature even more strongly among the newcomers, especially middle-aged couples who have migrated in anticipation of retirement. Most ex-urbanites have sufficient disposable income to purchase a superior residence and commute to the principal urban centres. Many in-migrants will have been fortunate enough to experience salary increases in real terms, possibly through possessing employment skills enabling them to enter highly valued professional, scientific and managerial occupations in the booming service sector of the national economy. In short, they are members of what is called the new service class; their lifestyles have become centred around urban-located production systems and rural-orientated patterns of consumption.

Summing up, urban–rural contrasts in environmental quality clearly play a major role in any comprehensive explanation of ex-urban growth. This leads to a paradox: although high environmental quality can be a 'pull' factor attracting population and industry into rural regions, the development it encourages may then threaten to erode those very characteristics which attracted growth initially. This reciprocal relationship will be the focus of attention through a case study in the next section.

18.4   COUNTERURBANIZATION IN SOUTH OXFORDSHIRE

18.4.1   The counterurbanization experience

As regional counterurbanization trends have been operating in the local authority district of South Oxfordshire for 30 years, they are discussed here as a case study.

South Oxfordshire lies close to the western fringes of the outer southeast region of England. Roughly one-third of the district lies within the Oxford green belt and a similar proportion within the Chilterns Area of Outstanding Natural Beauty (AONB) (Fig. 18.1). Functionally, South Oxfordshire forms part of the accessible or peri-urban or metropolitan countryside, being partly integrated into the orbit of London, but more closely linked to the major provincial towns of Oxford, Reading and High Wycombe.

The period 1961–90 saw a 45% increase in the district population (amounting to over 40 000 additional people in private households) and as many as 72% of parishes recorded net gains. Population dispersal was fairly widespread (Fig.18.2), particularly along the broad Didcot–Thame axis and into the Thames Valley south of Wallingford. Thus growth did not simply amount to suburbanization around Reading or Oxford or the main country towns. Few areas have been immune from population expansion, exceptions being mainly green belt parishes close to Oxford and in the Chilterns AONB.

The five principal country towns and other large villages increased their share of South Oxfordshire's population disproportionately, with relatively little 'trickle down' to places at the bottom end of the settlement hierarchy. Indeed, the smaller places have actually experienced a decline in the proportion of the district's population found in them. By 1991, more than one resident in three lived in the established country towns. Clearly, no inverse relationship between settlement size and rate of growth has evolved.

Thus, although *dispersal* of the district's additional inhabitants has occurred, *deconcentration* has been conspicuous by its absence. The spatially selective nature of these processes and patterns suggests that areas dominated by larger settlements are more likely to see changes in environmental quality than others. These will take on a number of forms, such as extra facilities and services (schools, shops, clinics, employment, mains drainage, road construction/widening). This brings benefits to some, such as those without regular access to a car, but may repel others who desire privacy and social exclusivity, which then becomes increasingly associated with village residential environments.

Counterurbanization also implies new housing development. Indeed, the growth of dwellings (from 28 000 in 1961 to 53 000 in 1990; a 90% increase) greatly exceeds the growth of population (see Chapter 7.4). This reflects societal trends; with more smaller households requiring self-contained dwellings, the outcome is substantial

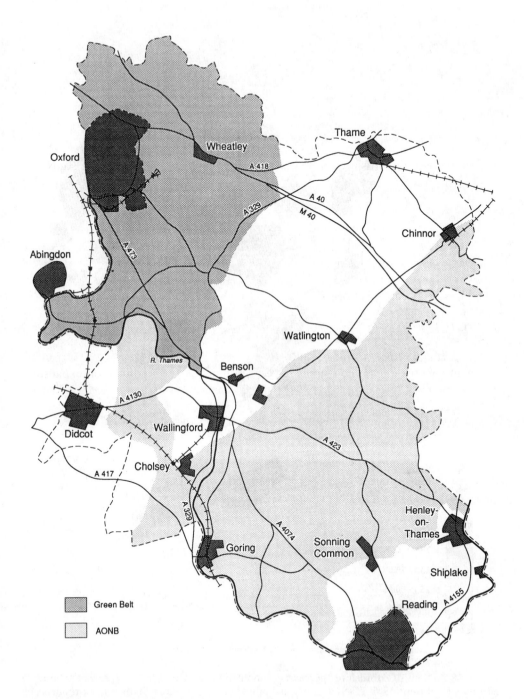

Figure 18.1 South Oxfordshire: constraints on development in relation to principal settlements and communications.

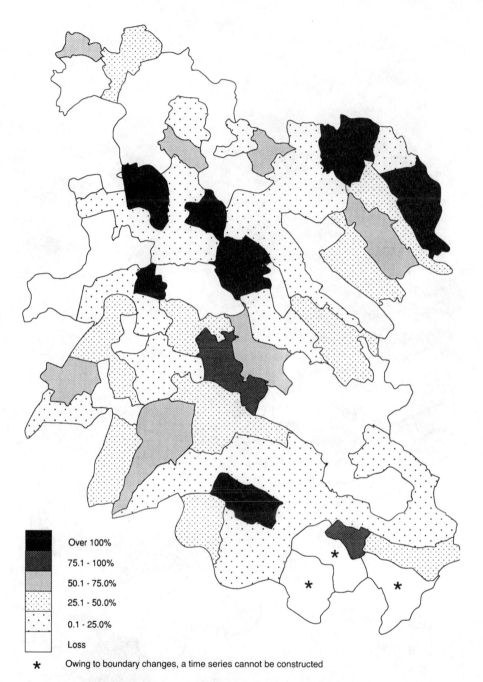

Figure 18.2   South Oxfordshire: percentage population change, 1961–90 (persons in private households). From: 1961–71–81 Census of Population, Small Area Statistics, reproduced by permission of the Office of Population Censuses and Surveys; 1990 Oxfordshire County Council Estimates.

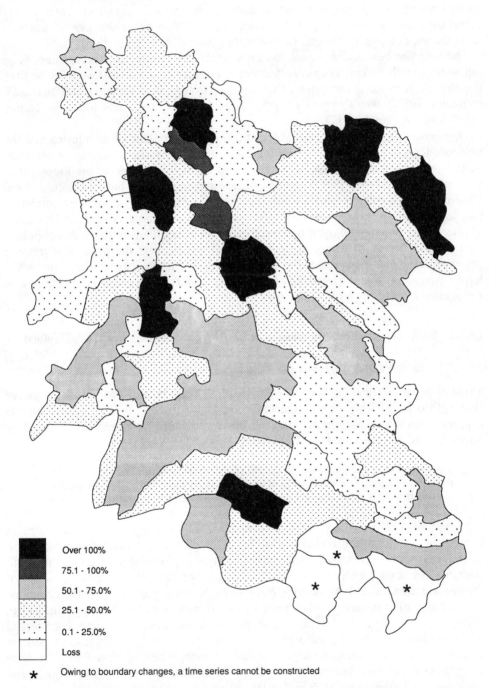

■	Over 100%
	75.1 - 100%
	50.1 - 75.0%
	25.1 - 50.0%
	0.1 - 25.0%
	Loss
✱	Owing to boundary changes, a time series cannot be constructed

Figure 18.3 South Oxfordshire: percentage change in the number of dwellings, 1961–91. From: 1961–71–81 Census of Population, Small Area Statistics, reproduced by permission of the Office of Population Censuses and Surveys; 1990 Oxfordshire County Council Estimates.

pressures for residential development which will affect environmental quality. In South Oxfordshire, as in other growing districts, the nature of environmental quality therefore stems not only from the impact of population expansion *per se*, but also from the implications of the creation of more households (see Fig.18.3).

The question now arises: where has the pressure for residential development been felt most acutely? With even less deconcentration of residential development than population, it is the country towns which have experienced very marked increases, averaging 105% over the period 1961–90, with the highest rates found in Thame (142%) and Wallingford (72%).

One explanation for this unequal redistribution of population and housing and the associated pressures on local environments is that it is the outcome of planning policies reflecting the values of a vociferous rural anti-development lobby. It is perhaps no coincidence that planning policies and regulations have protected visual environmental quality in and around the smaller settlements in the most attractive landscape settings, while steering most growth to the larger towns and villages. However, the benefits enjoyed by a small number of residents in these less populous communities (many of whom are middle class) has been at the price of increasing pressure in the historic cores and at the fringes of the market towns which, arguably, have environments equally worthy of preservation and enhancement, as is the case of the ancient market town of Wallingford.

18.4.2 Environmental quality issues in a 'pressurized' country town: Wallingford

18.4.2.1 Background: town and townscape

Situated in the Thames Valley between Reading and Oxford, the Saxon 'burgh' of Wallingford is arguably one of the most important historic towns in England. Its principal townscape features are outlined on Fig.18.4 and have been summarized by South Oxfordshire District Council (1990, p. 12):

> The prominent Saxon defences in the Bull Croft and the Kine Croft, the rectilinear pattern of its streets, the remains of its once magnificent castle, and its wealth of historic buildings present an eloquent physical testimony to its past.

By the mid-1970s, when the first local plan for Wallingford was drawn up, a number of locationally prominent historic buildings had lost their economic raison d'être and had deteriorated physically. There was therefore an urgent need to revitalize the town's economic base while simultaneously enhancing the qualities of the historic townscape. Fortunately, Wallingford began to participate in the economic and demographic upturn characteristic of all country towns in South Oxfordshire, with renewed demands for land and premises for residential and commercial activities. Significantly, the local planning authority was able to implement policies consistent with the spirit and purpose of sustainability (see Chapters 2 and 17). New investment was channelled into the re-use of the historic fabric, bringing national and international professional acclaim for the manner in which the townscape was thus conserved rather than redeveloped.

The question is, have these recent environmental achievements been enduring?

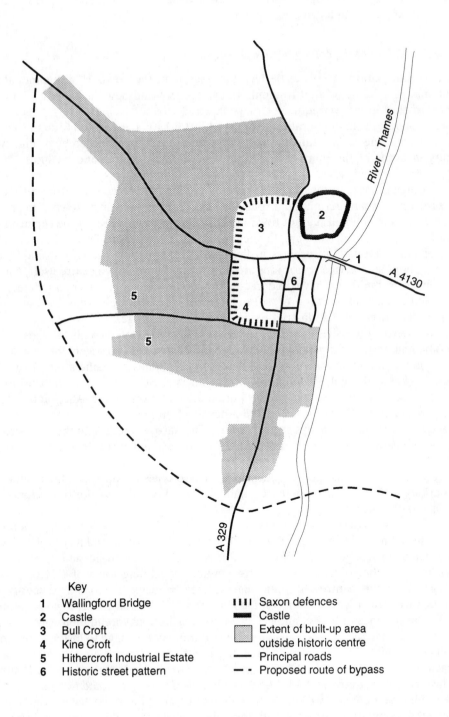

Key

1	Wallingford Bridge
2	Castle
3	Bull Croft
4	Kine Croft
5	Hithercroft Industrial Estate
6	Historic street pattern

IIII Saxon defences
▬ Castle
▨ Extent of built-up area
outside historic centre
— Principal roads
– – Proposed route of bypass

Figure 18.4   Wallingford: principal townscape features.

How has Wallingford coped with pressures relating to indigenous growth and counterurbanization-induced development?

18.4.2.2   Nature of the contemporary pressures

Some contemporary pressures reflect the fact that the town is accommodating additional people, associated residential development and new businesses. Many of these are the direct consequences of counterurbanization. The town's population grew, primarily through migration, from 4484 in 1961 to an estimated 6500 in 1990, a 45% increase. Inevitably, this has had a dramatic impact upon urban environmental quality in terms of the number and type of additional services and facilities which have been provided. Over the same period, the number of residences increased by 72%, bringing more intensive development within the town's historic core and pressures for urban peripheral expansion. As in many country towns, many of Wallingford's traditional market town industries have disappeared to be replaced by manufacturing activities and employment in the tertiary sector. This is illustrated by the arrival of Habitat Furnishings in the town, which is now the largest employer. Indeed, about 900 service sector jobs are now located in the town centre itself, where many historic buildings have undergone functional change to accommodate new office and retail uses. A total of 800 manufacturing and warehousing jobs are also found in the Hithercroft Industrial Estate on the urban fringe.

Cutting across the issues discussed so far are transport-related conflicts. These stem from the high levels of personal mobility commensurate with an acceptable quality of life for most inhabitants of rural localities. Large volumes of traffic attempt to gain access to or pass through Wallingford, which can cause severe congestion. This arises from the location of the town at the bottleneck junction of the A329 and A4130 roads. Wallingford Bridge, a scheduled Ancient Monument, also has very restricted capacity, causing major delays at peak hours. The narrow streets and their important historic and architectural qualities cannot be appreciated when traffic flows reach high volumes. Pavement widths are inadequate in the very areas where traffic flows are greatest, causing danger to pedestrians. There is also little space for additional car parking within the Saxon defences, and only very limited scope for redevelopment to alleviate the situation.

These pressures derive from a number of sources. First, traffic has been generated due to the economic buoyancy of South Oxfordshire as a whole and its country towns in particular. Through traffic therefore forms a substantial proportion of all flows, particularly as the principal route to the rapidly expanding town of Didcot passes through the town centre. Second, there is the growing economic prosperity of Wallingford itself. Developments on the Hithercroft Industrial Estate have generated susbtantial volumes of traffic, and heavy goods vehicles thus thread their way through Wallingford's narrow streets, causing noise pollution and damage to historic buildings through vibration. Third, some traffic represents tourist trips in response to the attractions discussed above. Fourth, residents of new dwellings constructed within the historic core require access to their own homes. Finally, it should not be forgotten that smaller settlements within the town's immediate hinterland have experienced substantial population growth, and they look towards Wallingford for services. Indeed, over the period 1961–90, the population of the four adjoining parishes grew from 6670 to 10 214. Given the paucity of local public transport, car access into Wallingford has become essential.

Overall, the net effect of increased vehicular flows creates a major conflict of interest between maintaining the historic townscape and accommodating road vehicles. Traffic problems arguably pose the greatest threat to improvements in the quality of the environment, which to date have come about largely through conservation-orientated policies.

18.4.2.3 Managing the pressures: local planning and environmental quality

In Wallingford, local planners have been confronted with the classic difficulty of how to incorporate some new development while simultaneously protecting environmental quality. In theory, a number of courses of action have been open to them: functional change, the reorganization of space within the Saxon defences, land use intensification and peripheral expansion. The 1990s are most likely to see the local authority implementing policies not unlike those dating back to 1976, which combine these options. The principal underlying aims of the 1990 Local Plan are to resist pressures for excessive and 'anyplace' development to protect agricultural interests and the special historic and environmental qualities of the town.

The quality of the residential environment in the town should be improved by adding to and rehabilitating the dwelling stock. To date, much new development (over 500 additional dwellings) has taken the form of sympathetically designed small units at high densities, largely by converting vacant industrial buildings and redeveloping a number of redundant spaces. This has helped to satisfy the growing need for easier access to good quality shelter for one and two person households (who make up over half of all Wallingford households) and for new development to respect the form and structure of the inherited physical fabric. Urban peripheral expansion for residential purposes will continue to be vigorously resisted as this, in particular, would undermine the Authority's conservation objectives by seriously eroding the natural environmental backcloth to the town and stimulating demand for more employment, services and commercial activities. These arguments have been accepted by the Department of Environment in dismissing an appeal against refusal to grant planning permission for a new housing estate on the western outskirts of the town.

Similar arguments have been used to restrain further business development in the town. Land and premises for further industrial development are to be allocated only to Wallingford based firms. Coupled with this, an industrial relocation policy within the central area has been pursued for over 15 years and will continue through the 1990s. It requires the resiting of a number of badly located businesses (from an aesthetic and operational point of view) to new premises at Hithercroft where they can operate more efficiently and cause less visual intrusion.

These long established policies implicitly aim to enhance the four interrelated components of urban environmental quality as discussed in Section 18.2, namely shelter, public safety, efficiency and aesthetics. The fact that this is to be achieved through the continuation of policies of restraint and by conservation indicates, however, that the aesthetic takes priority over other more tangible elements which might enhance the quality of life for most residents by providing better homes and more jobs. To date, the net outcome of local planning policies has been for a growing spatial separation of residential and industrial functions to enhance the physical quality of the built and unbuilt spaces which arguably give Wallingford its sense of place. It remains to be seen whether these policies will lead to gentrification within the Saxon defences: high quality environments can inflate property values which makes access difficult for low income groups. Planning policy thus generates a new series of social–environmental conflicts.

18.4.2.4 Environmental quality issues in the 1990s

Pressures on any environment are notoriously volatile and forecasts for the 1990s are difficult to make. In Wallingford, demographic related pressures might lead to a 5–10% population increase (the Oxfordshire estimate) over the next decade, probably exacerbated by further household fission (see Chapter 7.4). Furthermore, the local environment will almost certainly be adversely affected by road congestion. Traffic volumes are likely to grow at least in line with projected increases in the number of households. In rural areas, where public transport is poor, few households can afford to be without one (or more) cars, and railway closures means there is no alternative to road transport for freight.

A crucial factor in any forecast is the degree of urban–rural migration. Its fluctuations are closely linked to economic forces, the behaviour of which is difficult to predict. Of particular importance is the extent to which the economic buoyancy and prosperity associated with 'rural' localities in central southern England in the 1980s is eroded, and how rapidly recovery occurs. In the short to medium term, it is probable that the urban–rural shift of businesses and populations will abate, given the uncertain economic outlook. Wallingford might be spared much development pressure, at least until the mid-1990s.

What actually happens as regards environmental quality in the town and its environs will depend to a large degree on the manner in which these macro-scale forces are accommodated. This in turn depends on how far an alliance of central and local government planners, politicians and amenity groups possess sufficient power to bring about environmental sustainability in the face of market-generated growth. The crucial need here is development control: to adhere to targets for housing development, to restrict new job creation and to minimize the adverse impact of traffic.

A testing time for the effectiveness of local planning and political action will be the mid-1990s, when the environmental impact of the proposed bypass for the town becomes apparent. Given the environmental degradation caused by current traffic volumes, this appears at first sight to present a major opportunity to secure real environmental improvements. Indeed, the District Council is expecting reductions of between 40 and 65% in the volume of traffic using the town centre's narrow streets, together with a decrease in vehicles using residential streets to avoid the congested arterial roads. Through traffic should be virtually eliminated, while a link between the bypass and Hithercroft will channel heavy goods vehicles away from the central historic core. The net result should be the creation of an environment which is quieter, less polluted and safer for pedestrians and cyclists. It should be more conducive to appreciating Wallingford's townscape qualities. Owners of property on the main streets may then be more likely to maintain them in better condition, thereby contributing to an enhanced environmental quality commensurate with the objectives of urban conservation.

A number of important environmental gains are likely to occur in the short term. There are, however, doubts about their longevity. Although the local authority may be moving towards the solution to one set of environmental issues, they may be storing up other problems. The paradox is that although the new bypass will enable environmental improvements to be made, it will generate further pressures for access and new development. Wallingford is likely to be perceived as less congested, so that the demand for vehicular access to the town centre may well increase, thereby undermining any short-term benefits. The major difficulty concerns how far the bypass will be seen as a new outer boundary, thereby fuelling demands for the open space to be filled by housing development (Fig. 18.4). If all the land between the bypass and the town was developed, as many as 1500 additional dwellings could be constructed. This would increase the population of the town by around 70%. Clearly, this would pose a major threat to the conservation policies discussed above: the town centre will feel greater demands for car parking, additional retail floor space and other services.

The crucial issue here for the 1990s is, therefore, how far the local and structure planning systems possess sufficient power and political support to safeguard, and ideally enhance, the undoubted improvements which have occurred to the environmental quality of central Wallingford. This will require strenuous resistance to pressures for new development at the urban fringe. In this vein, local planners can argue that further housing would prejudice strategic targets for residential development in Oxfordshire, pre-empt development in other settlements and threaten the phased expansion of Didcot. The strength of these arguments may be tested locally and nationally. Much will depend on the attitude of central government towards two interrelated issues: first, attitudes towards the release of agricultural land and, second, how much support will be given to policies of constraint embedded in strategic and local land use plans.

In Britain, the late 1980s saw a fundamental re-appraisal of the need for agricultural land, given that mechanization and the application of new technologies has led to increased outputs from a smaller acreage. This has led to calls for a less protective attitude towards the release of lower quality land for development. Indeed, it has been calculated that between one and six million hectares of farmland and grazing land (representing 5–30% of the national total) will become 'surplus' to requirements by the year 2005. The implication here is that existing pressures to develop agricultural land around the edges of settlements will intensify, given the long-standing problems of urban fringe farming and the economic advantages of tacking additional development on to existing infrastructure. Unless local authorities, the Ministry of Agriculture Fisheries and Food and organizations aiming to protect the countryside can collectively oppose this, country towns such as Wallingford may expand peripherally. Population growth and the ancillary activities which come in its wake will intensify existing pressures on historic settlements.

In the case of Wallingford, it is fortunate that any fringe development would involve the loss of only 50 hectares of agricultural land of grades 1, 2 and 3a. However, even this would contravene government circular 16/87, which seeks to safeguard land of highest quality, even when national trends are towards reducing the amount of farmland. It therefore appears that on these grounds, national attitudes should help local planners to pursue policies of restraint.

In sensitive areas, the degree of constraint will also depend on how far the Department of the Environment adheres to agreed targets for new building and supports planners through the appeals system if conflicts with developers materialize. During the 1980s, the government was frequently accused of undermining the house-building policies of local authorities by covertly supporting the construction industry whenever a conflict of interest arose. Today, however, planners are less defensive, because central government has tended to support them. Many planners believe that the era of simply bowing to market pressures in the interests of national economic growth and restructuring, regardless of the environmental consequences, is over.

This change of heart may be because 'green' issues are now firmly on the political agenda: none of the principal parties can afford to advocate a reliance on unfettered market forces to bring about development without some consideration of its environmental impact. Indeed, it has become politically expedient for politicians to speak out against large-scale development to avoid alienating local supporters. Gradually, politicians have come to realize that all parties will have to be 'green' by the turn of the century. The question remains, how deep will be the shade of green?

In the early 1990s, the government appears to have taken a step in this direction by accepting that spatial and environmental issues can be promoted through planning. The Conservative Secretary of State for the Environment, Chris Patten, showed a willingness to take a more pragmatic view of strategic planning, thereby turning his back on established anti-collectivist and unfettered free-market philosophies. Indeed, Patten withdrew the controversial 1989 White Paper which would have dramatically diluted structure planning, and rejected appeals for permission to develop in sensitive locations. Nevertheless, critics point out that Patten's real political task was to successfully challenge vested financial and business interests, which he failed to do. The attitude of the Secretary of State appointed in November 1990, Michael Heseltine, is as yet unclear, but ministerial pronouncements indicate that Patten's course will probably be followed.

This change of emphasis, if genuine and permanent, augurs well for the environmental lobby. It should facilitate the continued conservation of historic Wallingford and the short-term enhancement of environmental quality by the provision of a bypass. If construction had commenced in the early to mid-1980s, it would have been much more difficult to resist subsequent development pressures. At that time, central government was determined to 'roll back the frontiers of the state' by removing planning constraints on the development industry. So far, the early 1990s have revealed a willingness to implement policies aimed at improving the quality of the environment in its widest sense, but provided that these do not threaten the status quo.

18.5 CONCLUSIONS

This chapter has focused on the interrelationships between social change, the urban–rural shift of population and development and the consequences for the quality of the environment in 'rural' areas. The principal argument has been that perceptions of environmental quality in rural settlements are an important ingredient in any comprehensive explanation for counterurbanization. Such areas are frequently seen very favourably. Quality thus stimulates counterurbanization and is, in turn, affected (positively or negatively) by the development which an influx of new residents and businesses brings in its wake. In theory, a stage might be reached when the environmental quality of rural settlements is itself eroded, so that the search for better environments for living and working becomes focused elsewhere. A re-appraisal of environmental quality in towns and cities might then lead to re-urbanization (van den Berg *et al.*, 1982), although in Britain current trends do not yet point in this direction.

The country town of Wallingford exemplifies these general arguments. Wallingford has experienced 'rounds' of population and employment growth, which initially provided opportunities for local planners to enhance environmental quality, particu-larly in terms of aesthetic improvements and a wider choice of housing opportunities and consumer services. However, the town's inherited historic fabric became incapable of accommodating more road vehicles, so that a deterioration of efficiency and public safety has occurred. These negative attributes of environmental quality are unlikely to stifle further development. Indeed, the local planning authority is still concerned about future growth and how to resist it. Its task might, paradoxically, be made more difficult by proposals to tackle the traffic issue through constructing a bypass. In such a historic town, it appears that the enhancement of some attributes of environmental quality can only occur at the expense of others. This stems principally from the difficulty of reconciling aesthetics with the need for efficiency and access to services.

Future levels of environmental quality in country towns such as Wallingford and the question of *whose* quality is enhanced will depend on the balance of power between local human agencies and macro-scale forces. One scenario is for a continuation of anti-urban trends, but interrupted by the recession of the early 1990s. This assumes the continuation of a market-orientated society with widening income inequalities, and a political ideology which merely pays lip service to 'green' issues. Those subgroups with most personal choice and freedom will search for residential exclusiveness. This might involve migration to a country town rich in aesthetic quality and with services orientated towards consumerist lifestyles. But as counterurbaniza-tion continues, small towns will fail to provide this. Attention will then be turned to villages, given that many have already become the expression of conspicuous consumption for the adventitious middle classes.

An alternative scenario envisages more radical social change, assuming that there is the political will to act now to secure long-term environmental benefits. It centres on the concept of sustainability, implying that development should meet the needs of the present without compromising the ability of future generations to meet their needs (see Chapter 2). If this gains acceptance, a much greater emphasis on environmental quality, energy efficiency and the quality of life for different social groups will be felt. Academics and planners will need to ascertain which settlement patterns are most likely to help achieve sustainability.

Research to date shows that there is no simple choice between compact city development, unfettered deconcentration, or dispersal with concentration (Chapter 17). There are, however, strong arguments that an energy conscious future demands some degree of dispersal into discrete settlements. Ironically, this corresponds to the pattern of development which has emerged in South Oxfordshire (and doubtless elsewhere) through planners' responses to counterurbanization. By default, resistance to deconcentration and the steering of growth into selected large villages and the country towns may have provided the skeletal framework for a system of settlements consistent with sustainability over a broad spatial scale. Interdependent communities of this type might operate efficiently, while also satisfying demands for higher environmental quality.

But who is likely to benefit? Given the present structure of society, this will depend very much on the lifestyles being sought by different individuals and households, and how far they can achieve them through competition in job and housing markets. The outstanding question then becomes, can a radical re-appraisal of attitudes towards the environment be extended to the socio-spatial structure of society, so that the improvements in environmental quality enjoyed by the 'haves' can be extended to the 'have nots'?

18.6   REFERENCES

Bolton, N. and Chalkley, B (1989) Counterurbanisation: disposing of the myths. *Town and Country Planning*, September, 249–250.

Burke, G. and Taylor, T. (1990) *Town Planning and the Surveyor*, 2nd edn. Reading: College of Estate Management.

Champion, A. G. (Ed.) (1989) *Counterurbanisation: The Changing Pace and Nature of Population Deconcentration*. London: Edward Arnold.

Fielding, A. J. (1989) Counterurbanisation. In Pacione, M. (Ed.). *Population Geography: Progress and Prospect*. London: Croom Helm, pp. 224–256.

Hudson, R. and Williams, A. (1986) *The United Kingdom*. London: Harper and Row.

Massey, D. B. (1984) *Spatial Divisions of Labour: Social Structures and the Geography of Production*. Basingstoke: Macmillan.

Perry, R., Dean, K. and Brown, B. (Eds). *Counterurbanisation: Case Studies of Urban to Rural Movement*. Norwich: Geo Books.

South Oxfordshire District Council (1990) *Wallingford Local Plan*. Crowmarsh Gifford: South Oxfordshire District Council.

van den Berg, L., Drewett, R., Klaasen, C. H., Rossi, A. and Vijvesburg, C. H. T. (1982) *Urban Europe: A Study of Growth and Decline*. Oxford: Pergamon Press.

18.7   FURTHER READING

Champion, A. G. and Townsend, A. R. (1990) *Contemporary Britain: A Geographical Perspective*. London: Edward Arnold.

Fielding, A. J. (1990) Counterurbanisation: threat or blessing? In: Pinder, D. (Ed.). *Western Europe: Challenge and Change*. London: Belhaven Press, pp. 226–240.

Gordon, I. (1988) Resurrecting counterurbanisation: housing market influences on migration fluctuations from London. *Built Environment*, **13** (4), 212–222.

CHAPTER 19

Must Tourism Destroy its Resource Base?

E. Cater and B. Goodall

19.1 TOURISM AND THE ENVIRONMENT

Tourism suggests holidays in the sun on sandy beaches and visits to distant places with grand scenery or alluring culture and history. Tourism has major economic, social and environmental impacts for destinations. Its growth creates problems, especially where fragile and remote environments are visited. Environmental degradation follows and the despoiled destinations become less attractive.

The exploitative approach of much tourism development neglects its resource dependency. The tourism industry may be brought into conflict with the environment, ultimately destroying its base in a particular destination. Research has focused upon impacts at destinations but tourism also contributes to global environmental change by its use of transport services. Tourism–environment interactions need to be better understood before recommendations can be made for a balanced relationship with the environment.

19.2 PATTERNS OF TOURISM

Tourism is an umbrella term for all the relationships and phenomena associated with people travelling. A tourist is defined by the World Tourism Organisation (WTO) as a visitor staying for more than 24 hours in a country visited for business or leisure purposes. Excursionists spend less than 24 hours in the foreign country.

Holidays may be categorized as sunlust or wanderlust. This reflects the different motivations for taking a holiday: the attraction of destinations with better climatic conditions than the holidaymaker's homebase and the desire to see new places and cultures, respectively. Activity holidays are a third type recognized.

Environmental Issues in the 1990s. Edited by A. M. Mannion and S. R. Bowlby
© 1992 John Wiley & Sons Ltd

19.2.1 International tourism

The WTO data provide a general synopsis of the global evolution of international tourism. In 1950 25.3×10^6 tourists were registered, rising by 1989 to 405.3×10^6. Numbers of tourists grew at an average annual rate of just below 4% over the previous decade during the 1980s. International tourism receipts (excluding fares) increased from US2.1×10^9 in 1950 to 209.2×10^9 in 1989, having grown at an average annual rate of 8.2% in current terms during the 1980s.

Tourists come from the more developed countries (MDCs), their demand fostered by high incomes (Chapter 1), increasing leisure time, high levels of education and new and cheaper forms of transport. In these countries fully integrated tourism companies, using sophisticated marketing techniques, have emerged to promote package tours and charter flights.

Table 19.1 shows that in 1988 the USA, Germany, Japan and the UK generated one-half of the total international tourism expenditure. Their combined populations

Table 19.1   International tourism expenditures and receipts for selected nations in 1988. Adapted from World Bank (1990), World Tourism Organization (1989)

Country	Gross national product per capita ($US)	International tourism expenditure (excluding transport) ($US millions)		International tourism receipts (excluding transport)	
		Total	Percentage	Total	Percentage
World		181 022	100	194 171	100
USA	19 840	32 122	17.7	29 202	15
West Germany	18 480	25 036	13.8	8449	4.3
Japan	21 020	18 682	10.3	2893	1.5
UK	12 810	14 650	8.1	11 023	5.7
Spain	7740	2440	1.3	16 686	8.6
Romania	—	56	0.03	176	0.09
Mexico	1760	3197	1.8	3994	2.1
Malaysia	1940	1324	0.7	766	0.4
Singapore	9070	930	0.5	2339	1.2
Kenya	370	25	0.01	410	0.2
Burkina Faso	210	15	0.01	6	0.003

constitute under 10% of the world total. The table also shows that while some MDCs are prime generators and attractors, e.g. the USA, others are predominantly net tourist attractors (Spain) or generators (West Germany). Among the less developed countries (LDCs) the richer, newly industrializing countries (NICs) are emerging as generators, for example, Malaysia, but on the whole the LDCs are net attractors, such as Kenya. Several, however, neither attract nor generate many tourists, e.g. Burkina Faso.

Tourists prefer certain destinations. Table 19.2 shows the dominance of Europe and the Americas, together accounting for almost three-quarters of tourist arrivals in

Table 19.2   International tourist arrivals by region in 1980 and 1989. Adapted from World Tourism Organization (1989)

Region	1980		1989		Average annual percentage change 1980–89
	Total arrivals (millions)	Share (%)	Total arrivals (millions)	Share (%)	
World	284.8	100	405.3	100	4.0
Africa	7.1	2.48	13.3	3.28	7.1
Americas	53.7	18.85	79.7	19.67	4.4
East Asia and Pacific	20.0	7.01	44.5	10.99	9.2
South Asia	2.2	0.80	3.2	0.79	4.2
Middle East	5.8	2.04	7.5	1.85	2.9
Europe	196.0	68.81	257.0	63.42	3.0

1989. However, these preferences change over time, with the fastest growing destination region during the 1980s being East Asia and the Pacific. The world share of arrivals in Europe and the Americas dropped from 96% in 1950 to 83% in 1989, while that of East Asia and the Pacific rose from 1 to 11%. The LDCs share of arrivals rose from 18.7% in 1980 to 22.1% in 1989.

Particular countries are outstandingly attractive as destinations. In 1985 50 countries recorded over one million tourist arrivals and are referred to as the 'tourist millionaires' (Bar-On, 1989). Once more the MDCs dominate. The top three destinations were France (36.7 million), Spain (27.5 million) and Italy (25 million). Only one LDC, Mexico, figures among the top ten destinations.

Tourists are also concentrated within destination countries, being drawn to the major resorts, e.g. in Spain to the Costas Brava, Blanca and del Sol. In Sri Lanka in 1987, 41.9% of foreign guest nights in graded establishments were located in south-west coastal resorts.

19.2.2   Domestic tourism

The WTO estimates that domestic tourism flows are ten times greater than international flows. Most people therefore holiday in their home country, although there are significant variations between MDCs, e.g. from 6% in Luxembourg to 93% for Greece. These reflect differing size of country, tourism resource endowment and level of affluence. Wealthy countries which are small, e.g. Luxembourg, or which lack tourism resources, especially coasts with favourable climate, e.g. Germany, have lower proportions of residents holidaying at home than countries with large land areas and varied resources, e.g. France and the USA. Social contacts, language difficulties, currency and documentation barriers and, perhaps, cost encourage the taking of domestic holidays. In developing countries domestic tourism is no more than an emerging or latent force.

Generally, domestic tourism is distributed throughout an MDC. Even so some regions are primarily origins and others destinations, e.g. the north and east of France are major source regions while the south and west (Brittany, Aquitaine, Languedoc-Roussillon and Provence) are the prime destinations of French holidaymakers. Domestic tourism often involves short-break holidays for which time and distance factors favour home country locations. Domestic tourists visit a wider range of destinations within a country than international visitors. The latter normally seek higher quality and more specific tourism resources, which are in shorter supply. Also the small number of international gateways used by inbound international tourists serves to increase their concentration. Domestic and international tourists therefore generate different forms and levels of impact.

19.3   ADVERSE ENVIRONMENTAL IMPACTS OF TOURISM

The tourist travels to the producer's location, i.e. destination, to consume the holiday product. Tourism is therefore dependent upon the attractive power of the destination's primary resources. Such resources are natural, e.g. beautiful scenery and wildlife, or people-made, either historical artefacts, cultural features, theme parks, or hallmark events such as the Olympic Games. Many primary resources are common property (see Chapter 6), for which no market exists, so making the avoidance of environmental damage more difficult. However, destinations, the tourism industry, and the tourists themselves share a common interest in ensuring these primary resources are sustained.

Tourists must travel to and within the destination and have somewhere to stay to experience these primary resources. Destinations therefore provide secondary resources, e.g. accommodation, transport facilities and service infrastructure. The provision of these secondary resources, along with created primary resources such as Disneyworlds, constitute destination tourism development. Such physical development restructures the destination environment by extending the built environment, e.g. French Mediterranean resorts developed since 1960 as part of the Languedoc-Roussillon project.

Tourist behaviour can have adverse consequences, both deliberate and unintentional, for the sustainability of the destination environment. The use of off-road vehicles in areas of sensitive ecosystems, such as dunes, is damaging but even wildlife observation can disturb the feeding and breeding of animals and birds. Furthermore, the presence of tourists generates extra waste. Litter is discarded by tourists at sites visited and sewage from tourist accommodation is dumped into the sea.

Restructuring of the physical environment and waste generation are compounded where tourism development attracts immigrants. The destination's attractiveness to tourists may also be reduced by non-tourism developments, e.g. industrial. Tourist travel has wide environmental consequences outside destinations, contributing to atmospheric pollution in tourist origin areas and intermediate regions traversed and thus to global warming.

Most tourism impact studies highlight the adverse consequences. Damage levels are recorded but the processes by which damage occurs are rarely identified. Usually a single environmental component in a destination is studied, ignoring interrelationships and interdependencies with the total environment. The research has also been topically uneven, e.g. more on vegetation trampling than on air pollution and on fragile coral reefs, coastal dunes and alpine meadows than other environments. Most research is reactionary, i.e. after-the-event analysis. This poses problems in distinguishing tourist impact from other effects because information is lacking on the original conditions of the destination environment.

Arising from such research is the idea that each destination has a tourism carrying capacity. This suggests a tolerance limit which, if exceeded, leads to an unacceptable degree of damage to the character and quality of the destination environment or the tourist's experience and satisfaction. This is a deceptively simple concept, fraught with difficulties in practical application because of the several interpretations: (a) as a physical carrying capacity, e.g. the hotel beds available in a destination; (b) as an economic carrying capacity which, interpreted narrowly, optimizes return on private investment but, in the wider destination context, maximizes net social benefit; (c) as a perceptual carrying capacity in which the quality of the tourist's experience depends upon the numbers of other tourists present; or (d) as an environmental carrying capacity which identifies the point of irreversible ecological damage to the natural environment. The latter appears most relevant to the sustainability of the destination's tourism resource base. However, the common property nature of many tourism resources poses destination tourism planners a problem. Negative externalities, i.e. environmental damage, arise from the overuse of primary resources available without charge. Even though management can influence environmental capacity, e.g. replacing ground vegetation susceptible to trampling damage by more resistant species, there remains the classic problem of how to make the 'polluter' pay. Any attempt to determine destination tourism carrying capacity must be site-specific, activity-specific and time-specific.

The adverse impact of tourism is illustrated in three environments—coastlines, mountains and built areas.

19.3.1 Coastlines

The coast is particularly alluring to the tourist seeking a sunlust holiday. Tourism is therefore a significant agent adding to the stresses already imposed upon fragile coastal resources. Consider, for example, tourism's impacts on coral reefs. The disruptive effects can be categorized as activities which pollute coastal waters and those which cause physical destruction. The effects of pollution on coral reefs have been discussed in Chapter 6. Tourism can add significantly to pollution via the discharge of partially treated sewage from hotels, by increasing turbidity in the construction phase of tourism development and by boating activities which add to oil and petrol spillages. Direct physical destruction results from the trampling of reefs at low tide, souvenir hunting, careless handling of scuba and boat equipment and the mining of coral for tourism construction purposes, e.g. in the Maldives a significant proportion of the coral mined from islands is used in resort construction.

The loss of the opportunity to explore a unique biological community is not the only adverse effect on future tourism prospects in such resorts. Coral reefs are important in terms of coastal protection and fisheries. Fringing reefs are self-repairing breakwaters, protecting beaches from destruction by wave action. Several tourist hotels along the Tanzanian coast are gradually collapsing due to beach erosion, partially attributable to destruction of the fringing reef. The Maldivian Islands are particularly vulnerable. Most of the islands lie, on average, only 1–1.5 m above mean sea level. The runway at Hulule airport, the tourist gateway to the Maldives, is only 0.5 m above the high water line. Coral reefs are highly productive fishing grounds. Any reduction in the productivity of coral reefs is therefore likely to affect the ability of the local fishing industry to supply tourist hotels.

Coastlines have historically dominated holiday tourism in the MDCs. Tourism development has permanently restructured the coastline, creating the traditional seaside resort, e.g. along the English south coast and the Belgian North Sea coast. Seafront promenades ensured public access to the beach, a feature not always retained in modern resort developments catering for mass sunlust tourism. Such tourism, controlled by tour operators, has produced a concrete sprawl of 'identikit' resorts along coastlines such as the Spanish Costas. Mass tourism leads to pollution where the resort infrastructure is overloaded and the local environment cannot absorb the wastes disposed. Most notable is the sewage pollution of bathing waters at older Mediterranean resorts, but the problem also affects modern resorts, e.g. Bondi Beach, Australia's third largest tourist attraction, is unsafe for bathing two days out of five. Resort development changes the shoreline configuration by dredging, the construction of groynes and sea walls and the building of marinas. It causes habitat loss, pollution by storm-water run-off, siltation problems and can disrupt marine food chains with consequences for inshore fishing, as in the Florida coast's mangrove belt (see Chapter 13).

Beaches are generally resistant to the pressures of large numbers of tourists. This is not so for any dune belt crossed to get to the beach. Where there are dunes there are good sandy beaches, e.g. the North Sea coasts of Belgium, The Netherlands, Germany and Denmark. The primary dunes are secured in place by grasses that are vulberable to trampling, as is the vegetation of the secondary dunes. The loss of stabilizing vegetation leads to linear blow-outs and the movement of sand smothers adjacent vegetation. Its loss from coastal dunes could reduce their natural protective role, increasing the risk of coastal flooding. To protect the dune belt controlled access paths are constructed, as along the Walcheren and Kennermerduinen coasts in The Netherlands or Studland Bay in the UK.

19.3.2 Mountains

Mountain ranges have distinctive ecosystems threatened by tourist pressures. Even the most remote mountains are at risk. In Nepal pressures from population growth have been exacerbated by a phenomenal increase in trekking tourism. It is estimated that the Annapurna region alone, home to about 40 000 mountain farmers, now receives 25 000 visitors annually. Additional pressures are placed on the soils, water and forests as well as on the local population. The average daily consumption of firewood has been estimated to be 6.4 kg per tourist. The average sized trekking group uses as much firewood in two weeks as a local family uses in six months. The disposal of human waste and litter has led to the route to Everest base-camp being dubbed as the 'Kleenex trail' and prompted an Environmental Expedition in 1990 to clean up the area. The scoring of paths across steep, unstable slopes has resulted in an increase in the incidence and scale of landslides and rockfalls. Among other detrimental effects are the trampling and partial destruction of natural vegetation cover and the virtual disappearance of many species of wildlife. Policy measures in Nepal are discussed later.

In MDCs much damage is attributable to winter tourism. Downhill skiing needs snowcover and steep slopes and ski resorts have been developed in areas of sparse population and difficult terrain. Examples are the high altitude integrated resorts of the French Alps, such as La Plagne, built well above normal settlement limits. Such development involves the construction of access roads, provision of accommodation, creation of ski-runs (pistes) and building of ski-lifts. The Alpine life zones, naturally squeezed into a few thousand vertical metres, are disrupted and further restricted; some may even be obliterated and animal migration routes severed.

Construction of roads and ski-lifts and the bulldozing of terrain to improve pistes can lead to slope erosion. Erosion is compounded by a loss of vegetation. Deforestation occurs when creating pistes and building lift systems. Ground vegetation is destroyed where soil compaction and drainage alterations result from the use of snow-compaction machines in preparing pistes and the actual downhill skiing itself. The risk of erosion is greater the steeper the pistes, the more frequent the run-off and the larger the catchment. In addition to erosion by soil loss and gullying there is an increased risk of avalanches and other mass movements, such as landslides and rockfalls. Drainage systems are altered, sometimes to ensure water supply to the resort. Run-off is increased because of the greater extent of the built environment and compacted slopes. The risk of flash flooding downstream may also be increased. Access to ski resorts is by roads which are often congested. The terrain often inhibits the dispersion of toxic fumes (carbon monoxide and lead) from vehicle emissions. These are higher in the mountains because of the lower vehicle speeds and the altitude. In the Rocky Mountains, Vail and Aspen have recorded pollution levels ten times greater than Denver on the edge of the Great Plains. There is also the problem of chemical run-off from 'salting' roads to keep them drivable throughout winter. Furthermore, such resorts with their access roads, hotels, ski-lifts, cable cars and power lines are often visually intrusive. Damage caused by winter tourism can reduce the area's attraction to summer tourists, a consequence acknowledged in Austria and in the Cairngorms (Scotland).

Hazard levels to tourists in mountainous areas rise with increasing tourism pressure. Many avalanches are triggered by skiers crossing the avalanche starting zone. Not surprisingly Austria and Switzerland have the highest per capita frequency of avalanche fatalities.

19.3.3   Historic and built environments

The built environment is also susceptible to tourism's adverse impacts. Architectural 'pollution' arises from the construction of large, often unaesthetic hotels and facilities, out of scale with the surroundings and failing to incorporate environmental considerations into their design. The development often flouts local environmental laws. The recently constructed Ramada Hotel Varca in southern Goa, India, violated both maximum height and minimum distance from high water mark criteria. The sprawl of buildings along coastlines may obstruct 'windows' to the sea and deny access to public beaches, e.g. along the south-west coast of Barbados. Similar obstructions to views may occur along scenic routes inland.

Excessive pressures are often placed on the infrastructure at LDC destinations. Tourist use per capita of utilities, such as water and electricity, is much higher than that of the resident population. Facilities such as air conditioning, swimming pools and en suite bathrooms in hotels place excessive demands on beleaguered supplies. Load shedding may result in frequent electricity blackouts and disruptions to water supply may occur. Holiday seasonality means that a destination has to scale up its infrastructure to cope with the period of maximum utilization. A costly option when for the remaining seven to eight months it will be under-utilized. This problem is particularly acute in small island economies where tourist numbers may outweigh resident populations.

The physical deterioration of historical buildings and artefacts is a further problem. The delicate tomb frescoes in the Valley of the Kings, Upper Egypt, are suffering from increased humidity levels due to the body moisture emanating from tourists. The use of camera flashes has been banned in an attempt to halt the progressive fading of the frescoes. Clambering over archeological sites can result in structural deterioration. The ruins at Karnak, Upper Egypt, are patrolled to prevent tourists climbing to higher vantage points. Stone steps up the side of Temple 1 in the Mayan ruins of Tikal, Guatemala are being reconstructed, not just for visual impression but also to harden the site for visitors wishing to climb to the top.

Similar impacts are evident in the MDCs. Heritage tourism can bring new life to buildings, from factories to stately homes, and may stimulate the conversion of other old buildings to tourist accommodation and cafes. It also leads to an expansion of the built environment, as in seaside resorts. The visual pollution of the high density concrete sprawl along Spain's Mediterranean coast to cater for cheap mass tourism in the 1960s is a prime example. Redevelopment of the built environment follows from pressures to meet tourist demands for central heating, air conditioning, car parking, en suite bathrooms and also more stringent government regulations pertaining to fire, health and safety. Increasing physical standardization of hotel accommodation results, but it can be tastefully integrated into the historic city as the Hotel Pullitzer in Amsterdam shows.

Within towns tourist flows, requirements for coach access, car parking and service access to tourist establishments can generate congestion. This is worst in historic cities where street patterns are not conducive to present day traffic volumes. The presence of large tourist numbers can degrade environments such as collegiate Oxford and Cambridge or the ecclesiastical enclaves of Canterbury and York. Indeed, the physical capacity to absorb tourists may be reached, necessitating at National Trust properties such as Chartwell (with its Churchillian associations) controlled access for a limited number of tourists via tickets with timed entry. Without control there are long queues, such as at ski resorts with restricted lift capacities, e.g. Chamonix, Kitzbühel and Verbier.

Again, historic artefacts are at risk from the feet (and hands) of too many tourists. Access to the Parthenon on the Acropolis in Athens and to Stonehenge has therefore been restricted (the latter not only because of tourist numbers but also because of vandalism). The National Trust reports having to renovate fabrics in a number of their properties because of people's irresistible urge to touch such artefacts.

19.4   THE NEED FOR SUSTAINABLE TOURISM

There is a circular and cumulative relationship between tourism development, the environment and socio-economic development. Most tourism development places additional pressures on the environmental resources upon which it is based, compromising the future prospects of the local population. The crucial issue of sustainability (defined in Chapter 2) applies to both hosts and guests as far as tourism is concerned. The destruction of tourism resources for short-term gain will deny the benefits to be gained from the mobilization of those resources in the future. Future generations of tourists will be denied the opportunity to experience environments very different from home. Host populations will be faced with environmental degradation and denied the tourism potential offered by the original attraction.

What are the prospects for sustainable tourism? Recently there has been increased emphasis on alternative forms of tourism, although none of these options constitute truly sustainable tourism. Even small group alternative tours can be damaging. In their search for 'unspoilt' locations, alternative tourists bring more and more remote communities, with delicately balanced environments and economies, into the locus of tourism. Enterprising travellers who penetrate new and as yet unspoilt areas unwittingly become the pioneers of tourism development. The principles of responsible tourism appear laudable, but are in danger of being co-opted by the tourist and the tourism industry to salve guilty consciences and promote seemingly more conscientious marketing. Green or eco-tourism may satisfy environmentalists, but unless the needs of the local population are also considered there will be no guarantee of sustainability.

Resource conservation policies may be regressive from the local population's viewpoint. The creation of National Parks to protect extensive tracts of land denies access to local populations for agriculture, the gathering of fuel, fodder and building materials. Simultaneously local populations rarely benefit from tourism. Such a conflict is evident in the National Parks of East Africa, where the pressures of a rapidly growing population and limited availability of land which can be cultivated conflict directly with wildlife tourism (Wyer and Towner, 1988). The failure to integrate local involvement in the tourism industry is exemplified by the Galapagos Islands. Here, in the face of the very rapid growth of tourism, the viability of existing conservationist strategies is in question. The maximum number of visitors to be admitted has been progressively revised upwards over recent years. Local tourism is now an established and increasingly important socio-economic factor due to the increased number of tourists staying on the islands. The Ecuadorean government has yet to formulate an overall plan for sustainable tourism which effectively involves Galapagan participation.

Nepal has been a pioneer in integrating the twin goals of conservation and local development to achieve sustainable tourism. The Annapurna Conservation Area utilizes a multiple land use approach, attempting to balance the needs of the local population, tourists and the natural environment. Local villagers continue to practise traditional methods of resource utilization. They participate in decision-making via the existing village assemblies, and also benefit directly from the funds generated by tourism.

What is to be gained from balancing the needs of tourists, of the local population and of the natural environment through policies of sustainable tourism? Eco-tourism is the fastest growing tourism sector. The largest nature tour agency in Costa Rica experienced a 46% growth in 1986. Recognition of the symbiotic relationship between tourism, the environment and development should stimulate conservation of the tourism attraction. Environmental destruction will not occur if it is possible to make more money by conservation. Tourism also motivates the conservation of historical sites and buildings, for example, the Buddhist temple at Borobudur, Indonesia and the Mayan site of Tikal, Guatemala. An ultimate example of such a symbiotic relationship in the built environment is Williamsburg, Virginia, where tourists stay in conserved colonial inns and taverns which are at the same time tourist attractions.

19.5   HOW TO ACHIEVE SUSTAINABLE TOURISM

19.5.1   Requirements of sustainable tourism

Sustainable tourism depends on:

(a) meeting the needs of the host population in terms of improved standards of living in the short and long term
(b) satisfying the demands of increasing tourist numbers and continuing to attract them to achieve this
(c) safeguarding the environment to achieve the two foregoing aims.

To accomplish these aims attitudes and policies will have to change. This involves four viewpoints: those of tourist destinations, the tourists themselves, tourism enterprises and, lastly, global considerations.

19.5.2   Viewpoint of the tourist destination

Local populations must be involved in tourism development if their needs are to be met. This involves four major policy considerations: ownership, scale, timing and location. The question of ownership is particularly pertinent in LDC destinations. Net foreign exchange earnings from tourism are considerably less than the gross receipts. Substantial leakages result from the repatriation of wages and profits, and imports. Transport carriers, hotel groups and tour operators based in the MDCs have all become increasingly transnational in their operations. Third World destinations receive only a small return for the exploitation of an increasingly scarce resource, their natural environment. Such destinations have to bear certain costs, both in terms of environmental degradation and in prospects for sustainability. Furthermore, due to their low level of development, they can rarely afford preventive and restorative measures. Any extra earnings which do accrue locally benefit a small commerical élite, more concerned with early profits than environmental considerations. It is therefore insufficient to advocate local as opposed to foreign ownership without considering distributional aspects and environmental accountability.

The scale of tourism development is a complex issue. Small-scale projects, locally controlled, can have a significant impact on raising living standards (Britton and Clarke, 1987), but are unlikely to meet the needs of large numbers of tourists. Some large-scale projects are inevitable, but it is important to consider the complementarity of large- and small-scale developments. As tourism development proceeds, indigenous firms and locals gain knowledge and experience. Government planners should co-ordinate investment in infrastructure with the needs of small-scale entrepreneurs and the needs of local communities, paying careful attention to the environmental component. Large-scale development is often the precursor to small-scale development. The growth in mass tourism and the building up of infrastructure in Senegal were essential prerequisites for the success of the Lower Casamance project in promoting rural development through tourism (Pearce, 1989). It is a two-way process. Large-scale developments benefit from the increased local skill base.

The location of tourism development is also crucial. Development concentrated in tourism enclaves, e.g. Nusa Dua in Bali, may minimize adverse impacts elsewhere, but does not constitute sustainable development. The local population may be denied continuance of their traditional practices as well as being excluded from any economic benefits of such development.

There appear, therefore, to be many contradictions concerning the ownership, scale, timing and location of tourism development at destinations. It is not as simple as resolving the issues of indigenous versus foreign, small versus large, gradual versus instantaneous and dispersed versus concentrated developments. It is more a question of ensuring complementarity between all these issues, so that tourism can contribute towards the development of an area while minimizing adverse environmental, social and economic effects to ensure sustainability.

19.5.3 The tourist's viewpoint

Are tourists aware that their actions may damage the natural environment and the culture of their destinations? Do they realize their contribution to environmental damage on a global scale?

Tourists know if their holiday is unsatisfactory, although despoiled destination environments may not be the only reason (e.g. flight delays, loss of baggage). Their response is to choose a different destination, hotel, or tour operator next time. Just as consumers in general have varied opinions on 'green' issues, so too with tourists. Some, e.g. allocentrics or explorers, are more environmentally aware of tourism's damaging potential than others, i.e. psychocentrics or mass tourists, who get enjoyment from holidaying in crowded situations and who seek developed ('theme park') destinations. Such facilities do not have to be located in fragile environments. Allocentrics are more of a problem, especially where demand exceeds supply. How can their tourism experiences be maintained? To spread them more widely implies more and distant destinations leading to an increase in travel consumption.

For the tourist, sustainable tourism offers the prospect of a guaranteed level of satisfaction whenever a destination is visited. But tourists' behaviour and attitudes must change—tourists must understand a destination's 'sense of place' if they are to respect its environment and culture. Tourists visiting less developed regions must forgo accustomed standards of comfort and convenience. Development using local materials and in scale with the vernacular may be more expensive. Who pays? Ultimately it is the tourist via entry charges and bed taxes. A danger is that sustainable tourism appears élitist—an up-market operation seeking longer stay tourists committed to quality crafts, organic foods, heritage, nature and quietness. It offers the richer tourist a means of preserving destinations from the masses with their noisy habits and bad tastes.

19.5.4 The tourism enterprise's viewpoint

Tourism enterprises are in the business for profit. If a hotel can minimize costs by discharging untreated sewage directly into the sea, because building regulations permit, it will do so. The hotel takes advantage of the sea as a common property resource (see Chapter 6) even though this imposes social costs, from pollution, on the destination as a whole. However, just as firms in retailing are responding to green consumerism, so too will tourism enterprises need to take account of environmental issues in their decision making. Indeed, some already do, for example, British Airways have an 'environmental policy' which covers all their operations.

Tourism enterprises may 'go green' operationally, i.e. act in an environmentally sound way. A 'green audit' of tourism enterprises would reveal the extent to which firms are conserving energy (insulation of hotels to reduce heating costs) and recycling wastes (British Airways and Consort Hotels recycle waste paper), minimizing pollution (Friendly Hotels use recycled paper products and 'ozone-friendly' toiletries and cleaning products, British Airways use lead-free petrol in their motor vehicles) and using local materials and produce (such as organic foods).

Tourism enterprises can encourage their customers to behave respectfully towards the environment. This educational role may serve as a basis for niche marketing, such as Consort Hotels' 'Go Green' weekends.

19.5.5 A global viewpoint

Even assuming that tourist destinations adopt sustainable tourism plans and tourism enterprises and tourists adopt more environmentally aware practices and behaviour, tourism will continue to contribute to global warming because of its heavy consumption of travel products. Can conventional tourism become more sustainable in this context? Currently tourist transport uses too much fuel. Public transport could be substituted for private transport, short-haul travel to nearer destinations for long-haul holiday journeys, high occupancy charter flights for scheduled flights and flight refuelling stops planned to minimize the need to tanker extra fuel. It is, however, questionable whether tourism can adopt measures which will reduce significantly emissions from the use of transport services and other energy sources.

Staying at home appears to be the 'greenest' way to holiday. Holographs have the potential to reproduce any environment artificially so, in the 21st century, it may be possible to holiday at home! The English Tourist Board has recently suggested the use of 'video scenics' in popular destinations such as the Lake District to dissuade tourists from visiting overcrowded locations. Alternatively, the development of holiday complexes, which provide artificial 'sun-warm water' environments (e.g. Center Parc villages), located at points of maximum market access, could be a way forward. What has to be faced at the global scale is that sustainable tourism could be more restrictive of holiday travel than conventional tourism.

Furthermore, global environmental change stemming from non-tourist activities will necessitate a response from the tourism industry. Global warming, for example, will alter the natural resource base and hence destinations' comparative advantages—new destinations will be developed for seaside holidays in the late 21st century.

19.6 FUTURE SCENARIOS

19.6.1 Sustained growth

Tourism is almost certain to become the biggest sector in international trade by the year 2000. Motivations to travel for business and leisure are very strong and all the indicators point towards the continued growth of tourism.

The 4% annual growth rate for 1980–89 produces a predicted rate of 637 million tourist arrivals by the year 2000. The outstanding destination region is likely to be Asia and Oceania, with its share of arrivals increasing to 21.9%, compared with 14.7% in 1989. The Americas and Africa are also likely to experience above average growth rates. All these are predicted to be at the expense of Europe, whose regional share is forecast to fall from 62% in 1989 to 53% by the year 2000.

Sustained rates of tourism growth will place primary tourism resources at greater risk. It is unlikely that policies and attitudes among destinations, tourists and tourism enterprises will change sufficiently to significantly reduce the pressures on destination environments. Indeed, the sustained growth scenario suggests that the relative increase in pressures could be greatest in the most vulnerable destinations in Asia, Africa and Oceania.

19.6.2 Accelerated growth

Estimates based on current trends are simply extrapolations of travel patterns by tourists largely from MDCs. Even within these nations rising real incomes, the extension of leisure time, changing employment patterns, the spread of higher education and the influence of the mass media coupled with sophisticated marketing techniques are all likely to contribute towards an increase in the propensity to travel. Estimates, cited by Pearce (1989), predict 784 million world tourist arrivals by 1995.

Furthermore, no account is taken of the changing economic fortunes of other countries. The opening up of the countries of Eastern Europe and their transition to market economies will undoubtedly result in a marked increase in travel abroad by their residents. Similarly, an increase in tourists from the better off countries among the LDCs is likely. Already the NICs are showing a marked increase in residents travelling abroad.

Accelerated rates of tourism growth have the most serious implication for further degradation of tourism's resource base. It may well threaten not only existing tourist destinations but also condition the rate and scale of tourism development in new destinations. The opportunity to capitalize on short-term gain by such developments in many developing countries may be taken at the expense of sustainable tourism.

19.6.3 Reduced rate of growth

A reduced rate of growth in world tourist arrivals is possible in the short term as the MDCs are approaching saturation level as tourist generators. A world recession exacerbated by oil price increases consequent upon the effects of the 1991 Gulf War is reducing growth rates and limiting the participation of tourists from NICs. In the MDCs where net departure rates exceed 50% further significant increases in the travel rate are unlikely. The demand for package holidays has plateaued, and is declining in the mass sunlust market, and, in the UK, recent increases in disposable income have not led to increased holiday-taking. Among European tourists there is evidence of shortening lengths of stay and a switch to domestic tourism has been noted in the UK and Germany.

Tourists also face higher holiday costs, not merely because they are demanding higher quality products. Fuel prices have a major bearing on travel costs and are notoriously unstable. Holiday prices are therefore variable and the effect of the Gulf War has added over 10% to the price of a package holiday. Further price increases in certain tourist generating countries will follow from the European Community's efforts to harmonize taxation procedures and improve consumer protection. The travel trade faces increased VAT liabilities, the loss of duty-free sales within the European community and increased liabilities for tourism products. In the short term, constraints to growth in Europe and Japan arising from airport congestion (primarily lack of runway space) and air traffic control limitations are unlikely to be removed.

Demand changes may also be expected. Consumers are responding to changing weather patterns, e.g. recent good summers in north-west Europe have encouraged domestic tourism and the demand for skiing has stagnated in Europe because of unreliable snow cover. Tourists' interest in long-haul travel may be dampened not only by price rises but also by reactions to uncomfortable travel on aircraft with limited leg-room per passenger to maximize loads. The more discerning tourists may be influenced by green issues and choose their holidays accordingly (although there is no incentive to delay visiting a destination if it is thought that the volume of tourists will spoil it). The decline in sunlust tourism may reflect increasing consumer awareness of medical evidence linking sunbathing with skin cancer.

Reduced rates of growth will restrict tourism's contribution to increasing global environmental pressures. Destinations have 'breathing space' to formulate sustainable tourism plans. However, tourist flows will be differentially affected with certain destinations coming under increased pressures, e.g. high altitude European ski resorts with good snow records. Generally the spatial patterns of change will be similar, although not as marked, as those outlined in the previous scenarios.

19.7 CONCLUSIONS

Tourism's future is undeniably linked to the destinations' environment and heritage. The quality of the tourist experience depends upon the quality of the destination environment. This can be achieved by promoting sustainable tourism (or green, responsible, alternative, sensitive, soft, ecological, conscious and post-industrial tourism). Claims that one or another of these forms of tourism is all things for all destinations are both naïve and unrealistic. Their common message is that tourism should adopt a general approach which ensures its activities are non-polluting and non-visually intrusive and which builds upon the character of a destination at a scale the host environment and culture can absorb and welcome.

Primary tourism resource conservation is essential and the prevention principle must be uppermost if irreversible environmental damage is to be avoided. Adopting good environmental practices will require some restructuring of the tourism industry. Ultimate responsibility must rest with the authorities closest to the resolution of environmental problems, i.e. the destinations. Sustainable tourism will mean higher prices for consumers and lower profits for commercial interests so supply-led sustainable tourism is unlikely. The more environmentally aware tourists become, the more likely they are to behave according to the customs of their destination and to travel in an ecologically responsible way. This will limit the damage to global and destination environments. Education of the tourist is the key and current tourism consumers' search for quality is a sign that sustainable tourism might eventually be demand-led.

19.8 REFERENCES

Bar-On, R. (1989) *Travel and Tourism Data*. London: Euromonitor.
Britton, S. and Clark, W. C. (Eds). (1987) *Ambiguous Alternative: Tourism in Small Developing Countries*. Fiji: University of South Pacific.
Pearce, D. (1989) *Tourist Development*. London: Longman.
World Bank (1990) *World Development Report*. Oxford: Oxford University Press.
World Tourism Organisation (1989) *Yearbook of Tourism Statistics*. Madrid: World Tourism Organisation.
Wyer, J. and Towner, J. (1988) *The UK and Third World Tourism*. Tonbridge: Ten Publications.

19.9 FURTHER READING

Krippendorf, J. (1987) *The Holiday Makers: Understanding the Impact of Leisure and Travel*. London: Heinemann.
Lea, J. (1988) *Tourism and Development in the Third World*. London: Routledge.
Mathieson, A. and Wall, G. (1982) *Tourism: Economic, Physical and Social Impacts*. London: Longman.
Murphy, P. E. (1985) *Tourism: A Community Approach*. London: Methuen.
Romeril, M. (1989) Tourism: the environmental dimension. In Cooper, C. P. (Ed.). *Progress in Tourism, Recreation and Hospitality Management*. Vol. 1. London: Belhaven Press, pp. 103–113.

SECTION IV

POSSIBILITIES

CHAPTER 20

Perspective and Prospect

S. R. Bowlby and A. M. Mannion

20.1 PREAMBLE

A number of themes emerge from the chapters in this book. Firstly, there are both certainties and uncertainties about the physical processes that underpin environmental change and about how these processes may be influenced by scientific and technological developments. Secondly, there are uncertainties about what future environmental policies will be needed and implemented. Here, the likely impacts of local, national and international power relations on the formulation of environmental policy are significant. A closely related issue concerns the nature of economic growth in the 1990s and beyond in the developed and developing worlds. All of these factors have a major role to play in the future identification and adoption of policies for sustainable development.

20.2 PERSPECTIVE

As individual chapters illustrate, there is both certainty and uncertainty about the physical processes and the direct and indirect human influences on them that have led to environmental change and hence to many of the issues on which this book focuses. Such disturbances to physical processes have been caused by the interplay of social, economic and political factors operating through human organization of production, consumption and social reproduction (see Fig. 1.3). All of these social factors will continue to operate in the 1990s and beyond (their interaction with physical processes highlights the reciprocity between people and environment). The world order, the product of power relationships between nations and economic interests, mainly dictates how these factors operate. Political pre-eminence will continue to play a major role in environmental issues in the future. In this context, it is important to examine the polarization between the developed and developing worlds.

Environmental Issues in the 1990s. Edited by A. M. Mannion and S. R. Bowlby
© 1992 John Wiley & Sons Ltd

20.2.1   Developed versus developing world

It is clear that many major environmental problems relate to the unequal divisions of economic and political power between the developing and developed world. At a global scale the principal damage to the environment through pollution has been created through the activities of the affluent nations of the developed world. The nations of the developing world have little power to prevent this or to force the developed nations to alter their systems and levels of production and consumption. They also lack the financial means to implement major programmes to restore or prevent environmental degradation.

- The implementation of measures to reduce global pollution thus depends upon 'enlightened self-interest' on the part of the wealthy nations.

Many people argue that the industrialized nations have an obligation to help those in the developing world as they have benefited greatly, and continue to benefit, from the transfer of resources from the developing world. They have improved their standard of living not only by consuming their own resource base, which they now seek to conserve, but also by importing a proportion of the resource base of the developing world. To help the world to move towards sustainable development the wealthy nations must not only persuade their own populations to accept changes in their ways of life, but must also be willing to give funds and expertise to the developing nations. Without such help it is unlikely that the developing countries will be able to improve the life-chances and well-being of their people and invest in the up-to-date technology which will allow increases in production and energy use and efficiency without high levels of damage to the environment (see Chapter 8).

- The creation of a co-operative relationship between the developed and developing worlds is a major political challenge.

Most poor people live in the developing world and it has been argued in the World Conservation Strategy (International Union for the Conservation of Nature, 1980) and the Brundtland Report (World Commission on Environment and Development, 1987) that, if the world's environmental problems are to be solved, the political and economic relationships between nations must be changed to reduce the dominance of the developed over the developing nations.

Many of the chapters in this book also highlight the impairment of particular fragile or ecologically or socially significant environments in the developing world (for example, arid lands, wetlands, some marine environments and some social environments) which stems from the way in which a capitalist market economy has affected peasant economies and societies.

- If humanity's damage to the environment is to be reduced it is essential to develop mechanisms whereby values other than those of pursuing narrowly defined economic growth and profit now can be brought to bear effectively on the decision making of the dominant economic and political agents in the world economy: this is an immensely difficult undertaking.

At present many people advocate the incorporation of the concept of sustainable development into such decision making as a means of ensuring that environmental costs and benefits and the interests of future generations are given appropriate consideration. In the last section of this chapter there is a brief discussion of the potential of policies, based on the concept of sustainability, to change current attitudes to the valuation and exploitation of the environment.

20.2.2 Power relations

An important theme of many of the chapters in this book concerns the relationship between the social and physical processes which create 'environmental problems'. It has been shown that many of the most difficult environmental problems are either a product of unequal power relationships or are harder to solve because of conflicts of interest between groups.

As many of the chapters have shown, poverty lies at the heart of the set of processes producing a large number of detrimental environmental changes such as soil erosion, desertification or the destruction of habitats, particularly in the developing world. Thirty-three per cent of the world's population is classified as being below a poverty line of $370 per capita per year (World Bank, 1990) and for many of these people their poverty means that to survive tomorrow they must take actions which will make their survival in the future more difficult. Examples of such situations have been given in the chapters on deforestation, population, soil erosion and desertification (Chapters 5, 7, 14 and 15). The poor thus find themselves in a situation whereby they reduce the productive potential of the very environment which they use to produce the food and materials on which they depend for their survival. Many poor people lack the ability to act in ways which would maintain or improve this environment. They lack this ability because they do not have the resources of time, equipment and money to invest in, for example, environmentally benign methods of which they are often well aware, e.g. alternative farming methods or new ways of producing and marketing goods. Moreover, their poverty means that they lack the political power to influence others to help them achieve appropriate investment.

- Poor people are forced to trade off long-term sustainability against short-term survival.

Furthermore, the resulting damage to the environment, by creating further poverty, means that people's own potential to live with dignity and to improve and sustain the environment is impaired.

- The degradation of environmental resources leads to damage to human resources: a vicious circle operates.

- Environmental degradation is often, simultaneously, a *result* of underdevelopment, a *symptom* of underdevelopment and a *cause* of underdevelopment (Blaikie, 1985).

Inequality of wealth and political power is also an important influence on environmental problems other than those directly connected with poverty. Two questions arise. Firstly, who decides what is an environmental problem and, secondly, how can the competing interests of different groups in society be balanced when selecting and implementing policies to deal with these problems?

It is significant that there is often a dispute over whether something is an environmental problem and for whom. Consider acidification (Chapter 11). After its identification in the 1960s many industries and governments argued, using scientific evidence, either that emissions causing acid deposition did not constitute an environmental problem or that it was not their responsibility. Nor were they willing, until the mid-1980s, to acknowledge that mitigation of the problem lay in their hands through the implementation of emission reduction strategies which required financial investment. Ironically, given the fears of the polluters, one country that did address acidification, Germany, now finds itself at a technological advantage.

- The root of such conflicts lies in the short-term financial considerations of individual firms, governments and government agencies who are not obliged to include a proper assessment of long-term environmental costs to society (both within and beyond their national boundaries) in their calculations of the benefits and disbenefits of taking action to reduce pollution.

Again, in situations where some groups argue for the preservation of an environment on aesthetic or ecological grounds, other groups will often argue that the environment in question has little aesthetic or ecological value.

- The economic and political framework within which problems are identified is vital to understanding which environmental problems society will recognize.

Many environmental problems affect a variety of groups with different interests, all of whom will argue for different policies. In some instances, groups will be differentiated by location—urban versus rural, centre versus periphery, one country versus another. In most such situations these locational divisions will be cross-cut by divisions based on income and class or by consumption pattern—house-owners versus tenants, or car-owners versus those without cars, or land-owners versus the landless.

- Conflicts between commercial, residential and business interests often influence the definition of an environmental problem and the policies which are adopted to deal with it.

The conflicting interests which are involved in the 'environmental problem' of the growth of population in the smaller towns and villages in the south-east of Britain (discussed in Chapter 18) illustrate these points. The counterurbanization of population can be seen (a) as conflicting with the interests of people already living in the south-east by damaging the beauty of the countryside and the historic environment of small towns and villages; (b) as being against the interests of all of the population because of this damage; and (c) as spoiling these resources for future generations. However, others argue that to preserve this environment is simply to preserve the interests of the better-off inhabitants of the area. They suggest that such preservation will damage the interests of poorer groups living in the countryside by blocking the growth of much needed jobs, and will prevent people finding and enjoying a better quality of life outside the older cities. Moreover, the countryside so conserved will be available only for the enjoyment of those who already live there and those who have the means to visit it using the motor car. It is suggested that those people who are dependent on public transport—the poor, the old and many women—will be excluded from such enjoyment.

Commercial and business interests also enter the debate. For example, house-builders, who would benefit from a building programme in the south-east of Britain are keen to argue that there is plenty of land available and that building need not detract from the pleasures of the countryside and small towns. Retailers are also eager to argue that to develop new facilities will merely enhance the living standards of the local population without damaging the environment. Established local businesses may take a different view of the threat of competition and raise environmental objections to further development. Clearly some of the groups involved have greater political and economic influence than others.

- It is difficult to ensure that the decisions reached are in the long-term interests of the whole of the population—assuming that it is possible to decide what these are and that conflicts between them can be resolved.

20.2.3 Certainties and uncertainties

For many issues the causes of environmental degradation are well established. In most instances social and economic factors conspire to alter environmental processes and culminate in the consumption, directly and indirectly, of resource capital at an accelerating rate. This undermines both inter- and intra-generational security. Even where mitigating measures are available, inertia in decision-making and implementation are often contributory factors to environmental deterioration—too little, too late. This is most apposite to the developed world where profligate resource use is the norm. The situation in the developing world is made even more acute by the scarcity of appropriate technology and the reluctance of developed nations and their companies to assist in the transfer of such technology.

Acidification and cultural eutrophication are two environmental problems which involve processes that are fairly well understood. These relate to disturbances of the biogeochemical cycles of sulphur, nitrogen and phosphorus, which result in high accumulations of compounds which thus become pollutants. The activities causing these disruptions include energy consumption, intensive fertilizer use in agriculture and sewage and waste water effluents. Mitigating strategies are available and fall into two categories: short-term solutions that address the symptoms and long-term solutions that tackle the underlying causes.

- In terms of resource use and conservation the most appropriate mitigating strategies are those relating to the underlying causes.

The environmental processes involved in soil erosion, desertification, deforestation and wetland destruction are also well, or sufficiently well, understood to facilitate conservation strategies. Again, economic and social factors tend to dominate these issues in both the developed and developing worlds. For example, in Europe and North America the increased use of artificial fertilizers has continued to raise productivity and thus masks, to a certain extent, the impact of soil erosion. Wetland destruction continues apace because of the continuing demand for garden peat-based products and the grants available for drainage to provide agricultural land. Economic factors are also contributing to soil erosion, desertification and deforestation in the developing world and relate in part to the need to produce cash crops for export in return for foreign currency. Social factors are also important, notably the growing number of land-poor and landless people (Section 20.2.2) who are migrating into fragile environments in the search for survival. Some governments also operate resettlement programmes which are exacerbating already existing problems, especially tropical deforestation.

The interests of the developed world are also reflected in the environmental problems of the developing world. The expansion of large-scale ranching in Central and South America, for example, reflects the increasing demand for cheap beef in Europe and North America. Likewise commercial logging enterprises cater for expanding western markets for tropical wood and wood products. These examples are a reminder that even where there is some agreement about the causes of environmental degradation it can be very difficult to identify solutions that people are willing to implement.

- The social and economic pressures that sustain environmental degradation in the developing world are more urgent and basic than those of the developed world due to the extreme poverty of some of the population and their consequent aspirations for development.

- At the same time the developed world provides a ready market, and thus a major stimulus, for environmental exploitation in the developing world.

- While the developed world is willing to buy resources, the developing world has few alternatives but to supply them.

- Thus, although there is a conflict of interest between the haves and have-nots there is also a mutual interdependence that creates a vicious circle.

There are other issues, however, which are surrounded by scientific uncertainty because the physics, chemistry and biology are still not well understood. This is particularly true of the marine environment and the effects of pollution and waste disposal on its food chains and biogeochemical cycles. Further uncertainties surround the potential impact of biotechnology and, especially, genetic engineering. Most important, however, is the uncertainty that surrounds climatic change, particularly in relation to the enhanced greenhouse effect. This issue is, in turn, related to uncertainties about future trends in fossil fuel consumption.

The question of the capacity of the oceans to absorb waste products continually has recently been highlighted by the problems created by oil releases into the Persian Gulf during and following the Gulf War of early 1991. Losses of wildlife and problems of oil-slick dispersion are cases in point.

- The role of the oceans in regulating the earth's life-support systems is thus an important issue that warrants further research.

Moreover, the impairment of marine resources has its parallels with deforestation and habitat destruction; all represent a reduction in genetic resources which means that intra-generational opportunities are being curtailed as the resource base is diminished. This, in turn, imposes constraints on sustainable development in the future. It is ironic, in view of recent scientific developments, that species extinction is occurring at unprecedented rates. Just as society is learning to manipulate genetic resources they are declining and thus limiting future opportunities. This is not to imply that genetic engineering will solve the world's environmental problems. Little is so far known about how transgenic organisms—plant or animal—will affect existing ecosystems, agricultural ecosystems and environmental processes. Although there is much optimism in relation to improved crops, waste water treatment, recycling, and so on, it is Utopian to expect that such developments will be entirely environmentally benign. The absence, to date, of internationally accepted guidelines for the field testing of such organisms is worrying, especially as there are already instances of the export of transgenic organisms to developing countries for testing. Although science has clearly much to offer in this context, notably the possibility of tailoring crops to match the environment rather than vice versa (the traditional approach), caution must be exercised. Moreover, these opportunities are more relevant to the food production problems of the developing world and their adoption and regulation will be affected by power relations within and between countries. In particular, their use will depend on the relationships between the developed and developing world as technology transfer will be needed. Who is to pay the bill?

• Diminishing gene pools, as extinction proceeds in tandem with habitat destruction, are reducing opportunities for the future.

• Genetic engineering, and its relevance to environmental problems, is surrounded by scientific and political controversy.

The possibility of transgenic organisms being used as agents of warfare may be a further threat to inter- and intra-generational security. In 1991 it became clear that existing environmental resources can be used as agents of warfare. The firing and release of Kuwaiti oil reserves during the Gulf War is a case in point. This has created a new and unanticipated environmental threat. Such deliberate destruction of natural resources to gain political momentum and strategic advantage constitutes another potent agent of degradational environmental change. It is also important to note that conventional warfare inflicts, directly and indirectly, major changes on the environment (Shaw, 1990).

• Warfare in all its actual and potential forms is thus a threat to global environmental security.

There remains then, the most vexed question of climatic change. Much uncertainty surrounds the physical processes involved, which means that prediction is enigmatic and speculative. Equally speculative are the possible repercussions, in terms of environmental processes, of economic, social and political factors. The crux of the debate is whether and, if so, how fast, enhanced greenhouse warming is occurring in response to increased atmospheric concentrations of heat-trapping gases, notably carbon dioxide. The evidence to date is not definitive. The most widely accepted data are those compiled by the Intergovernmental Panel on Climatic Change (IPCC, 1990), which states that over the last century the global mean surface temperature has increased by 0.3–0.6°C. This, however, is within the range of natural climatic variability and the IPCC suggests that it will be at least another 10 years before sufficient data are available to make unequivocal decisions. Nevertheless, as increasing atmospheric carbon dioxide concentrations, of a magnitude similar to those which have occurred since the Industrial Revolution of the mid 18th century, reinforced global warming as the last ice age ended it seems unlikely that a further 25% increase over the last 200 years will have no repercussions.

If global warming occurs there will be substantial changes in the distribution and nature of the world's natural ecosystems and agricultural systems. Thus, the resource base will alter and environmental change will occur. This will open up new possibilities as well as curtailing present activities. How such changes will alter social, economic and political configurations is a matter for much debate. They will certainly affect patterns of global food production and trade. Those nations which currently enjoy food security may no longer do so. Undoubtedly, sea-levels will rise. IPCC (1990) states that over the last century global sea-levels have risen by 10–20 cm, although the rise has not been temporally or spatially uniform. Much of the rise is the result of thermal expansion and some melting of mountain glaciers and the margins of the Greenland ice-cap. Global warming would accentuate this rise and although the rise may not be as large as some predictions suggest because of increased snow accumulation in the polar regions, it would have a significant effect on nations with low-lying areas such as Bangladesh and Egypt and on island nations. Even a small rise in sea-level could have major implications for their populations and force mass migrations. Food production and tourism would also be affected. For example, irrigation systems are likely to be adversely affected by rising groundwater tables and agricultural land will be lost to the sea.

It could be a brave new world indeed, although rather different to that envisaged by Aldous Huxley and for very different reasons. Then again, the enhanced greenhouse effect could turn out to be just a puff of hot air! The global climatic future is, thus, uncertain, and this makes the formulation and implementation of strategies for sustainable development much more difficult. What the earth needs is a sustainable climatic environment. Nature seems to offer no guarantee of this.

The issue of global warming exemplifies many of the problems of formulating and implementing policies to prevent environmental degradation and unsustainable development. Nations are reluctant to pay when there is uncertainty over the likely extent and impact of global warming. However, they recognize that some action is required now should the more gloomy predictions prove correct. The dilemma for policy makers, therefore, is to choose actions whose costs are not likely to be greater than the future benefits from averted global warming. Moreover, there are major conflicts of interest between the developed and developing worlds over policy. The latter see no reason to cut back their development programmes to reduce the production of greenhouse gases or to pay for expensive technology to reduce their emissions when they consider the problem to have been produced by the industrial growth of the developed world. They therefore look to the richer nations to compensate them and to shoulder the principal burden of controlling emissions.

20.3   FUTURE PROSPECTS: SUSTAINABLE DEVELOPMENT

The concept of sustainable development is intended to provide a framework within which people can begin to develop more sensible ways of organizing global, national and local economic decision making. It has lent weight to arguments for proactive rather than reactive environmental policies (and global warming illustrates the kind of problem for which proactive policy making could be vital). As Chapter 2 describes, sustainable development has been adopted enthusiastically by many major global agencies such as the World Bank, and the use of its terminology is the current fashion among politicians when discussing environmental issues. Chapter 2 critically examines the merits and implications of the concept and illustrates some of the difficulties in deciding how to implement it. Here it is appropriate to endorse the value of the idea of sustainable development but to stress that it does not of itself solve the issues of uncertainty and economic and political power (Sections 20.2.2 and 20.2.3). It can, however, help policy makers to examine these issues from a new perspective. Indeed, much of the value of sustainable development as an approach to environmental problem solving is dependent on how far it is possible to use it to think afresh about these issues. The danger, of course, is that sustainable development becomes a slogan without substance.

To adopt the principles of inter- and intra-generational equity and of preserving or even enhancing the 'capital' that one generation passes on to the next—principles that lie at the heart of sustainable development—it is necessary to value resources both now and for the future. This involves devising new methods of assessing the assets and actions of both nations and individual producers and consumers. For example, current techniques of national accounting would need to be modified and extended to take account of damage to natural environments and the consumption of natural capital (Pearce *et al.*, 1989; Pearce, 1991). Such a valuation is made more difficult (a) in a technical sense, by uncertainty about the future pattern of society and technological developments and (b) in a practical or political sense, by the disparities of power between groups with alternative views about the value of particular resources. Furthermore, selecting appropriate policies to ensure that valued resources are, indeed, preserved or enhanced will depend on a reasonably accurate understanding of the physical and social interrelationships which lead to environmental degradation and damage. The chapters in this book illustrate the difficulties involved in formulating policies to sustain even a particular environment, let alone the global environment. Given the uncertainties which exist about many of the most environmentally significant relationships, a focus on the issue of sustainability may be a useful way of deciding on research priorities.

The implementation of policies for sustainable development will be a difficult task as it involves persuading members of current generations to suffer in the short term for the (uncertain) long-term benefit of future generations. It also involves persuading some powerful groups in society to accept an unpalatable analysis of the causes of particular environmental problems. Above all, particularly in relation to environmental degradation in the developing world, it will involve changes in the distribution of wealth and power. The history of humanity does not suggest that this will be easy. The environmental degradation of Eastern Europe is a timely reminder that socialist planned economies are no more effective than capitalist market economies at preserving environmental quality or facilitating a move towards sustainable development. The new political thinking that will be necessary to achieve sustainable development must transcend the traditional divisions between East and West as well as those between North and South.

There has been a gradual shift in society's attitudes towards the view that human beings are part of the global ecosystem rather than separate from and, in some sense, independent of it. The popularity of the concept of sustainable development is evidence of this shift and also suggests that it could be possible, albeit slowly and uncertainly, to move towards a way of using the resources of the globe for the benefit of all societies and generations. Otherwise, 'for those who come late only the bones are left'.

20.4 REFERENCES

Blaikie, P. (1985) *The Political Economy of Soil Erosion in Developing Countries*. Harlow: Longman.

Intergovernmental Panel on Climate Change (1990) *Climatic Change: The IPCC Scientific Assessment*. Cambridge: Cambridge University Press.

International Union for the Conservation of Nature (IUCN)/WWF/UNEP (1980) *World Conservation Strategy*. Switzerland, Gland: IUCN.

Pearce, D. W., Markardy, A. and Barbier, E. B. (1989) *Blueprint for a Green Economy*. London: Earthscan.

Pearce, D. W. (Ed.) (1991) *Blueprint 2: Greening the World Economy*. London: Earthscan.

Shaw, R. P. (1990) Rapid population growth and environmental degradation: ultimate versus proximate factors. *Environmental Conservation*, **16** (3), 199–208.

World Bank (1990) *World Development Report 1990*. Oxford: Oxford University Press.

World Commission on Environment and Development (1987) *Our Common Future*. Oxford: Oxford University Press.

20.5 FURTHER READING

McCarta, R. (1990) *The Gaia Atlas of Future Worlds: Challenge and Opportunity in an Age of Change*. London: Gaia Books.

Mungall, C. and McLaren D. J. (Eds) (1990) *Planet Under Stress*. Oxford: Oxford University Press.

Turner, B. L. II., Clark, W. C., Kates, R. W., Richards, J. F., Mathews, J. T. and Meyer, W. B. (Eds) (1990) *The Earth as Transformed by Human Action*. Cambridge: Cambridge University Press with Clark University.

Worldwatch Institute (1990) *State of the World 1990: A Worldwatch Institute Report on Progress Towards a Sustainable Society*. New York and London: W. W. Norton.

Index